MASS SPECTROMETRY

WILEY-INTERSCIENCE SERIES IN MASS SPECTROMETRY

Series Editors

Dominic M. Desiderio
Departments of Neurology and Biochemistry
University of Tennessee Health Science Center

Nico M. M. Nibbering
Vrije Universiteit Amsterdam, The Netherlands

A complete list of the titles in this series appears at the end of this volume.

MASS SPECTROMETRY

Instrumentation, Interpretation, and Applications

Edited by

Rolf Ekman
Jerzy Silberring
Ann Westman-Brinkmalm
Agnieszka Kraj

WILEY

A JOHN WILEY & SONS, INC., PUBLICATION

Published by John Wiley & Sons, Inc., Hoboken, New Jersey
Published simultaneously in Canada

Library of Congress Cataloging-in-Publication Data

Mass spectrometry : instrumentation, interpretation, and applications / edited by
Rolf Ekman ... [et al.].
p. cm.
Includes index.
ISBN 978-0-471-71395-1 (cloth)
1. Mass spectrometry. I. Ekman, Rolf, 1938–
QD96.M3M345 2008
543′.65—dc22

2008041505

Printed in the United States of America

10 9 8 7 6 5 4 3 2 1

CONTENTS

FOREWORD xiii

CONTRIBUTORS xv

PART I INSTRUMENTATION 1

1 DEFINITIONS AND EXPLANATIONS 3
Ann Westman-Brinkmalm and Gunnar Brinkmalm
References 13

2 A MASS SPECTROMETER'S BUILDING BLOCKS 15
Ann Westman-Brinkmalm and Gunnar Brinkmalm
2.1. Ion Sources 15
2.1.1. Gas Discharge 16
2.1.2. Thermal Ionization 16
2.1.3. Spark Source 19
2.1.4. Glow Discharge 20
2.1.5. Inductively Coupled Plasma 21
2.1.6. Electron Ionization 23
2.1.7. Chemical Ionization 24
2.1.8. Atmospheric Pressure Chemical Ionization 24
2.1.9. Photoionization 25
2.1.10. Multiphoton Ionization 25
2.1.11. Atmospheric Pressure Photoionization 26
2.1.12. Field Ionization 26
2.1.13. Field Desorption 27
2.1.14. Thermospray Ionization 27
2.1.15. Electrospray Ionization 27
2.1.16. Desorption Electrospray Ionization 29
2.1.17. Direct Analysis in Real Time 30
2.1.18. Secondary Ion Mass Spectrometry 31

2.1.19. Fast Atom Bombardment 33
2.1.20. Plasma Desorption 34
2.1.21. Laser Desorption/Ionization 34
2.1.22. Matrix-Assisted Laser Desorption/Ionization 35
2.1.23. Atmospheric Pressure Matrix-Assisted Laser Desorption/Ionization 37
2.2. Mass Analyzers 38
2.2.1. Time-of-Flight 40
2.2.2. Magnetic/Electric Sector 45
2.2.3. Quadrupole Mass Filter 49
2.2.4. Quadrupole Ion Trap 51
2.2.5. Orbitrap 55
2.2.6. Fourier Transform Ion Cyclotron Resonance 58
2.2.7. Accelerator Mass Spectrometry 62
2.3. Detectors 65
2.3.1. Photoplate Detector 65
2.3.2. Faraday Detector 67
2.3.3. Electron Multipliers 67
2.3.4. Focal Plane Detector 69
2.3.5. Scintillation Detector 69
2.3.6. Cryogenic Detector 70
2.3.7. Solid-State Detector 70
2.3.8. Image Current Detection 70
References 71

3 TANDEM MASS SPECTROMETRY 89

Ann Westman-Brinkmalm and Gunnar Brinkmalm

3.1. Tandem MS Analyzer Combinations 91
3.1.1. Tandem-in-Space 91
3.1.2. Tandem-in-Time 95
3.1.3. Other Tandem MS Configurations 97
3.2. Ion Activation Methods 97
3.2.1. In-Source Decay 97
3.2.2. Post-Source Decay 98
3.2.3. Collision Induced/Activated Dissociation 98
3.2.4. Photodissociation 100
3.2.5. Blackbody Infrared Radiative Dissociation 100
3.2.6. Electron Capture Dissociation 101

3.2.7. Electron Transfer Dissociation 101
3.2.8. Surface-Induced Dissociation 101
References 102

4 SEPARATION METHODS **105**
Ann Westman-Brinkmalm, Jerzy Silberring, and Gunnar Brinkmalm
4.1. Chromatography 106
4.1.1. Gas Chromatography 106
4.1.2. Liquid Chromatography 107
4.1.3. Supercritical Fluid Chromatography 109
4.2. Electric-Field Driven Separations 110
4.2.1. Ion Mobility 110
4.2.2. Electrophoresis 111
References 113

PART II INTERPRETATION **117**

5 INTRODUCTION TO MASS SPECTRA INTERPRETATION: ORGANIC CHEMISTRY **119**
Albert T. Lebedev
5.1. Basic Concepts 119
5.2. Inlet Systems 121
5.2.1. Direct Inlet 121
5.2.2. Chromatography-Mass Spectrometry 121
5.3. Physical Bases of Mass Spectrometry 128
5.3.1. Electron Ionization 129
5.3.2. Basics of Fragmentation Processes in Mass Spectrometry 130
5.3.3. Metastable Ions 135
5.4. Theoretical Rules and Approaches to Interpret Mass Spectra 137
5.4.1. Stability of Charged and Neutral Particles 137
5.4.2. The Concept of Charge and Unpaired Electron Localization 148
5.4.3. Charge Remote Fragmentation 151
5.5. Practical Approaches to Interpret Mass Spectra 152
5.5.1. Molecular Ion 152
5.5.2. High Resolution Mass Spectrometry 155

5.5.3. Determination of the Elemental Composition of Ions on the Basis of Isotopic Peaks 158
5.5.4. The Nitrogen Rule 164
5.5.5. Establishing the ^{13}C Isotope Content in Natural Samples 166
5.5.6. Calculation of the Isotopic Purity of Samples 166
5.5.7. Fragment Ions 168
5.5.8. Mass Spectral Libraries 173
5.5.9. Additional Mass Spectral Information 173
5.5.10. Fragmentation Scheme 175
References 177

6 SEQUENCING OF PEPTIDES AND PROTEINS 179

Marek Noga, Tomasz Dylag, and Jerzy Silberring

6.1. Basic Concepts 179
6.2. Tandem Mass Spectrometry of Peptides and Proteins 181
6.3. Peptide Fragmentation Nomenclature 183
6.3.1. Roepstorff's Nomenclature 183
6.3.2. Biemann's Nomenclature 185
6.3.3. Cyclic Peptides 187
6.4. Technical Aspects and Fragmentation Rules 188
6.5. Why Peptide Sequencing? 190
6.6. De Novo Sequencing 192
6.6.1. Data Acquisition 193
6.6.2. Sequencing Procedure Examples 194
6.6.3. Tips and Tricks 205
6.7. Peptide Derivatization Prior to Fragmentation 207
6.7.1. Simplification of Fragmentation Patterns 208
6.7.2. Stable Isotopes Labeling 209
Acknowledgments 210
References 210
Online Tutorials 210

7 OPTIMIZING SENSITIVITY AND SPECIFICITY IN MASS SPECTROMETRIC PROTEOME ANALYSIS 211

Jan Eriksson and David Fenyö

7.1. Quantitation 212
7.2. Peptide and Protein Identification 213

7.3. Success Rate and Relative Dynamic Range 218
7.4. Summary 220
References 220

PART III APPLICATIONS **223**

8 DOPING CONTROL **225**
Graham Trout
References 233

9 OCEANOGRAPHY **235**
R. Timothy Short, Robert H. Byrne, David Hollander, Johan Schijf, Strawn K. Toler, and Edward S. VanVleet
References 241

10 "OMICS" APPLICATIONS **243**
Simone König
10.1. Introduction 243
10.2. Genomics and Transcriptomics 246
10.3. Proteomics 248
10.4. Metabolomics 251

11 SPACE SCIENCES **253**
Robert Sheldon
11.1. Introduction 253
11.2. Origins 254
11.3. Dynamics 256
11.4. The Space MS Paradox 257
11.5. A Brief History of Space MS 259
11.5.1. Beginnings 259
11.5.2. Linear TOF-MS 260
11.5.3. Isochronous TOF-MS 262
11.6. GENESIS and the Future 264
References 264

12 BIOTERRORISM **267**

Vito G. DelVecchio and Cesar V. Mujer

12.1. What is Bioterrorism? 267

12.2. Some Historical Accounts of Bioterrorism 267

12.3. Geneva Protocol of 1925 and Biological Weapons Convention of 1972 268

12.4. Categories of Biothreat Agents 268

12.5. Challenges 269

12.6. MS Identification of Biomarker Proteins 270

12.7. Development of New Therapeutics and Vaccines Using Immunoproteomics 271

References 272

13 IMAGING OF SMALL MOLECULES **275**

Małgorzata Iwona Szynkowska

13.1. SIMS Imaging 277

13.2. Biological Applications (Cells, Tissues, and Pharmaceuticals) 278

13.3. Catalysis 280

13.4. Forensics 281

13.5. Semiconductors 282

13.6. The Future 283

References 285

14 UTILIZATION OF MASS SPECTROMETRY IN CLINICAL CHEMISTRY **287**

Donald H. Chace

14.1. Introduction 287

14.2. Where are Mass Spectrometers Utilized in Clinical Applications? 288

14.3. Most Common Analytes Detected by Mass Spectrometers 288

14.4. Multianalyte Detection of Clinical Biomarkers, The Real Success Story 289

14.5. Quantitative Profiling 291

14.6. A Clinical Example of the Use of Mass Spectrometry 292

14.7. Demonstrations of Concepts of Quantification in Clinical Chemistry 294

14.7.1. Tandem Mass Spectrometry and Sorting (Pocket Change) 294
14.7.2. Isotope Dilution and Quantification (the Jelly Bean Experiment) 295

15 POLYMERS **299**

Maurizio S. Montaudo

15.1. Introduction 299
15.2. Instrumentation, Sample Preparation, and Matrices 300
15.3. Analysis of Ultrapure Polymer Samples 301
15.4. Analysis of Polymer Samples in which all Chains Possess the Same Backbone 301
15.5. Analysis of Polymer Mixtures with Different Backbones 303
15.6. Determination of Average Molar Masses 303
References 306

16 FORENSIC SCIENCES **309**

Maria Kala

16.1. Introduction 309
16.2. Materials Examined and Goals of Analysis 311
16.3. Sample Preparation 312
16.4. Systematic Toxicological Analysis 312
16.4.1. GC-MS Procedures 315
16.4.2. LC-MS Procedures 315
16.5. Quantitative Analysis 317
16.6. Identification of Arsons 319
References 319

17 NEW APPROACHES TO NEUROCHEMISTRY **321**

Jonas Bergquist, Jerzy Silberring, and Rolf Ekman

17.1. Introduction 321
17.2. Why is there so Little Research in this Area? 322
17.3. Proteomics and Neurochemistry 323
17.3.1. The Synapse 324
17.3.2. Learning and Memory 324

17.3.3. The Brain and the Immune System 325
17.3.4. Stress and Anxiety 327
17.3.5. Psychiatric Diseases and Disorders 329
17.3.6. Chronic Fatigue Syndrome 329
17.3.7. Addiction 330
17.3.8. Pain 331
17.3.9. Neurodegenerative Diseases 331
17.4. Conclusions 333
Acknowledgments 333
References 334

PART IV APPENDIX **337**

INDEX **353**

FOREWORD

Over the last two decades mass spectrometry has become one of the central techniques in analytical chemistry, and the analysis of biological (macro)molecules in particular. Its importance is now comparable to that of the more traditional electrophoresis and liquid separations techniques, and it is often used in conjunction with them as so-called "hyphenated" techniques, such as LC-MS.

This development was originally triggered by the discovery of novel techniques to generate stable ions of the molecules of interest and the development of associated ion sources. Such a technique has to meet two basic requirements: first the molecules, usually existing in the liquid or solid condensed state, have to be transferred into the gas phase and eventually into the vacuum of a mass analyzer; second, the neutral molecules have to acquire one or several charges to be separated and detectable in the mass analyzer. Both steps had traditionally been prone to internal excitation of the molecules leading to fragmentation and loss of analytical information. The two techniques that evolved as the frontrunners and nowadays dominate mass spectrometry are electrospray ionization (ESI) and matrix-assisted laser desorption/ionization (MALDI). Even though these two techniques solve the problem of transfer from the condensed to gas phase as well as the ionization in very different ways and were developed completely independently, their breakthrough happened concurrently in 1988. This concurrent development was, most probably, not a shear coincidence. The basics of the macromolecular structure and function of biological systems, the role of DNA and proteins in particular, had evolved over the three decades before and it had become apparent at least to a small group of scientists that to unravel the details of their structure required a leap in the development of more sensitive and more specific analytical techniques. Mass spectrometry held at least the promise for such a leap, even though most of the "experts" thought it impossible. This might suggest that in science, as in other fields of human development it holds that, "where is a need, there is a way." It is also important to realize in this context, that both ESI and MALDI make use of principles developed in the years before, such as field desorption and desorption by particle beams, as well as chemical ionization in the gas phase.

The novel ionization mechanisms have early on induced the revival of some mass analyzer principles such as the (axial ion extraction) time-of-flight (TOF) instruments, which had been written off as having too low a performance earlier. More recently a whole plethora of new mass spectrometers have been marketed, combining both ESI and MALDI with high performance spectrometers such as the orthogonal extraction TOFs, Fourier transform ion cyclotron (FT-ICR) and orbitrap instruments. These developments have been largely introduced by the instrument manufacturers. The parallel development of high speed digital signal processing, data analysis, and data banking has also played a major role in the development of the field.

Mass spectrometry has meanwhile become an important part of academic education in analytical chemistry. It can be found in the curricula of most undergraduate as well as graduate courses in the field. The publication of this dedicated textbook is, therefore, a timely undertaking and the editors and authors are to be complimented for the effort to put the book together.

How much detail does a student need to know and how much detail should a textbook then contain? This is an almost unsolvable problem because of the diversity of students and their analytical needs. The majority of students will eventually move on into special fields in (bio)chemistry, molecular or systems biology or polymer chemistry. For them mass spectrometry will "only" be one of the commodities to help them solve their problems, which are defined by their field of activity, not the analytical technique. How much of the basics in mass spectrometry will they need to know? Again, this depends on the problem at hand. For many a routine application of commercial instruments and the manufacturers' manuals will suffice. However, if the problem is not routine the analytical technique cannot be either. Mass spectrometry is and, most probably, will remain a rather complex technique. To fully exploit its tremendous potential, but, equally important, to avoid its many pitfalls, a deeper understanding of the mechanisms and the technology will be mandatory. This book will, hopefully, help students to lay the basis for this expertise and, once the need arises, allow them to go back to the more specialized literature at a later time. It is in this sense that I hope this book will be a real help to many of them.

FRANZ HILLENKAMP

Münster, Germany
August 2008

CONTRIBUTORS

Jonas Bergquist, Department of Physical and Analytical Chemistry, Uppsala University, Uppsala, Sweden.

Gunnar Brinkmalm, Institute of Neuroscience and Physiology, The Sahlgrenska Academy, University of Gothenburg, Molndal, Sweden.

Robert H. Byrne, College of Marine Science, University of South Florida, St. Petersburg, Florida.

Donald H. Chace, Pediatrix Analytical, Bridgeville, Pennsylvania.

Vito G. DelVecchio, Vital Probes Inc., Mayfield, Pennsylvania.

Tomasz Dylag, Poland Faculty of Chemistry and Regional Laboratory, Jagiellonian University, Krakow, Poland.

Rolf Ekman, Institute of Neuroscience and Physiology, The Sahlgrenska Academy, University of Gothenburg, Molndal, Sweden.

Jan Eriksson, Department of Chemistry, Swedish University of Agricultural Sciences, Uppsala, Sweden.

David Fenyö, The Rockefeller University, New York, New York.

Franz Hillenkamp, Institute for Medical Physics and Biophysics University of Muenster, Muenster, Germany.

David Hollander, College of Marine Science, University of South Florida, St. Petersburg, Florida.

Justyna Jarzebinska, Faculty of Chemistry and Regional Laboratory, Jagiellonian University, Krakow, Poland.

Maria Kala, Institute of Forensic Research, Krakow, Poland.

Agnieszka Kraj, Poland Faculty of Chemistry and Regional Laboratory, Jagiellonian University, Krakow, Poland.

Simone König, Integrated Functional Genomics, Interdisciplinary Center for Clinical Research, University of Münster, Münster, Germany.

Albert T. Lebedev, Organic Chemistry Department, Moscow State University, Russia.

Maurizio S. Montaudo, Italian National Research Council (CNR), Institute of Chemistry and Technology of Polymers, Catania, Italy.

Cesar V. Mujer, Calvert Laboratories, Olyphant, Pennsylvania.

Marek Noga, Poland Faculty of Chemistry and Regional Laboratory, Jagiellonian University, Krakow, Poland.

Hana Raoof, Faculty of Chemistry and Regional Laboratory, Jagiellonian University, Krakow, Poland.

Johan Schijf, Aquatic Environmental Geochemistry, UMCES/Chesapeake Biological Laboratory, Solomons, Maryland.

Robert Sheldon, National Space Science and Technology Center, Huntsville, Alabama, USA.

R. Timothy Short, SRI International, St. Petersburg, Florida.

Jerzy Silberring, Faculty of Chemistry and Regional Laboratory, Jagiellonian University, Krakow, Poland.

Filip Sucharski, Faculty of Chemistry and Regional Laboratory, Jagiellonian University, Krakow, Poland.

Małgorzata Iwona Szynkowska, Institute of General and Ecological Chemistry, Technical University of Lodz, Lodz, Poland.

Strawn K. Toler, SRI International, St. Petersburg, Florida.

Graham Trout, National Measurement Institute, Pymble, Australia.

Edward S. VanVleet, College of Marine Science, University of South Florida, St. Petersburg, Florida.

Ann Westman-Brinkmalm, Institute of Neuroscience and Physiology, The Sahlgrenska Academy, University of Gothenburg, Molndal, Sweden.

PART I

INSTRUMENTATION

INTRODUCTION

The first part of this book is dedicated to a discussion of mass spectrometry (MS) instrumentation. We start with a list of basic definitions and explanations (Chapter 1). Chapter 2 is devoted to the mass spectrometer and its building blocks. In this chapter we describe in relative detail the most common ion sources, mass analyzers, and detectors. Some of the techniques are not extensively used today, but they are often cited in the MS literature, and are important contributions to the history of MS instrumentation. In Chapter 3 we describe both different fragmentation methods and several typical tandem MS analyzer configurations. Chapter 4 is somewhat of an outsider. Separation methods is certainly too vast a topic to do full justice in less than twenty pages. However, some separation methods are used in such close alliance with MS that the two techniques are always referred to as one combined analytical tool, for example, GC-MS and LC-MS. In effect, it is almost impossible to study the MS literature without coming across at least one separation method. Our main goal with Chapter 4 is, therefore, to facilitate an introduction to the MS literature for the reader by providing a short summary of the basic principles of some of the most common separation methods that have been used in conjunction with mass spectrometry.

Mass Spectrometry. Edited by Ekman, Silberring, Westman-Brinkmalm, and Kraj

1

DEFINITIONS AND EXPLANATIONS

Ann Westman-Brinkmalm and Gunnar Brinkmalm

The objective of this chapter is to provide the reader with definitions or brief explanations of some key terms used in mass spectrometry (MS). As in many other scientific fields there exist in the MS community (sometimes heated) debates over what terms and definitions are correct and what the everyday MS terminology really stands for. Maybe this is an inevitable phenomenon in any multidisciplined, highly active, and fast evolving branch of science. However, in this chapter we will try to keep out of harms way by providing the reader mainly with definitions based on suggestions by the current IUPAC project "Standard Definitions of Terms Relating to Mass Spectrometry" [1], see also "Mass Spectrometry Terms and Definitions Project Page" [2]. This project is currently in its final stage and will be officially published in the near future. However, in some cases we could not refrain from adding some contrary opinions. See also Chapters 5 and 6 for more detailed explanations of some of the basic concepts of MS. When studying the different chapters in this book the reader will notice that all authors (including ourselves) have not adhered strictly to the list of recommended definitions found in this chapter. This is a realistic reflection of the MS literature in general and the reader should not allow herself or himself to be too confused or discouraged. Our general advice is, "when in Rome, do as the Romans do." However, to aid the reader, the editors have when possible provided alternative or additional terms in concordance with the IUPAC definitions.

Mass Spectrometry. Edited by Ekman, Silberring, Westman-Brinkmalm, and Kraj

Accurate Mass An experimentally determined mass of an ion that is used to determine an elemental formula. For ions containing combinations of the elements C, H, N, O, P, S, and the halogens, with mass less than 200 Da, a measurement with 5 ppm uncertainty is sufficient to uniquely determine the elemental composition. See also related entries on: *average mass; dalton; molar mass; monoisotopic mass; nominal mass; unified atomic mass unit.*

Atomic Mass Unit See *unified atomic mass unit.*

Average Mass The mass of an ion, atom, or molecule calculated using the masses of all isotopes of each element weighted for their natural isotopic abundance. See also related entries on: *accurate mass; dalton; molar mass; monoisotopic mass; nominal mass; unified atomic mass unit.*

Dalton (Da) A non-SI unit of mass (symbol Da) that is equal to the unified atomic mass unit. See also related entries on: *accurate mass; average mass; molar mass; molecular weight; monoisotopic mass; nominal mass; unified atomic mass unit.*

Daughter Ion See *product ion.*

Dimeric Ion An ion formed by ionization of a dimer or by the association of an ion with its neutral counterpart such as $[M_2]^{+\bullet}$ or $[M\text{-}H\text{-}M]^+$.

Electron Volt (eV) A non-SI unit of energy defined as the energy acquired by a particle containing one unit of charge through a potential difference of one volt, $1\ \text{eV} \approx 1.6 \cdot 10^{-19}\ \text{J}$.

Extracted Ion Chromatogram A chromatogram created by plotting the intensity of the signal observed at a chosen m/z value or series of values in a series of mass spectra recorded as a function of retention time. See also related entry on: *total ion current chromatogram.*

Field Free Region Any region of a mass spectrometer where the ions are not dispersed by a magnetic or electric field.

Fragment Ion See *product ion.*

Ionization Efficiency Ratio of the number of ions formed to the number of atoms or molecules consumed in the ion source.

Isotope Dilution Mass Spectrometry (IDMS) A quantitative mass spectrometry technique in which an isotopically enriched compound is used as an internal standard. See Chapter 14 for a more detailed explanation.

Isotope Ratio Mass Spectrometry (IRMS) The measurement of the relative quantity of the different isotopes of an element in a material using a mass spectrometer.

m/z The three-character symbol m/z is used to denote the dimensionless quantity formed by dividing the mass of an ion in unified atomic mass units by its charge number (regardless of sign). The symbol is written in italicized lower case letters with no spaces. Note 1: The term mass-to-charge ratio is deprecated. Mass-to-charge ratio has been used for the abscissa of a mass spectrum, although the quantity measured is not the quotient of the ion's mass to its electric charge. The three-character symbol m/z is recommended for the dimensionless quantity that is the independent variable in a mass spectrum. Note 2: The proposed unit thomson (Th) is deprecated [1].

Comment: Here the authors feel obliged to state that a mass analyzer *does* separate gas-phase ions according to their mass-to-charge ratio (m/q, see formulas below) and neither mass nor charge are dimensionless quantities. z, being the number of charges, is dimensionless, leading to the fact that the unit for m/z is u or Da. The SI unit for m/q is kilogram/coulomb (kg/C), but is not practical because of the actual numbers involved. Alternative units for m/q would be atomic units. Historically u/e has been used, where e equals the elementary charge. Unfortunately e is constant (the value of the charge of the proton and electron), not a unit—there is presently no accepted atomic unit for charge. Therefore such a unit has been suggested—millikan (Mi). A unit for m/q has also been suggested—thomson (Th), where Th = u/Mi or Da/Mi, all being atomic units. All this can seem like nit-picking, but it is very impractical not to have an accepted unit for the very thing we measure in mass spectrometry.

Time-of-flight $t_{TOF} = \frac{L}{v} = L\sqrt{\frac{m}{2qU_a}}$

Magnetic sector $\frac{m}{q} = \frac{B^2r^2}{2U_a}$

FTICR $f_c = \frac{qB}{2\pi \cdot m}$

Mass See entries on: *accurate mass; average mass; dalton; molar mass; molecular weight; monoisotopic mass; nominal mass; unified atomic mass unit.*

Mass Accuracy Difference between measured and actual mass [3]. Can be expressed either in absolute or relative terms.

Mass Calibration (time-of-flight) A means of determining m/z values from their times of detection relative to initiation of acquisition of a mass spectrum. Most commonly this is accomplished using a computer-based data system and a calibration file obtained from a mass spectrum of a compound that produces ions whose m/z values are known.

Mass Defect Difference between exact and nominal mass [3].

"Mass" Limit The m/z value above or below which ions cannot be detected in a mass spectrometer.

Mass Number The sum of the protons and neutrons in an atom, molecule, or ion.

Mass Peak Width ($\Delta m_{50\%}$) The full width of a mass spectral peak at half-maximum peak height [3].

Mass Precision Root-mean-square (RMS) deviation in a large number of repeated measurements [3].

Mass Range The range of m/z over which a mass spectrometer can detect ions or is operated to record a mass spectrum.

Mass Resolution The smallest mass difference Δm (Δm in Da or $\Delta m/m$ in, e.g., ppm) between two equal magnitude peaks such that the valley between them is a specified fraction of the peak height [3].

Ten Percent Valley Definition

Let two peaks of equal height in a mass spectrum at masses m and m, Δm, be separated by a valley which at its lowest point is just 10% of the height of either peak. For similar peaks at a mass exceeding m, let the height of the valley at its lowest point be more (by any amount) than 10% of either peak. Then the "resolution" (10% valley definition) is $m/\Delta m$. The ratio $m/\Delta m$ should be given for a number of values of m [4].

Comment: This is a typical example of the confusion regarding the definition of the term resolution. Here resolution is used instead of the more appropriate phrase mass resolving power (which is the inverse of resolution).

Peak Width Definition

For a single peak made up of singly charged ions at mass m in a mass spectrum, the "resolution" may be expressed as $m/\Delta m$, where Δm is the width of the peak at a height that is a specified fraction of the maximum peak height. It is recommended that one of three values 50%, 5%, or 0.5% should always be used. (Note that for an isolated symmetrical peak recorded with a system that is linear in the range between 5% and 10% levels of the peak, the 5% peak width definition is equivalent to the 10% valley definition). A common standard is the definition of resolution based upon Δm being the full width of the peak at half its maximum (FWHM) height [4]. See Fig. 1.1 and also Chapter 5.

Comment: See comment for *Ten percent valley definition* above.

Mass Resolving Power ($m/\Delta m$) In a mass spectrum, the observed mass divided by the difference between two masses that can be separated, $m/\Delta m$. The method by which Δm was obtained and the mass at which the measurement was made should be reported.

Mass Spectrometer An instrument that measures the m/z values and relative abundances of ions. See also discussion in entry m/z.

Mass Spectrometry Branch of science that deals with all aspects of mass spectrometers and the results obtained with these instruments.

MS/MS The acquisition and study of the spectra of the electrically charged products or precursors of m/z selected ion or ions, or of precursor ions of a selected neutral mass loss. Also termed tandem mass spectrometry.

Comment: There are two different opinions of what MS/MS is an abbreviation of. One is mass spectrometry/mass spectrometry [1]. The other is mass selection/mass separation.

Mass Spectrum A plot of the detected intensities of ions as a function of their m/z values. See discussion in entry m/z.

Mass-to-Charge Ratio or Mass/Charge See discussion in entry m/z.

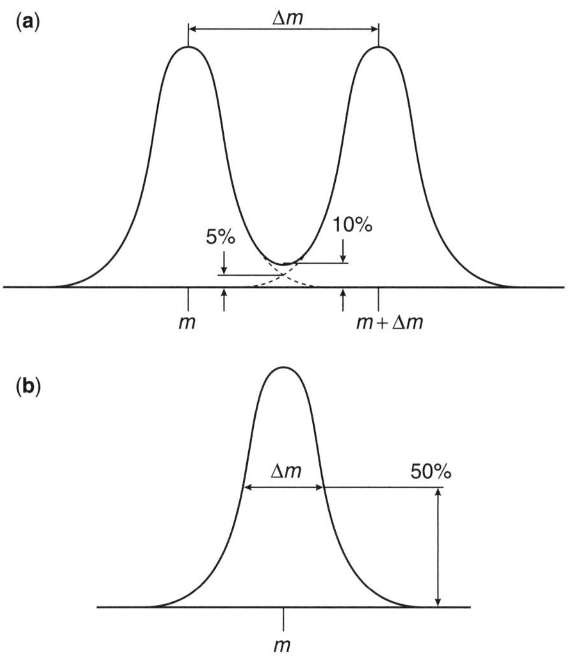

Figure 1.1. The two different ways of establishing mass resolution or mass resolving power. (a) The 10% valley definition. The peak separation Δm is defined as the distance between the centers of two peaks of equal height when the valley bottom between them is 10% of their height. If the peaks are symmetric this will also in theory correspond to the peak width at 5% peak height. This is a true peak separation definition but is usually problematic to establish because of the difficulty of finding two peaks of equal height properly separated in a mass spectrum. (b) The full width at half-maximum (FWHM) definition. Here the peak width Δm is determined at 50% of the peak height. This number is easy to obtain since just a clearly separated peak is required, but it is instead not directly addressing peak separation capability. For Gaussian peak shapes the FWHM definition will yield a mass resolving power number roughly twice that of the 10% valley definition.

Metastable Ion An ion that is formed with internal energy higher than the threshold for dissociation but with a lifetime long enough to allow it to exit the ion source and enter the mass spectrometer where it dissociates before detection.

Molar Mass Mass of one mole ($\approx 6 \cdot 10^{23}$ atoms or molecules) of a compound. Note: The use of the term molecular weight is urged against because "weight" is the gravitational force on an object, which varies with geographical location. Historically the term has been used to denote the molar mass calculated using isotope-averaged atomic masses for the constituent elements.

Molecular Ion An ion formed by the removal of one or more electrons to form a positive ion or the addition of one or more electrons to form a negative ion.

Molecular Weight See *molar mass.*

Monoisotopic Mass Exact mass of an ion or molecule calculated using the mass of the most abundant isotope of each element [1]. This recommendation refers to a somewhat unfortunate statement by Yergey et al. [5], who have contributed with an otherwise enlightening paper on the subject. In the paper, only elements for which their most abundant isotopes are also their lightest, are considered. The practical problem arises when this is not the case (e.g., for Fe or B). Here the authors instead prefer the definition "the mass of an ion or molecule calculated using the mass of the lightest isotope of each element." For molecules containing the most common elements, such as C, H, N, O, S where the lightest isotope also is the most abundant the two suggested definitions give the same end result, but this is not the case for B, Fe, and many other elements. Cytochrome c includes one Fe; with the definition that the monoisotopic peak is the one containing only the most abundant isotopes of the elements, the result is that one of the isobars of the second isotopic peak would be the monoisotopic. The second isotopic peak also includes the isobars of one ^{2}H, or one ^{13}C, or one ^{15}N, or one ^{17}O, or one ^{33}S, in sum there are six isobars of which only one is the true monoisotopic peak. With the "lightest isotope" definition, the first isotopic peak does not have isobars and is therefore well defined.

Multiple Reaction Monitoring (MRM) See *selected reaction monitoring.*

Multiple-Stage Mass Spectrometry (MS^n) Multiple stages of precursor ion m/z selection followed by product ion detection for successive progeny ions.

Neutral Loss Loss of an uncharged species from an ion during either a rearrangement process or direct dissociation.

Nominal Mass Mass of an ion or molecule calculated using the mass of the most abundant isotope of each element rounded to the nearest integer value and equivalent to the sum of the mass numbers of all constituent atoms [1].

Example: The nominal mass of an ion is calculated by adding the integer masses of the lightest isotopes of all elements contributing to the molecule, for example, the nominal mass of H_2O is $(2 \cdot 1) + 16$ Da $= 18$ Da.

Comment: The same problem as for monoisotopic mass immediately arises for compounds containing elements such as Fe or B. See discussion in entry *monoisotopic mass.*

Peak A localized region of a visible ion signal in a mass spectrum. Although peaks are often associated with particular ions, the terms peak and ion should not be used interchangeably.

Peak Intensity The height or area of a peak in a mass spectrum.

A word of caution from the authors: The peak height and peak area are not interchangeable quantities. Consider for example how the height-to-area relationship depends on the resolving power of the mass analyzer or the response time of the detector.

Precursor Ion Ion that reacts to form particular product ions. The reaction can be unimolecular dissociation, ion/molecule reaction, isomerization, or change in charge state. The term parent ion is deprecated (but still very much in use).

Product Ion An ion formed as the product of a reaction involving a particular precursor ion. The reaction can be unimolecular dissociation to form fragment ions, an ion/molecule reaction, or simply involve a change in the number of charges. The terms fragment ion and daughter ion are deprecated (but still very much in use).

Progeny Ions Charged products of a series of consecutive reactions that includes product ions, first generation product ions, second generation product ions, etc.

Protonated Molecule An ion formed by interaction of a molecule with a proton, and represented by the symbol $[M+H]^+$. The term protonated molecular ion is deprecated; this would correspond to a species carrying two charges. The terms pseudo-molecular ion and quasi-molecular ion are deprecated; a specific term such as protonated molecule, or a chemical description such as $[M+Na]^+$, $[M-H]^-$, etc., should be used [1].

Selected Ion Monitoring (SIM) Operation of a mass spectrometer in which the abundances of one or several ions of specific m/z values are recorded rather than the entire mass spectrum.

Selected Reaction Monitoring (SRM) Data acquired from specific product ions corresponding to m/z selected precursor ions recorded via two or more stages of mass spectrometry. Selected reaction monitoring can be preformed as tandem mass spectrometry in time or tandem mass spectrometry in space. The term multiple reaction monitoring is deprecated [1].

Space-Charge Effect Result of mutual repulsion of particles of like charge that limits the current in a charged-particle beam or packet and causes some ion motion in addition to that caused by external fields.

Tandem Mass Spectrometry See MS/MS.

Thomson (Th) See discussion in entry m/z.

Torr Non-SI unit for pressure, 1 torr = 1 mmHg = 1.33322 mbar = 133.322 Pa.

Total Ion Current (TIC) Sum of all the separate ion currents carried by the different ions contributing to a mass spectrum.

Total Ion Current Chromatogram Chromatogram obtained by plotting the total ion current detected in each of a series of mass spectra recorded as a function of retention time. See related entry on *extracted ion chromatogram.*

Transmission The ratio of the number of ions leaving a region of a mass spectrometer to the number entering that region.

Unified Atomic Mass Unit (u) A non-SI unit of mass defined as one twelfth of the mass of one atom of ^{12}C in its ground state and $\approx 1.66 \times 10^{-27}$ kg. The term atomic mass unit (amu) is not recommended to use since it is ambiguous. It has been used to denote atomic masses measured relative to a single atom of ^{16}O, or to the isotope-averaged mass of an oxygen atom, or to a single atom of ^{12}C.

Abbreviations and Units

2-DGE	two-dimensional gel electrophoresis
a	atto, 10^{-18}
AC	alternating current
AMS	accelerator mass spectrometry
APCI	atmospheric pressure chemical ionization
API	atmospheric pressure ionization
AP-MALDI	atmospheric pressure matrix-assisted laser desorption/ionization
APPI	atmospheric pressure photoionization
ASAP	atmospheric-pressure solids analysis probe
BIRD	blackbody infrared radiative dissociation
c	centi, 10^{-2}
CAD	collision-activated dissociation
CE	capillary electrophoresis
CF	continuous flow
CF-FAB	continuous flow fast atom bombardment
CI	chemical ionization
CID	collision-induced dissociation
cw	continuous wave
CZE	capillary zone electrophoresis
Da	dalton
DAPCI	desorption atmospheric pressure chemical ionization
DART	direct analysis in real time
DC	direct current
DE	delayed extraction
DESI	desorption electrospray ionization
DIOS	desorption/ionization on silicon
DTIMS	drift tube ion mobility spectrometry
EC	electrochromatography
ECD	electron capture dissociation
EI	electron ionization
ELDI	electrospray-assisted laser desorption/ionization
EM	electron multiplier
ESI	electrospray ionization
ETD	electron transfer dissociation
eV	electron volt
f	femto, 10^{-15}
FAB	fast atom bombardment
FAIMS	field asymmetric waveform ion mobility spectrometry
FD	field desorption
FI	field ionization
FT	Fourier transform
FTICR	Fourier transform ion cyclotron resonance

FWHM	full width at half maximum
GC	gas chromatography
GD	glow discharge
GE	gel electrophoresis
GLC	gas-liquid chromatography
GPC	gel permeation chromatography
GSC	gas-solid chromatography
HIC	hydrophobic interaction chromatography
HPLC	high performance liquid chromatography
ICP	inductively coupled plasma
ICR	ion cyclotron resonance
IEC	ion-exchange chromatography
IEF	isoelectric focusing
IMS	ion mobility spectrometry
IR	infrared
IRMPD	infrared multiphoton dissociation
ITP	isotachophoresis
k	kilo, 10^3
LA	laser ablation
LC	liquid chromatography
LDI	laser desorption/ionization
LMIG	liquid metal ion gun
l-QIT	linear quadrupole ion trap
LSIMS	liquid secondary ion mass spectrometry
m	milli, 10^{-3}
m/z	mass-to-charge ratio
μ	micro, 10^{-6}
M	mega, 10^6
MALDESI	matrix-assisted laser desorption electrospray ionization
MALDI	matrix-assisted laser desorption/ionization
MCP	microchannel plate
MECA	multiple excitation collisional activation
MEKC	micellar electrokinetic chromatography
MIKES	mass-analyzed ion kinetic energy spectrometry
MPI	multiphoton ionization
MRM	multireaction monitoring
MS	mass spectrometry
MS/MS	mass selection/mass separation or mass spectrometry/mass spectrometry
MS^n	MS/MS of higher generations
n	nano, 10^{-9}
NSD	nozzle-skimmer dissociation
NPC	normal-phase chromatography
oa	orthogonal acceleration

p	pico, 10^{-12}
PD	plasma desorption; photodissociation
pI	isoelectric point
PI	photoionization
PLOT	porous layer open tubular
ppb	part per billion, 10^{-9}
ppm	part per million, 10^{-6}
ppt	part per trillion, 10^{-12}
PSD	post-source decay
q	quadrupole (or hexapole/octapole) used as collision chamber
Q	quadrupole, quadrupole filter
QIT	quadrupole ion trap (Paul trap)
QqQ	triple quadrupole
Qq-TOF	quadrupole–time-of-flight
REMPI	resonance-enhanced multiphoton ionization
RF	radio frequency, here used instead of AC (alternating current) to imply a high frequency
RI	resonance ionization
RPC	reversed-phase chromatography
RTOF	reflector time-of-flight
SDS	sodium dodecyl sulfate
SEC	size exclusion chromatography
SELDI	surface-enhanced laser desorption/ionization
SEM	scanning electron microscopy
SFC	supercritical fluid chromatography
SID	surface-induced dissociation
SIMS	secondary ion mass spectrometry
SNMS	secondary neutral mass spectrometry
SORI	sustained off-resonance irradiation
SRM	selected reaction monitoring
SS	spark source
SSD	solid state detector
STJ	superconducting tunnel junction
SWIFT	stored waveform inverse Fourier transform
TDC	time-to-digital converter
TI	thermal ionization
TOF	time-of-flight
TOF-TOF	tandem time-of-flight
TSI	thermospray ionization
UPLC	ultra performance liquid chromatography
UV	ultraviolet
VLE	very low-energy
WCOT	wall coated open tubular
z	zepto, 10^{-21}
ZE	zone electrophoresis

REFERENCES

1. IUPAC. Standard Definitions of Terms Relating to Mass Spectrometry—Provisional Recommendations. 2006. Available at http://www.iupac.org/web/ins/2003-056-2-500.
2. IUPAC. Mass Spectrometry Terms and Definitions Project Page. 2007. Available at http://www.msterms.com/wiki/index.php?title=Main_Page.
3. A. G. Marshall, C.L. Hendrickson, and S. D. Shi. Scaling MS Plateaus with High-Resolution FT-ICRMS. *Anal. Chem.*, **74**(2002): 252A–259A.
4. J. Inczédy, T. Lengyel, A. M. Urc, A. Gelencsér, and A. Hulanicki. *Compendium of Analytical Nomenclature (The Orange Book).* 1997. Available at http://www.iupac.org/publications/analytical_compendium/.
5. J. Yergey, D. Heller, G. Hansen, R. J. Cotter, and C. Fenselau. Isotopic Distributions in Mass Spectra of Large Molecules. *Anal. Chem.*, **55**(1983): 353–356.

2

A MASS SPECTROMETER'S BUILDING BLOCKS

Ann Westman-Brinkmalm and Gunnar Brinkmalm

A mass spectrometer consists of three major parts: the ion source, the mass analyzer, and the detector. Since the mass analyzer and the detector (and many of the ion sources) require low pressure for operation the instrument also needs a pumping system. Moreover, another system is required to record the signal registered by the detector. In the early days, such a system could often be a photographic plate but for almost all modern instruments a computer-based system is used. Computers are also utilized to process much of the acquired data, for example, database searches for protein identification. Finally, additional separation devices are often used in conjunction with mass spectrometers, for example, chromatographic separation can be performed prior to the mass spectrometric analysis, either online or offline. This chapter will neither describe the vacuum systems nor the data recording systems, which nevertheless are vital components of a mass spectrometer. Many of today's analysis techniques would be impossible to use or be severely limited without computers—perhaps the most striking example is chemical imaging.

2.1. ION SOURCES

In a mass spectrometer the role of the ion source is to create gas phase ions. Analyte atoms, molecules, or clusters are transferred into gas phase and ionized either

Mass Spectrometry. Edited by Ekman, Silberring, Westman-Brinkmalm, and Kraj

concurrently (as in electrospray ionization) or through separate processes (as in the glow discharge). The choice of ion source depends heavily on the application. So-called soft ion sources can produce intact ions of large fragile molecules such as proteins, nucleic acids, or even noncovalently bound complexes. Other ion sources, such as the glow discharge and inductively coupled plasma, atomize the analyte but can provide very accurate quantitative data and are used for example in isotope ratio measurements, which deal with very small differences in ratio that may reflect geographical origin. The number of available ion sources is quite large, not to mention all variations within a general type, and we have attempted to cover the main types, even if the description is brief in some cases. A few of the sources described are not commonly used today, but are of historical interest. Some sources are very specialized while others have a broad area of application. It is somewhat difficult to order them into categories since there are always overlaps. However, some main categories can be distinguished. Table 2.1 provides an overview of ion generation methods described in this chapter. The reader should not be discouraged by the complexity of the subject, but instead read selectively those portions that may be of particular interest at the moment.

2.1.1. Gas Discharge

Although their nature was not understood at the time, the first record of ions produced is from 1886, when Goldstein discovered that for gas discharge tubes with a perforated cathode a glow could be observed at the cathode end [1]. Wien [2] and Thomson [3] both explored the gas discharge further and eventually Thomson constructed the first mass spectrograph using a gas discharge ion source [4]. At the time the typical gas discharge source consisted of an anode and a cathode located in a glass tube filled with gas at a pressure of about 1 torr or less. When applying a high voltage potential between the electrodes a discharge is obtained and the gas is ionized. Hence, both rays of electrons and gaseous ions are produced and can be detected with proper arrangement. Glow discharge (GD) and inductively couple plasma (ICP) sources are, in principle, developments of the gas discharge source (see Sections 2.1.4 and 2.1.5).

2.1.2. Thermal Ionization

In thermal ionization mass spectrometry (TIMS, also referred to as surface ionization) ions are created by electrical heating of one or more metal filaments. TI is one of the earliest ionization techniques and dates back to 1906 when Gehrcke and Reichenheim produced Na^+ ions by heating sodium salt anodes in a discharge tube [5]. It was first used in mass spectrometry by Dempster in his scanning magnetic sector instrument 1918 [6]. In 1953 Inghram and Chupka introduced a triple filament source [7], a still common arrangment (Fig. 2.1). Here the sample is deposited onto one of the outer filaments, which is heated to produce a neutral vapor that is directed towards the much hotter central filament. Upon impact onto the hotter filament the analyte is ionized. This way the evaporation and ionization are decoupled, which allows for better-controlled experiments. The ionization efficiency may be increased several magnitudes compared to a single filament source. Thus, elements with ionization potential greater

TABLE 2.1. Overview of the Ion Generation Methods Described in this Chapter

Method	Acronym	Category	Ion type	Applications[a]
Gas discharge	–	Discharge	Atomic ions	First ionization mechanism to be used in MS
Thermal ionization	TI	Ionization by heating	Atomic ions	Isotope ratio, Trace analysis; Solid samples
Spark source	SS	Discharge	Atomic ions	Trace analysis in solid samples
Glow discharge	GD	Plasma source	Atomic ions	Trace analysis
Inductively coupled plasma	ICP	Plasma source	Atomic ions	Isotope ratio; Trace analysis
Electron ionization	EI	Electron induced ionization	Volatile molecular ions	Smaller molecules; GC-MS; Extensive libraries
Chemical ionization	CI	Electron induced ionization	Volatile molecular ions	GC-MS
Atmospheric pressure chemical ionization	APCI	Electron induced ionization	Nonvolatile molecular ions	Smaller molecules; LC-MS
Photoionization	PI	Photoionization	Volatile molecular ions	Smaller molecules; GC-MS
Multiphoton ionization	MPI	Photoionization	Atomic and molecular ions	Resonance-enhanced MPI is highly selective; Trace analysis
Atmospheric pressure photoionization	APPI	Photoionization	Nonvolatile molecular ions	LC-MS; Nonpolar compounds
Field ionization	FI	Ionization by strong electric field	Volatile molecular ions	Molecular compounds
Field desorption	FD	Desorption/ionization by strong electric field	Nonvolatile molecular ions	First soft method; Large molecules
Thermospray ionization	TSI	Spray	Nonvolatile molecular ions	LC-MS
Electrospray ionization	ESI	Spray	Nonvolatile molecular ions	Soft method, LC-MS; Large molecules

(*Continued*)

TABLE 2.1. *Continued*

Method	Acronym	Category	Ion type	Applications[a]
Desorption electrospray ionization	DESI	Spray	Nonvolatile molecular ions	Direct, preparation-free analysis of samples
Direct analysis in real time	DART	Discharge	Nonvolatile molecular ions	Direct, preparation-free analysis of samples
Secondary ion (mass spectrometry)	SIMS	Particle induced desorption/ ionization	Nonvolatile molecular ions	Semiconductors; Surface analysis; Imaging
Fast atom bombardment	FAB	Particle induced desorption/ ionization	Nonvolatile molecular ions	Soft method; Large molecules
Plasma desorption	PD	Particle induced desorption/ ionization	Nonvolatile molecular ions	Soft method; Large molecules
Laser desorption/ ionization	LDI	Photon induced desorption/ ionization	Nonvolatile atomic and molecular ions	Isotope ratio; Trace analysis
Matrix-assisted laser desorption/ ionization	MALDI	Photon induced desorption/ ionization	Nonvolatile molecular ions	Soft method; Large molecules
Atmospheric pressure matrix-assisted laser desorption/ ionization	AP-MALDI	Photon induced desorption/ ionization	Nonvolatile molecular ions	Soft method; Large molecules

[a]The "applications" column does not cover all applications, but some examples.

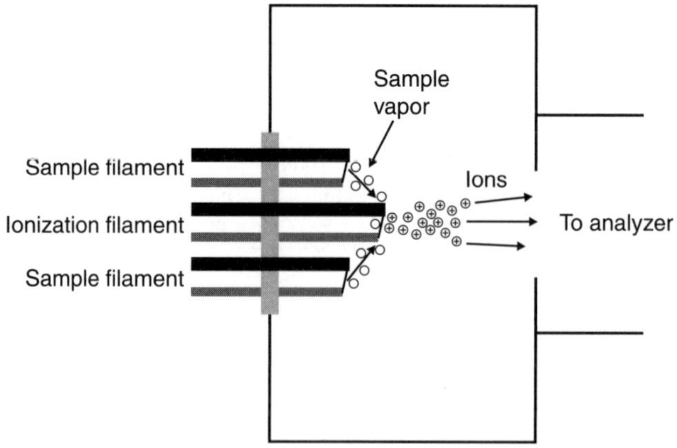

Figure 2.1. Schematic of a thermal ionization (TI) source. Each filament consists of two pins connected by a wire.

than the work function can also be effeciently ionized. The ionization efficiency depends strongly on the chemical and physical properties of the filament surface and it is important to minimize contaminations. Both positive and negative ions can be formed, depending on the analyte.

TI is a very precise and accurate method in stable isotope ratio measurements and quantification of inorganic elements, for example, by isotope dilution mass spectrometry [8]. Because TI is a continuous ion source, it could be coupled to any analyzer that is suitable for such sources. However, because the strength of TI lies in the quantitative precision and accuracy, sector analyzers are preferred to ensure maximum quality.

Since about 1990, however, inductively coupled plasma (ICP, see Section 2.1.5) has become increasingly popular at the expense of TI in this area of application [9]. Although TI can provide better results for some analyses, ICP is more versatile and requires less sample preparation effort. Moreover, the advantage of better precision for TI is often compromised by the sample, for example, sample inhomogeneity. Nevertheless, there are still many examples where TI is used, such as for isotope analysis [10–13] and geochronology [14].

2.1.3. Spark Source

In the 1930s Dempster introduced the spark source (SS, also referred to as spark ionization or vacuum spark) to analyze isotopes of metals, a class of compounds that could not be ionized by TI [15, 16]. The ions produced in a spark source have wide initial energy distributions (several kiloelectronvolts), so double focusing (see Section 2.2.2) is necessary to obtain sufficient resolution. The spark also produces an ion current that fluctuates heavily, making operation difficult. Gorman et al. [17] introduced an electrical recording device that included a monitoring collector located prior to the analyzer entrance. This way the measured intensity after m/z separation could be intensity calibrated, which allowed for quantitative analysis. The take-off for spark source mass spectrometry (SSMS) came in 1954 when Hannay introduced an instrument for semiconductor analysis capable of detecting sub-parts per million levels of impurities [18–20].

The SS is useful for analysis of solid samples. In the most common type of SS a vacuum spark discharge is generated between two electrodes by a pulsed high voltage (some tens of kilovolts) radio frequency (RF) potential. The electrode ends constitute the sample to be analyzed (Fig. 2.2). If the sample consists, for example, of a powder, an electrode can be formed by pressing it to the desired shape. Electron impact processes in the discharge plasma causes evaporation, atomization, and ionization of the sample material. Mainly singly charged atomic ions are produced. The ions are accelerated through a high voltage field towards the exit aperture of the source and then into the analyzing region of the mass spectrometer. Applications for SSMS include multielement trace analysis of conducting and semiconducting materials and of nonconducting materials, such as geological samples. Being relatively demanding, the spark source has generally been succeeded by plasma ionization methods (see Sections 2.1.4 and 2.1.5). For a comprehensive review of the SS and several other ion sources used in MS analysis of inorganic samples, see Reference 21.

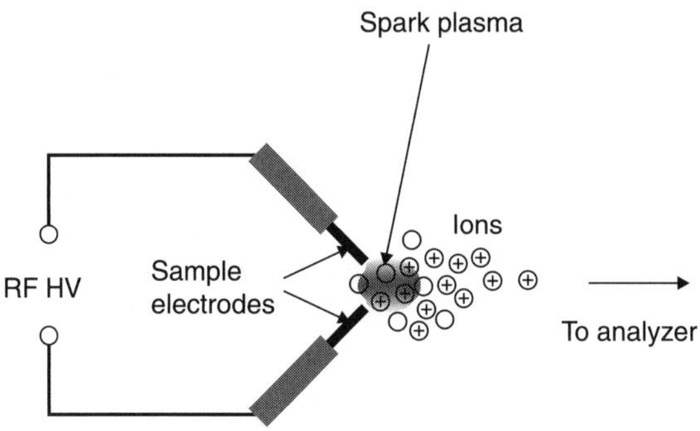

Figure 2.2. Schematic of a spark ion source.

2.1.4. Glow Discharge

Glow discharge mass spectrometry (GDMS) was introduced in 1974 by Harrison and Magee [22]. A GD source can be used also by analysis techniques other than MS, for example, atomic absorption, atomic emission, and atomic fluorescence [23]. The GD ion source consists of a metal pin cathode, containing the sample, mounted on a removable probe, and a stainless steel anode that surrounds the sample cathode like a housing (Fig. 2.3). The anode and cathode are separated by an insulator. The discharge gas, typically high purity argon, enters the source through a needle valve in the anode, and the pressure is kept at about 1 torr. The most commonly used GD source is the direct current (DC) type. When the applied voltage between the electrodes is high enough, a discharge occurs. Under certain conditions of pressure, voltage, and current, a brilliant discharge is obtained. In the case of Ar as discharge gas, Ar^+ ions are formed in the glow

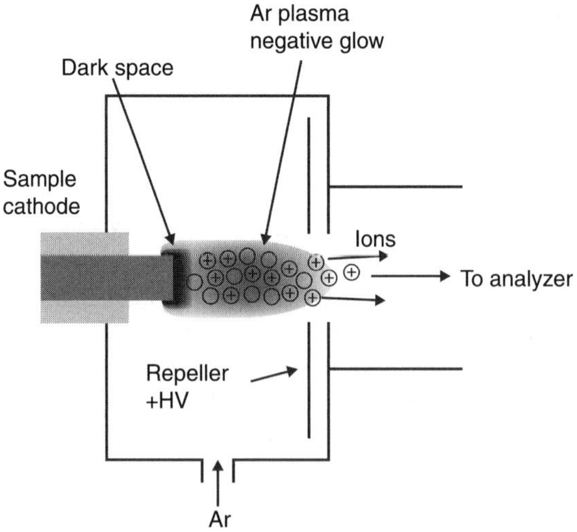

Figure 2.3. Schematic of a glow discharge (GD) ion source.

discharge. The Ar^+ ions are accelerated towards the sample cathode where they cause sputtering of the sample material. This sputtering process is similar to that of secondary ion mass spectrometry (SIMS) and fast atom bombardment (FAB; see Sections 2.1.18 and 2.1.19). Positively charged analyte species emitted from the cathode surface will immediately return (the negative species are accelerated towards the anode). Hence, only neutral species will diffuse out into the negative glow region, where they will be subject to ionization processes. Penning ionization is the principal mechanism, but also electron ionization and charge exchange processes are involved. The positively charged analyte ions exiting the source are those formed close to the exit aperture that are swept out by the flowing gas.

In many elemental mass spectrometric methods atomization and ionization occur in one step (e.g., in thermal ionization [TI], laser desorption/ionization [LDI], and SIMS). This leads to severe matrix effects, limiting the quantitative information extractable [24]. In the GD the atomization occurs already at the surface of the cathode while the ionization occurs in the negative glow region. This decoupling decreases the matrix effects and the composition of the cathode surface will be better reflected (secondary neutral mass spectrometry is another technique where vaporization and ionization are decoupled, see Section 2.1.18). Disadvantages with GD are that several minutes are required to achieve equilibrium so sample throughput may be low, and that significant contamination of the instrument may occur, necessitating extensive cleaning for successful trace analysis.

GD is a continuous source and suitable analyzers are magnetic sectors and quadrupoles, although trapping analyzers and time-of-flights (TOFs) can also be utilized. Its main application is multielement trace analysis of high purity solid conductors and semiconductors. Detection of sub-parts per billion levels are possible. Nonconducting solids can also be analyzed but with more difficulty, because of charge build-up in the DC GD source. Such samples can be mixed with conducting material like in SSMS or an RF GD source can be employed. Because other ion sources, such as inductively coupled plasma (ICP; see Section 2.1.5), offer less experimental complexity the use of GD is presently relatively limited. For a comprehensive review of GD, see References 23 and 25, and for GD as well as several other ion sources used in MS analysis of inorganic samples, see Reference 21.

2.1.5. Inductively Coupled Plasma

Analytically useful inductively couple plasma (ICP) mass spectra were obtained by Houk et al. in 1978 [26]. Prior to this achievement Gray had shown in 1974 that mass spectra could be obtained from a DC plasma [27]. Like the GD, ICP was first coupled to analyzing techniques other than MS, namely atomic emission spectrometry. At present ICP-MS is the most popular mass spectrometric technique for analysis of inorganic compounds, and applications include isotope ratio measurements as well as trace element analysis. The techique is sensitive and versatile. Detection limits are in the 1 to 100 pg/L range and down to sub-picogram amounts can be analyzed. An atmospheric pressure interface allows for easy interfacing with, for example, liquid chromatography. The precision is also high, making it an excellent method for obtaining quantitative data.

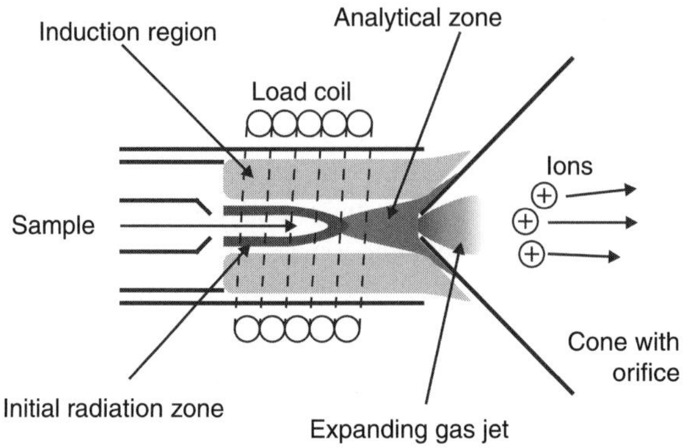

Figure 2.4. Schematic of an inductively coupled plasma (ICP) ion source.

A typical ICP source consists of a torch and load coil into which the analyte aerosol enters (Fig. 2.4). The plasma is fed by RF energy applied to the load coil. The energy is transferred by the electromagnetic field to the outer portion of the plasma, which has a toroidal shape. The analyte is carried via an argon flow in the axial direction through the central part of the torus, which is heated by energy from the outer region. The separation of the domain through which the analyte is flowing and the domain where the energy is added means that the type of analyte has little effect on the processes sustaining the plasma. Moreover, since no electrodes are in contact with the plasma no (or little) contamination of those will occur. The challenge with ICP-MS is to transfer the analyte ions from the 5000 K plasma at atmospheric pressure to the low pressure mass spectrometric analyzer. This is achieved by letting the plasma flow around the tip of a water-cooled metal cone with an orifice of less than 1 mm diameter. Behind is a skimmer with another orifice, located so that as many analyte ions as possible can enter the second vacuum chamber. Here the pressure is low enough to allow the use of steering and focusing devices so the beam can be transported to the analyzer with minimum intensity loss.

The analyte spends about 2 ms in the plasma and is efficiently atomized and, in most cases, ionized. The ionization efficiency can be estimated and more than 50 elements are ionized with more than 90% efficiency [28]. The main part of the ions are singly charged, although a few elements will not be ionized since they have higher ionization potentials than argon, and another few will to some extent be doubly charged due to their low second ionization potentials. For some elements oxide ions are also produced. Argon can be replaced by helium to provide better ionization efficiency for elements with high ionization potential, such as halogens [29]. Being a continuous ion source the preferred analyzers for ICP are magnetic sectors and quadrupoles.

There are several sample introduction methods that are used in conjunction with ICP, including nebulization, electrothermal evaporation, gas chromatography, hydride generation, and laser ablation [30]. Laser ablation combined with ICP (LA-ICP) is useful for analysis of solids. In such a source the sample is positioned in a chamber prior to the ICP source, the ablation cell. Argon gas at atmosperic pressure flows through the cell towards the ICP source. The sample is irradiated by a laser beam and the ablated material is transported with the argon stream to the plasma, where the analyte species are ionized and subsequently separated by the mass analyzer. For a comprehensive review of ICP and several other ion sources used in MS analysis of inorganic samples, see Reference 21. See also Chapter 9 for another application.

2.1.6. Electron Ionization

Electron ionization (EI) was introduced in 1921 by Dempster, who used it to measure lithium and magnesium isotopes [31]. Modern EI sources are, however, based on the design by Bleakney [32] and Nier [33, 34], who both worked in Prof. J. T. Tate's laboratory. In EI ions are produced by directing an electron beam into a low pressure vapor of analyte molecules.

The EI source consists of a chamber with some openings (Fig. 2.5). Analyte molecules are introduced directly into the source. The electron beam is created by heating a filament, and the beam is directed through the source and afterwards collected in a trap. Magnets provide for a spiral motion of the electrons so that the path is increased and thus the chance for electron–molecule reactions. Because the electron mass is significantly smaller than any ion mass the latter will not be affected by the magnetic field strength required. Upon interaction one (occasionally more) electron is removed from the analyte creating typically positive radical molecular ions. If the analyte has high electron affinity negative ions can be formed through electron attachment, although the utility in negative ion mode is limited. For positive ions the electron energy is in most cases set to

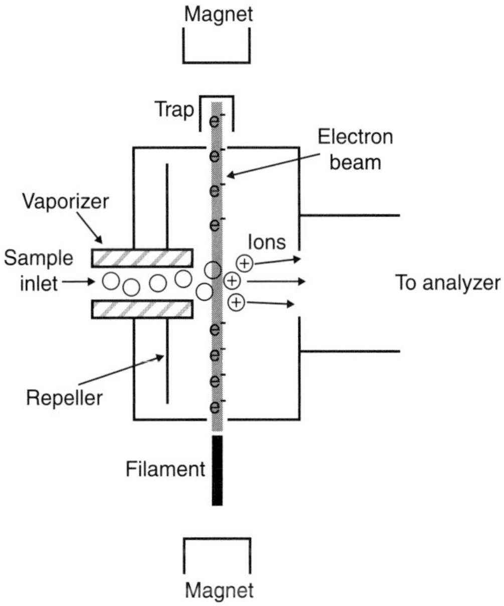

Figure 2.5. Schematic of an electron ionization (EI) source.

70 eV, a value that is quite close to the maximum cross section for most molecules [35]. The molecular ion intensity will be close to maximum and at the same time fairly intense fragment ions are obtained that can give structural information.

Normally the sample input is regulated such that ion–molecule reactions are kept at a minimum. Even if extensive fragmentation often is undesired, an advantage is that compound fingerprints can be recorded. Moreover, since the EI fragment patterns are relatively reproducible and instrument independent, large spectral libraries have been created that can be used for compound identification. Fragmentation can be more or less eliminated by lowering the electron energy, but the ionization efficiency is also reduced severely. EI is a continuous source and is therefore suitable with analyzers such as quadrupoles and magnetic sectors, but other analyzers are used as well. EI is commonly utilized in GC-MS analysis of organic compounds. See Chapter 5 for a more detailed description and Chapters 8 and 16 for additional application examples.

2.1.7. Chemical Ionization

Chemical ionization (CI) was introduced by Munson and Field in 1966 by allowing a reagent gas into an EI source [36–38]. The pressure is typically ~1 torr and the electron beam is more energetic than in EI, typically 500 to 1000 eV. While in EI the analyte is directly ionized by the electron beam, CI is a two-step process, where first reagent gas molecules are ionized by the electron beam and thereafter the reagent gas ions transfer charge to the analyte. Originally CI was employed for positive ion analysis, before negative ion analysis was introduced by Hunt et al. in 1976 [39]. The reaction gas is of a type that is nonreactive with itself, but undergoes an exoergic reaction with the sample molecules. Initially methane was used to produce stable reagent ions, which in turn react with the analyte forming different types of molecular or fragment ions. The choice of reaction gas depends on application (and ion analysis mode). The advantage over EI is that it is softer so intact molecular ions are easier to obtain.

Although most analytes can form positive ions, negative ion formation is selective and can be very efficient, providing very high sensitivity. This is exploited, for example, in detection of trace amounts of strong electrophores by GC-negative ion-CI-MS. A disadvantage with CI compared to EI is that the ion source requires cleaning more often. For a comprehensive review of CI-MS see Reference 40 and for analysis of negative ions see Reference 41. CI is also assumed to play a major role in the plume chemistry and analyte ion formation in MALDI (see Section 2.1.22). The same type of analyzer choice as for EI applies for CI.

2.1.8. Atmospheric Pressure Chemical Ionization

Atmospheric pressure chemical ionization (APCI) was introduced in 1973 by Horning et al. [38, 42, 43] and coupled to GC. This is also the introduction of atmospheric pressure ionization (API) in general. The next year corona discharge was introduced for ion generation as well as successful coupling to LC [44, 45]. In APCI of a liquid, a pneumatic nebulizer induces the flow of liquid to form a spray at atmospheric pressure. The spray droplets pass a corona discharge electrode situated close to the orifice, which is the inlet to the vacuum region of the mass spectrometer. Being a continuous ion source it is readily combined with mass analyzers, such as magnetic sectors, quadrupoles, trapping analyzers, and oa-TOFs. APCI has to a great extent replaced low pressure CI and is frequently available with instruments that provide ESI. APCI is better suited than ESI for compounds that are less polar and have lower mass (see Section 2.1.15). Moreover, APCI allegedly has better tolerance to salts and buffer compounds than ESI. See, for example, Chapter 8 for an application of APCI.

2.1.9. Photoionization

Photoionization (PI) is the term used for the photoelectric effect when it occurs in the gas phase. The photoelectric effect was discovered by Hertz [46] and later explained by Einstein [47]. The phenomenon is, for example, the cause of aurora borealis (the northern light). A PI source is in principle an EI source, but with a photon beam instead of the electron beam. The first experiments with one-step photoionization in combination with mass analysis were peformed by Ditchburn and Arnot [48], who ionized potassium using light produced by an iron arc. By employing a monochromator to the light generated by a discharge lamp, a monoenergetic UV photon beam could be generated in vacuum [49]. Thus, it became possible to measure ionization yield as a function of photon energy. PI was introduced as a detection method for GC in the mid-1970s [50]. It has also been utilized with liquid chromatography [51] and ion mobility spectrometry [52]. Selective ionization of the analyte is obtained if the ionization energy of the analyte is lower than that of the carrier medium and the lamp is chosen so the photon energy is in between. A recent application is analysis of flame chemistry using a synchrotron as a tunable photon source [53].

2.1.10. Multiphoton Ionization

By employing a laser for the photoionization (not to be confused with laser desorption/ ionization, where a laser is irradiating a surface, see Section 2.1.21) both sensitivity and selectivity are considerably enhanced. In 1970 the first mass spectrometric analysis of laser photoionized molecular species, namely H_2, was performed [54]. Two years later selective two-step photoionization was used to ionize rubidium [55]. Multiphoton ionization mass spectrometry (MPI-MS) was demonstrated in the late 1970s [56–58]. The combination of tunable lasers and MS into a multidimensional analysis tool proved to be a very useful way to investigate excitation and dissociation processes, as well as to obtain mass spectrometric data [59–62]. Because of the pulsed nature of most MPI sources TOF analyzers are preferred, but in combination with continuous wave lasers quadrupole analyzers have been utilized [63]. MPI is performed on species already in the gas phase. The analyte delivery system depends on the application and can be, for example, a GC interface, thermal evaporation from a surface, secondary neutrals from a particle impact event (see Section 2.1.18), or molecular beams that are introduced through a spray interface. There is a multitude of different source geometries.

In resonance-enhanced multiphoton ionization (REMPI, also commonly referred to as resonance ionization—RI) near-UV photons can be used for ionization [60]. When the first photon is absorbed the molecule will be excited to a real resonant intermediate eigenstate and the second photon will cause ionization. This process not only leads to higher ionization efficiency than for nonresonant excitation, but it is also highly selective, allowing for spectroscopic studies of neutral molecules and clusters. See References 63 and 64 for reviews of REMPI.

One method to study energy-selected ions is threshold ionization, in which ions with precisely defined energy contents are produced. These ions can then be used to study unimolecular fragmentation, ion–molecule reactions, van der Waals clusters, and hydrogen-bonded clusters [62].

2.1.11. Atmospheric Pressure Photoionization

After the initial development by Revel'skii et al. [65, 66], atmospheric pressure photoionization (APPI) was introduced generally by Robb et al. in 2000 [67], which is also the first example of photoionization in conjunction with LC-MS. APPI is useful particularly for nonpolar compounds, which may be practically impossible to ionize by ESI or APCI. Both positive and negative ions can be analyzed. Two types of APPI are being practiced, direct ionization [68] and ionization through a photoionizeable dopant, such as toluene or acetone, that is added in amounts greatly exceeding that of the analyte [67, 69]. Ionization of analyte molecules in LC-APPI-MS is quite inefficient since the solvent tends to deplete the emitted photons. The dopant increases the efficiency by first becoming photoionized itself and then transferring the charge to the analyte. The effect is apparently dependent on the illuminance of the lamp—with greater illuminance the improvement caused by adding dopant is reduced [68]. APPI sources are available as an alternative ion source for instruments designed for ESI and APCI analysis. For a review of APPI see Reference 70.

2.1.12. Field Ionization

Field ionization (FI) was discovered by E. W. Müller, who observed the generation of positive ions of gas phase atoms or molecules near a metal surface in a high electrostatic field and exploited the phenomenon to construct the field ion microscope [71]. In 1954 Inghram and Gomer coupled an FI source to a mass spectrometer [72]. In FI ionization is achieved by exposing analyte molecules to a very high electrostatic field. Typical field strengths are in the 10^7 to 10^8 V/cm range, which is at least 1000 times higher than the typical field strength of, for example, a SIMS or MALDI ion source. This is accomplished by using, for example, a thin tungsten wire that has been activated by benzonitrile [73, 74]. The activated wire has protrusions that increase the field strength close to the wire. Typically FI is considerably less sensitive than EI, but produces a larger relative amount of intact molecular ions, that is, it is a "soft" ionization technique. Hence, the EI and FI techniques are complementary, and some instruments have been equipped with both ionization techniques. Coupling of an FI source with GC was made by Damico and Barren [75], as well as Beckey et al. Since FI is a continuous ion source it is readily combined with quadrupole or magnetic sector analyzers.

2.1.13. Field Desorption

Field desorption (FD) was introduced by Beckey in 1969 [76]. FD was the first "soft" ionization method that could generate intact ions from nonvolatile compounds, such as small peptides [77]. The principal difference between FD and FI is the sample injection. Rather than being in the gas phase as in FI, analytes in FD are placed onto the emitter and desorbed from its surface. Application of the analyte onto the emitter can be performed by just dipping the activated emitter in a solution. The emitter is then introduced into the ion source of the spectrometer. The positioning of the emitter is crucial for a successful experiment, and so is the temperature setting. In general, FI and FD are now replaced by more efficient ionization methods, such as MALDI and ESI. For a description of FD (and FI), see Reference 78.

2.1.14. Thermospray Ionization

In 1983 thermospray ionization (TSI) was introduced by Blakley and Vestal as a means to couple LC at conventional flow rates ($\sim$1 mL/min) to mass spectrometry [79]. Contrary to earlier LC-MS methods, TSI was successful in forming gas-phase ions of nonvolatile molecules. The liquid is sprayed when emerging from a heated metal capillary and subsequently entering a low pressure chamber with heated gas. Due to the hot gas, the small droplets in the resulting supersonic jet continue to evaporate and gas-phase ions are created. The method requires charged or polar analyte compounds as well as volatile buffers. The vaporizer temperature is critical and dependent on the solvent composition. Being a continuous ion source, TSI is used with quadrupole mass filter and magnetic sector analyzers. TSI is now replaced by more stable and more sensitive ionization techniques such as ESI and APCI. Extensive reviews of the technique and applications are given in References 80 and 81.

2.1.15. Electrospray Ionization

Electrospray ionization (ESI) was first introduced by Dole and coworkers in 1968 [82] and coupled to MS in 1984 by Yamashita and Fenn [83]. In ESI the sample is dissolved in a polar, volatile solvent, and transported through a needle placed at high positive or negative potential (relative to a nozzle surface) [83–85]. The high electric potential (1 to 4 kV) between the needle and nozzle causes the fluid to form a Taylor cone, which is enriched with positive or negative ions at the tip. A spray of charged droplets is ejected from the Taylor cone by the electric field. The droplets shrink through evaporation, assisted by a warm flow of nitrogen gas passing across the front of the ionization source (Fig. 2.6). Ions are formed at atmospheric pressure and pass through a cone-shaped orifice, into an intermediate vacuum region, and from there through a small aperture into the high vacuum of the mass analyzer. ESI has been used in conjunction with all common mass analyzers. The exact mechanism of ion formation from charged droplets has still not been fully elucidated and there are different theories proposed [82, 86, 87].

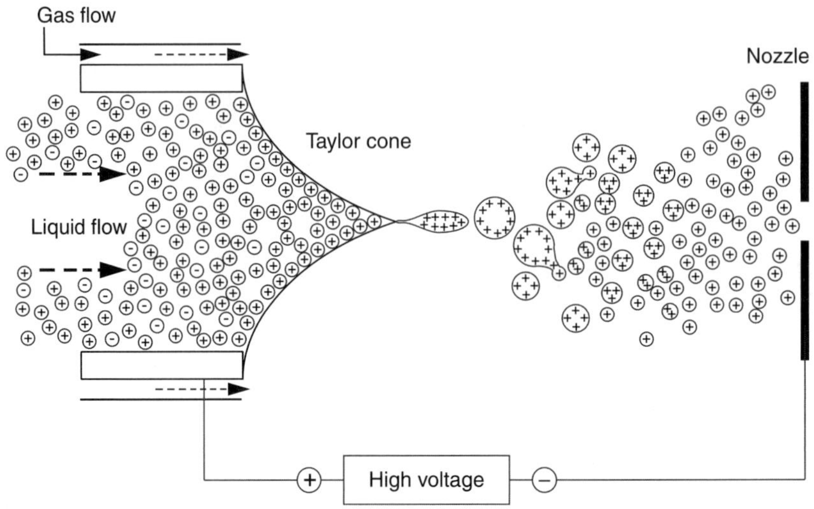

Figure 2.6. Schematic of an electrospray ionization (ESI) source. Reprinted from A. Westman-Brinkmalm and G. Brinkmalm (2002). In *Mass Spectrometry and Hyphenated Techniques in Neuropeptide Research*, J. Silberring and R. Ekman (eds.) New York: John Wiley & Sons, 47–105. With permission of John Wiley & Sons, Inc.

Sample preparation requires only dissolution of the sample to a suitable concentration in a mixture of water and organic solvent, commonly methanol, isopropanol, or acetonitrile. A trace of formic acid or acetic acid is often added to aid protonation of the analyte molecules in the positive ionization mode. In negative ionization mode ammonia solution or a volatile amine is added to aid deprotonation of the analyte molecules.

The sensitivity of ESI-MS is good, with low femtomole or attomole detection levels for many peptides. However, the sensitivity of ESI is a function of the concentration of the injected sample. High flow rates, that is, 1 to 1000 μL/min in conventional ESI-MS, result in high sample consumption. It is therefore advantageous to use the lowest possible flow rate. Nano-ESI (or nanospray) is a low-flow-rate (20 to 200 nL/min) version of ESI [88, 89], with lower sample consumption and considerably higher sensitivity. Nano-ESI has also been shown to be more tolerant to salts than conventional ESI [88]. Subattomole levels of analyte have been detected by coupling capillary electrophoresis (CE) to ESI (see Section 2.2.6). ESI-MS can be used for analysis of polar molecules ranging from less than 100 Da to a whole virus with mass >2 MDa [90] and even a 100 MDa single DNA ion [91]. An important feature of ESI is the capability to generate a distribution of multiply charged ions, which allows the analysis of very large proteins using mass analyzers with limited m/z range. Relatively small changes in analysis conditions, such as pH, solvent composition, salt concentration, and partial denaturation of the analyte molecules, can alter the charge state distribution of a large molecule [92, 93]. The complex pattern of multiple-charged ions makes interpretation of ESI spectra from complex mixtures difficult and in practice computers are used to transform the charge state envelopes to single peaks at the respective molecular mass (or zero charge state).

ESI is an extremely gentle ionization method, accompanied by very little fragmentation of the formed molecular ions. Consequently, weak bonds are often preserved and analysis of intact post-translationally modified peptides/proteins and noncovalently bound complexes, such as protein–ligand complexes, can be successfully performed with ESI-MS [94–99]. Even though fragments are seldom produced in ESI, the ions generated are especially favorable for collision induced dissociation (CID) (see Chapter 3), because the high charge state of the molecular ions increases the energy available for the collision event [100]. Analyte signal suppression caused by charge competition between electrolytes and, for example, other analytes, is a major problem in ESI and may in practice prevent thorough analysis of complex mixtures if chromatographic prefractionation is not applied. These charge competition phenomena as well as the analyte signal's strong dependence on experimental conditions, such as pH, solvent composition, and salt concentration, make it risky to draw quantitative conclusions from ESI-MS data. However, as in MALDI-MS, quantification can be achieved, within a limited concentration range, by using a carefully chosen internal calibrant of known quantity and close chemical resemblance to the peptide/protein of interest [101]. The combination of ionization at atmospheric pressure and the continuous flow of solvent used in ESI allow direct coupling with separation techniques, such as liquid chromatography (LC; see Chapter 4) and capillary electrophoresis (CE; see Chapter 4). See also Chapters 8, 10, and 12 for some examples of applications.

2.1.16. Desorption Electrospray Ionization

Desorption ESI (DESI) was introduced by Takátz et al. [102]. The phenomenon actually was observed earlier but was discarded as a nuisance (e.g., an analyte or calibration mixture that coated the entrance of the transfer capillary and contributed to undesired peaks in the spectra). The idea of using the electrospray for desorption is as clever as it is simple. The method is sensitive and large species such as proteins can be detected. The ions observed are more or less the same as with regular ESI.

A DESI source consists of a spray capillary and a coaxial capillary providing the nebulizer gas. High voltage is applied to the spray needle, which is directed towards the target surface (Fig. 2.7). Sample species are then desorbed and will subsequently enter the orifice to the mass spectrometer. Normal distances between the spray needle, sample, and orifice range from some millimeters to several centimeters. The optimum geometry depends on the sample and on the size of the desired sampling area. The advantages with DESI are that the target can be in principle any type of surface and that the analysis time often can be very short, on the order of seconds. This means that rapid analyses can be performed without the need for sample preparation. A sample, such as a dollar bill, tomato, or tablet can be placed close to the spectrometer inlet and after a few seconds of spraying a spectrum is recorded. There are numerous applications, including high-throughput analysis, screening for trace levels of drugs, explosives, pesticides, and contaminations. DESI also has a

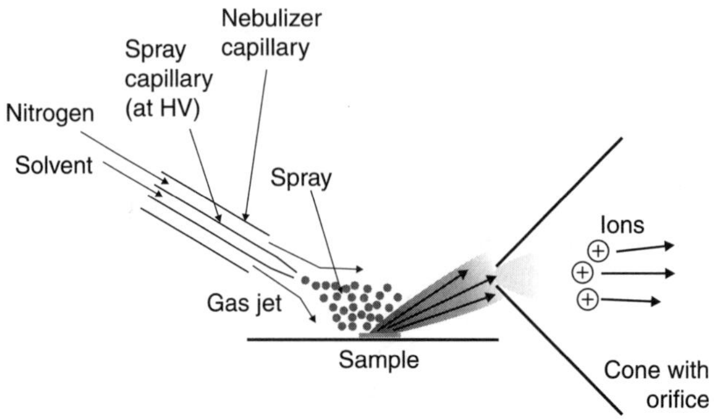

Figure 2.7. Schematic of a desorption electrospray ionization (DESI) source.

potential for chemical imaging [103]. See References 104 and 105 for a comprehensive description of DESI.

There are some variants that have emerged in the wake of DESI. By replacing the electrospray emitter by a metal needle and allowing solvent vapor into the coaxial gas flow desorption APCI (DAPCI) can be performed [106]. Other versions are atmospheric-pressure solids analysis probe (ASAP) where a heated gas jet desorbs the analyte, which is subsequently ionized by a corona discharge [107], and electrospray-assisted laser desorption/ionization (ELDI) where a laser ablates the analyte and charged droplets from an electrospray postionizes the desorbed neutrals [108].

2.1.17. Direct Analysis in Real Time

Direct analysis in real time (DART) was introduced in 2005 by Cody et al. [109]. Similarly to DESI samples can be analyzed directly without preparation. While analytes in the small protein range can be analyzed by DESI, DART is in practice limited to analysis of ions in the region below 1 kDa or slightly above. However, as in SIMS, for example, fragment spectra of larger componds can be acquired, which can provide useful information, although the molecular ion cannot be observed.

In DART source gas, for example, helium or nitrogen, enters a discharge chamber. The discharge is initiated by applying a high voltage between the electrodes in the chamber and a plasma is created containing charged and excited species. The stream can then be manipulated by electrodes, for example, to remove unwanted species, and heated if desired. After exiting the source the flow can be directed towards a sample and desorbed ions will enter the mass spectrometer through an orifice. Typically the distance between the source outlet, the sample, and the orifice is about 5 to 25 mm but ions have been observed with the source as far away as 1 m, so the positioning is not critical. Being a continuous source, DART is suitable for coupling to the same analyzers as ESI, that is, quadrupoles, magnetic sectors, Qq-TOFs, and trapping analyzers. The applications are more or less the same as for DESI (see Section 2.1.16). See Reference 110 for a comparison of DART to DESI and DAPCI.

2.1.18. Secondary Ion Mass Spectrometry

Secondary ion mass spectrometry (SIMS) has quite a long historical record, although the term SIMS was not coined until 1970. Ejection of neutrals and ions from a surface as a result of ion bombardment was first observed by J. J. Thomson as early as 1910 [111]. In 1949 Herzog and Viehböck constructed an ion source to a secondary ion mass spectrometer [112], while the first practical instrument was developed by Honig in the 1950s [113–115]. In the early 1960s Castaing and Slodzian introduced a secondary ion microscope [116]. Further development in the area was carried out by Liebl, who designed an ion microprobe in 1967 [117]. In the mid-1970s Benninghoven et al. obtained the first secondary ion mass spectra of amino acids [118]. Since then the technique has continued to develop with new primary ion sources and refined components into today's powerful chemical imaging, depth profiling, and surface analysis technique. It is even possible to analyze compounds up to about 10 kDa, although the sensitivity is normally poor for species >1 kDa. For most applications the mass range of interest is rather <300 Da. Larger species can nevertheless be probed by fragment signatures, which often provide sufficient information.

The primary ion impact is believed to start a cascade of collisions between the impacting particle and the atomic nucleii in the sample, resulting in ejection of neutral molecules and ions through so-called sputtering (Fig. 2.8) [119]. The SIMS

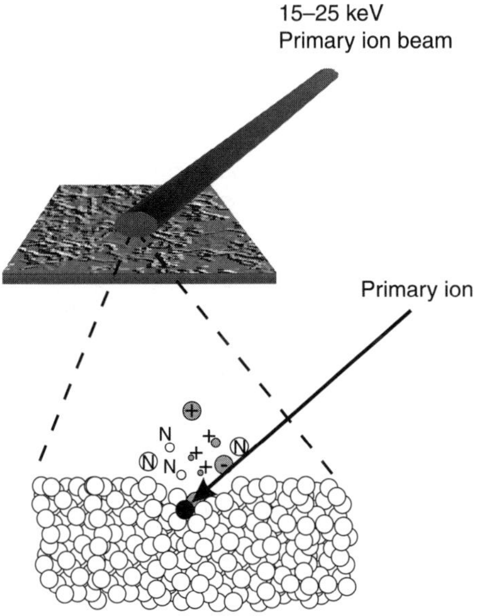

Figure 2.8. Schematic of a sputtering event in secondary ion mass spectrometry (SIMS). Mainly neutral species are ejected, but also some positively charged and negatively charged ions. For samples containing analyte with relatively low masses, intact molecular ions can be desorbed. The greater portion of ejected compounds is, however, fragments.

technique can be divided into two major areas of application, static SIMS where the objective is to analyze the surface of a sample, and dynamic SIMS where the sample depth profile is anayzed. In static SIMS the irradiation of the sample is kept at a low level (below a total dose of $\sim 10^{13}$ ions/cm^2) so that surface damage can be neglected. The TOF analyzer has proven to be well suited for static SIMS, especially for imaging, a technique that can offer submicrometer spatial resolution [120]; see also Chapter 13. Dynamic SIMS, on the other hand, exploits the fact that material is desorbed only from the surface. Here, a pit is literary dug in the sample in order to analyze the material as a function of depth. In dynamic SIMS most molecular information is lost, but atomic and small molecular species can be detected. Since dynamic SIMS is considerably faster with a continuous primary ion beam than with a pulsed ion beam, quadrupoles and magnetic sectors are the analyzers of choice. Major applications of static SIMS include microelectronics, materials science, polymers, particles, and life sciences, while applications for dynamic SIMS include semiconductors, metallurgy, geochronology, and biology. Comprehensive reviews of SIMS, including applications, are given in References 121–123. The development of SIMS is described in References 114 and 115.

In SIMS the analyte ions (i.e., the secondary ions) are created by letting a pulsed or continuous beam of primary ions impinge on the sample surface (Fig. 2.8). The primary ions are typically singly charged with kinetic energy in the range of a few to tens of kiloelectronvolts, but there are ion sources that operate below 1 keV. Commercial instruments are often supplied with power sources that can generate a potential of 25 to 30 kV, which defines the upper limit on the available particle energy. Generally the secondary ion yield increases with increasing primary ion mass, which means better sensitivity and shorter analysis time. The drawback can be technical difficulties, compromised lateral and mass resolution. TOF and magnetic sector analyzers can offer mass resolving powers exceeding 10,000, which is sufficient to resolve many isobaric compounds below 100 Da. The best lateral resolution attainable is <50 nm but is highly sample dependent. Usually the best results are obtained from flat conductors or semiconductors while organic substrates are more difficult.

There are a number of different primary ion beams to select from, depending on applications and demands. In liquid metal ion guns (LMIG) Ga^+ and In^+ are typical primary ions. They do not give high secondary yields but can be focused to a narrow diameter, which is desirable in chemical imaging. Cs^+ ions give higher yield, especially for electronegative compounds, and is common for sputter cleaning and depth profiling of negative species. O_2^+ ions enhance the yield significantly for electropositive compounds, for example, many metals. Fairly recently molecular and cluster beams have been introduced. Primary ion beams such as SF_5^+, Bi_n^+, Au_n^+, and C_{60}^+ all give enhanced yield compared to monoatomic primary ions, especially in the higher mass range. This is important in analysis of biological samples. Moreover, the damage inflicted on the sample surface is comparatively low. For example, C_{60}^+ is promising for molecular depth profiling [124].

Utilizing a TOF analyzer poses particular considerations since the primary ion beam must be pulsed. The actual sensitivity is not decreased but since the target is not subject to irradiation most of the time the analysis is relatively time consuming. Moreover, in a

TOF analyzer it is critical for the mass resolution that the secondary ions are ejected at a precisely defined time. This means that the primary ion pulse should be as narrow in time as possible, preferably <1 ns. At the same time maximum lateral resolution is desired. Unfortunately, there is a trade-off between these two parameters if the primary ion intensity is not to be sacrificed [122]. Therefore, TOF-SIMS instruments have two modes of operation, high mass resolution and high lateral resolution. An advantage with the pulsed source is that an electron flood gun can be allowed to operate when the primary ion gun is inoperative. Thus, charge-compensation is effectively applied when analyzing insulating materials.

Dynamic SIMS is used for depth profile analysis of mainly inorganic samples. The objective is to measure the distribution of a certain compound as a function of depth. At best the resolution in this direction is <1 nm, that is, considerably better than the lateral resolution. Depth profiling of semiconductors is used, for example, to monitor trace level elements or to measure the sharpness of the interface between two layers of different composition. For glass it is of interest to investigate slow processes such as corrosion, and small particle analyses include environmental samples contaminated by radioisotopes and isotope characterization in extraterrestrial dust.

Finally, secondary neutral mass spectrometry (SNMS) deserves mentioning. SNMS was introduced in the early 1970s by Oechsner and Gerhard [125]. In SIMS the ionization is very selective so the signal typically does not reflect the true sample composition. Most ejected species in a sputtering event, however, are neutral. Post-ionization of these with an appropriate technique allows for better quantitative analysis. This can be achieved in different ways, for example, by using an electron beam, plasma, or laser for the ionization. The typical application of SNMS is depth profiling of inorganic samples such as semiconductors, but trace element analysis of biological samples has also been performed. For reviews of SNMS, see References 126 and 127.

2.1.19. Fast Atom Bombardment

A technique very closely related to SIMS is fast atom bombardment (FAB), where a liquid sample is bombarded with energetic atoms (typically Ar or Xe atoms of ~10 keV kinetic energy) instead of ions [128, 129]. There is actually a technique named liquid SIMS (LSIMS) were the liquid sample is bombarded with energetic ions (see Section 2.1.18) [118, 130]. In principle there is no difference in the sputtering mechanism whether the primary particles are ions or atoms. The original reason to use a neutral beam was to avoid influence from the comparatively high accelerating voltage in the source of the magnetic sector instrument where it was introduced [131]. Charging of insulating targets is also significantly reduced. The FAB ion source can be combined with many different mass analyzers, but is most widely used together with quadrupole and magnetic sector instruments. A disadvantage with FAB is the rapid contamination of the ion source region, so frequent cleaning is required. In order to keep the sample in the liquid state when it enters the high vacuum ion source the sample is usually dissolved in a viscous solvent with low vapor pressure and freezing point, such as glycerol [132–136]. The matrix also shields the sample molecules from damage caused by the impinging high-energy particles [137]. In continuous-flow FAB (CF-FAB) sample solution is continuously delivered to the target, thereby making it possible to provide FAB-MS with online coupling to LC and CE [138, 139]. Less organic matrix ($\sim$5% glycerol instead of $>$90% glycerol) is necessary to keep the sample liquid in CF-FAB, resulting in greatly reduced chemical background. The reduced background and the constant refreshment of the surface layers makes CF-FAB more sensitive than conventional FAB [140].

FAB mass spectra are dominated by singly charged molecular ions, although doubly charged molecular ions and dimers are also occasionally observed. Prompt fragmentation of the analyte ions often gives partial sequence information, but is usually not of sufficient intensity to fully sequence a peptide of unknown structure. However, fragmentation readily occurs with CID (see Chapter 3) and provides further structural information. One major disadvantage with FAB is the intense chemical background due to matrix cluster ions and matrix fragment ions. The matrix can also react directly with the sample molecules, forming radical anions or causing reduction of the analyte. The FAB ion source is not as "soft" and sensitive as ESI or MALDI, the techniques that presently are preferred instead, and the upper mass limit, which is strongly sample dependent, seldom exceeds 10 kDa.

2.1.20. Plasma Desorption

The plasma desorption (PD) ionization method was the most successful way to transfer large molecules into gas phase before MALDI and ESI. It was discovered by Torgerson et al. in 1974 [141], who showed that $\sim$100 megaelectronvolt ^{252}Cf fission fragments could desorb amino acids from a thin foil. The very high energy of the primary ions distinguishes this method from SIMS. Apart from californium, ion beams from an accelerator were also used in some laboratories [142–145]. The accelerator beams allowed for defined primary ions and for experiments that studied the effects of varying types of ion and charge. PDMS was capable of producing molecular ions up to 45 kDa [146]. In the early 1990s, however, PDMS as a tool for mass spectrometry was more or less outrun by the softer, more efficient ionization techniques of MALDI and ESI. The geometry and principle of a typical ion source is quite similar to that of SIMS and MALDI and will not be described in more detail here. References 147 to 150 contain comprehensive descriptions of the method and its development.

2.1.21. Laser Desorption/Ionization

Mass spectrometric measurements of ions desorbed/ionized from a surface by a laser beam was first performed in 1963 by Honig and Woolston [151], who utilized a pulsed ruby laser with 50 μs pulse length. Hillenkamp et al. used microscope optics to focus the laser beam diameter to $\sim$0.5 μm [152], allowing for surface analysis with high spatial resolution. In 1978 Posthumus et al. [153] demonstrated that laser desorption/ionization (LDI, also commonly referred to as laser ionization or laser ablation) could produce spectra of nonvolatile compounds with mass $>$1 kDa. For a detailed review of the early development of LDI, see Reference 154. There is no principal difference between an LDI source and a MALDI source, which is described in detail in Section 2.1.22 In LDI no particular sample preparation is required (contrary to

MALDI, which is LDI utilizing a particular sample preparation). Although the performance of MALDI is superior to LDI in the analysis of many groups of compounds, LDI is still the perferred choice in some important applications, including crude oil analysis [155], fullerene detection in rocks [156], atmospheric aerosol analysis [157], semiconductors, and surface analysis [158]. Reference 21 is a comprehensive review of the use of LDI (and several other ion sources) in analysis of inorganics.

There is a special case of LDI worth mentioning, desorption/ionization on silicon (DIOS) [159, 160], in which analyte compounds are deposited on a surface of etched silicon. With this substrate the mass range can be extended to a few kilodaltons, allowing for analysis of, for example, small peptides without the involvement of a matrix.

2.1.22. Matrix-Assisted Laser Desorption/Ionization

Matrix-assisted laser desorption/ionization (MALDI) was developed by Karas, Hillenkamp, and coworkers in the late 1980s [161–163]. At the same time, a related technique was introduced by Tanaka et al. and involved mixing of the analyte with a very finely ground metal powder [164].

In principle MALDI is a special case of LDI, in which a particular type of sample preparation is employed (the success of the method warrants its own section, though). In the most widespread form of MALDI-MS the sample consists of analyte molecules dilutely embedded in a matrix of highly light absorbing, low-mass molecules. The matrix molecules are resonantly excited by an ultraviolet (UV)—by far the most common case—or infrared (IR) laser pulse of typically nanosecond duration, and the absorbed energy causes an explosive breakup of the sample and ionization of a fraction of the analyte molecules. Consequently, a volume of the matrix and the trapped analyte molecules are ejected into the gas phase. The ejected material contains both neutral and charged species that interact with each other during the expansion of the plume in the ion source (Fig. 2.9). The standard MALDI ion source (which is the same as an LDI source except for the matrix preparation) usually operates under high vacuum conditions, typically a pressure $<10^{-6}$ torr, and is combined with a TOF mass analyzer with axial extraction. When coupling to other analyzers, such as trapping analyzers or Qq-TOFs, it is for practical technical reasons also common with low vacuum, $\sim$1 torr. For AP-MALDI, see Section 2.1.23.

The matrix is the key part of the MALDI method. Most of the UV-absorbing matrices found so far are low-mass aromatic compounds such as the cinnamic acid derivatives, but several other compounds, including H_2O ice, glycerol, and urea for IR-MALDI have also been successfully applied as matrices [165–171]. Different matrix compounds are suitable for different classes of analytes. So far, the discovery of new matrices has to a large extent been a matter of trial and error experimentation, and the properties that separate most promising candidates from the few that actually work as MALDI matrices remain somewhat obscure. Briefly, a matrix has to be able to absorb strongly at the appropriate wavelength and to rapidly transfer into gas phase and also ionize the embedded analyte molecules without heating them too much. In vacuum-MALDI it is also important that the matrix is relatively stable, that is, does not readily evaporate, under high vacuum conditions. Other key factors for obtaining

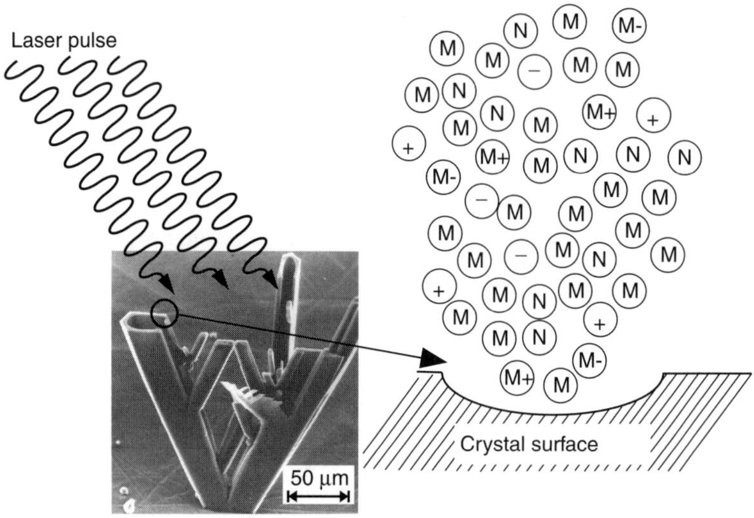

Figure 2.9. Schematic of a matrix-assisted laser desorption/ionization (MALDI) event. The SEM micrograph depicts sinapinic acid–equine myoglobin crystal from a sample prepared according to the dried drop sample preparation method. In the desorption event neutral matrix molecules (M), positive matrix ions (M+), negative matrix ions (M−), neutral analyte molecules (N), positive analyte ions (+), and negative analyte ions (–) are created and/or transferred to the gas phase. Reprinted from A. Westman-Brinkmalm and G. Brinkmalm (2002). In *Mass Spectrometry and Hyphenated Techniques in Neuropeptide Research*, J. Silberring and R. Ekman (eds.) New York: John Wiley & Sons, 47–105. With permission of John Wiley & Sons, Inc.

useful MALDI spectra are the morphology of the matrix/analyte film and the nature of the molecular incorporation in the matrix crystals [172–174]. The relative analyte concentration and the local morphology at different points on the sample can differ considerably. Many MALDI users have reported that it is necessary to search for "sweet" spots on the sample and the MALDI sample preparation is still widely considered to be something of an art. However, the understanding of the crystallization and phase separation phenomena involved is improving and several methods to prepare more homogeneous and reproducible samples have also been described [175–178]. Other types of preparation include utilization of anchor target plates [179] or solvent-free praparation for insoluble analytes [180]. Reference 181 contains a number of useful MALDI sample preparation protocols including comments. Most MALDI mass spectrometers are also equipped with a video camera to inspect the sample during data acquisition.

One feature of MALDI-MS, making it especially promising for analysis of biological samples, is the ability to detect biomolecules in complex mixtures in the presence of relatively large concentration of salts, buffers, and other species. Because of this ability, MALDI-MS has been utilized to study proteins and peptides in serum, cerebrospinal fluid, blood, tissue extracts, and whole cells [176, 182–185]. Recently imaging of biological samples has emerged as a promising technique [120, 186]. However, sample contamination often disturbs the crystallization of the matrix/analyte, and causes peak broadening by fragmentation and adduct formation, thereby reducing sensitivity and mass accuracy. MALDI-MS is extremely sensitive, allowing detection of peptites in the attomole range [187]. The mass range of MALDI-MS is strongly dependent on the mass analyzer, but proteins with masses up to 1 MDa have been successfully analyzed with MALDI-TOF-MS [188]. Because of the high background generated by the matrix MALDI-MS analysis of small peptides (<600 Da) is often difficult.

In both positive and negative ion mode MALDI mainly generates singly charged ions (somewhat dependent on the matrix), but multiply charged ions are common, especially for high mass proteins, and dimers occur as well. Because of the soft nature of the MALDI method, it is possible to desorb, ionize, and detect large, intact proteins. In many cases it is also possible to detect proteins or peptides intact with post-translational modifications, for example, phosphorylation and glycosylation [189–191]. Only a few examples of the detection of intact noncovalent complexes, for example, protein–ligand complexes or protein–metal-ion complexes, have been reported [192].

The molecular ions produced in the MALDI process have relatively high initial velocities, which can cause reduction in mass resolving power and transmission, primarily for TOF analyzers with axial ion extraction (see Section 2.2.1). Hence, the MALDI-MS mass resolving power depends strongly on laser fluence and is highest when the laser fluence is close to the threshold level.

The reasons behind the selectivity of the MALDI process have not yet been fully explained, but are probably a combination of varying ionization efficiencies and analyte/matrix incorporation efficiencies. Replacing the matrix or modifying the solvent system has been shown to alter the relative intensities between different components in mixtures [173]. However, within a limited concentration range, quantification can be achieved by using known quantities of the analyte to determine the concentration dependence of the MALDI-MS signal intensity [193]. Several research groups have also shown that quantitative data can be obtained with MALDI-MS by using a carefully chosen internal calibrant of known quantity and close chemical resemblance to the analyte of interest [184, 194]. See Reference 195 for a description of the early development of MALDI and Chapters 10, 12, and 15 for additional examples of applications for MALDI.

Also for MALDI, there is a special case worth mentioning. Surface-enhanced laser desorption/ionization (SELDI) is a technique that utilizes special sample plates [196, 197]. These have different modified surfaces, for example, hydrophobic, anionic, or antibody treated. Which type of surface to select depends on the application. After application of analyte the surface is washed according to a protocol leaving only the desired components on the target. Finally, a MALDI matrix is applied before analysis in the spectrometer. See Chapter 12 for an application example of SELDI.

2.1.23. Atmospheric Pressure Matrix-Assisted Laser Desorption/Ionization

Atmospheric pressure MALDI (AP-MALDI) was introduced in 1998 by Laiko et al. [198, 199]. The advantage with the AP version of MALDI is the possibility of coupling to analyzers other than TOF, such as a quadrupole ion trap (QIT) or Fourier transform ion cyclotron resonance (FTICR) cell, which are well suited for coupling to pulsed ion sources. This opens up the possibility of using instruments that can have both ESI and MALDI sources, which can be beneficiary. Moreover, low-energy CID as well as other "slow" fragmentation techniques can thus be employed with ions created with MALDI. Higher-order fragmentation, such as MS^3, is also possible to perform with coupling to trapping analyzers.

In AP-MALDI the target is positioned close to the inlet to the spectrometer. Several target arrangements have been utilized, including a target surface perpendicular to the spectrometer axis, a tilted target surface (almost parallel to the spectrometer axis), and transmission geometries where the laser light comes from behind through a quartz glass [199]. After desorption part of the ejected material will enter the vacuum region through the orifice. This can, like in ESI, be assisted by a gas flow. The ion injection in AP-MALDI resembles that of ESI, for example, and since there is no axial TOF the problem with broad initial velocity distributions is considerably smaller.

Still there are some limitations when coupling MALDI with analyzers other than TOF, such as the rather limited m/z range. With pulsed dynamic focusing [200] the ion transmission is increased about one order of magnitude compared to a static operation. Moreover, the target alignment sensitivity is also significantly reduced. The sensitivity of AP-MALDI is high, although the sensitivity of intermediate-vacuum MALDI can still be somewhat higher for the same instrument [201]. The sensitivity also depends on the analyzer and will therefore be higher for vacuum MALDI coupled to a regular axial TOF analyzer.

There is a recent hybrid between AP-MALDI and ESI, matrix-assisted laser desorption electrospray ionization (MALDESI) [202], where species desorbed from a MALDI target are subjected to an electrospray before entering the mass spectrometer. The method is similar to ELDI except that the analyte is embedded in a matrix as in MALDI.

2.2. MASS ANALYZERS

A mass analyzer is a device that can separate species, that is, atoms, molecules, or clusters, according to their mass. The separation should also be independent of the chemical conformation of the species. All mass analyzers presently in use are based on electromagnetism so ions are required to obtain separation. Therefore, an ion source has to be coupled to the analyzer. The analyzer will then separate ions coming from the source according to their m/z. There are several types of mass analyzers used in mass spectrometric research and they can be divided into different categories, such as magnetic or pure electric, scanning or nonscanning (pulse based), and trapping or nontrapping analyzers. In this section, the following mass analyzers are described: time-of-flight (TOF), magnetic/electric sector, quadrupole mass filter (Q), quadrupole ion trap (QIT), orbitrap, Fourier transform ion cyclotron resonance (FTICR), and also the technique of accelerator mass spectrometry (AMS). Table 2.2 contains a brief overview of the analyzers. There is a research field that deals with extremely accurate mass determinations of

TABLE 2.2. A Rough Comparison of Some Features of the Mass Analyzers

Analyzer	TOF	Sector	Q filter	QIT	Orbitrap	FTICR	Accelerator
Resolution	Low–high	Very high	Low–medium	Low–high	Very high	Highest	Very high
Mass accuracy	High	Very high	Low	Low–medium	Very high	Very high	Very high
m/z range	Very high	Medium	Low	Low–medium	Low[a]	Medium	Very low
Sensitivity	High	High	High	High	Medium	Medium	High
Dynamic range	Medium	Very high	High	Low–medium	Medium	Medium	Very high
Quantification	Medium–good	Very good	Good–very good	Poor	Medium	Medium	Very good
Speed	Fast	Slow	Medium–fast	Medium–fast	Slow–medium	Slow–medium	Slow
Ion-source	Pulsed/ continuous	Continuous	Continuous	Pulsed/ continuous	Pulsed/ continuous	Pulsed/ continuous	Continuous
Handling	Easy–medium	Medium–demanding	Easy	Easy	Medium	Demanding	Very demanding

[a]There is presently only one instrument type available, and its maximum m/z is set to 4000 Th.

atomic species (mass accuracy about 10 ppt). These determinations demand highly specialized equipment, such as storage rings and other specialized trapping analyzers, and will not be covered here [203].

2.2.1. Time-of-Flight

A time-of-flight (TOF) mass analyzer separates ions according to the time difference between a start signal and the pulse generated when an ion hits the detector, that is, the time of flight.

2.2.1.1. Principle. The principle of the TOF analyzer was first published by Stephens in 1946 [204], and a schematic is shown in Fig. 2.10. A must in TOF-MS is a well-defined start signal. TOF analyzers are therefore very well suited for the pulsed ion sources such as MALDI where the laser pulse, which creates ions in the gas phase, can be used to start the time measurement. Normally the sample plate is floated on a high positive or negative potential, typically 5 to 30 kV, and as the ions enter the gas phase they are accelerated towards ground potential. When leaving the acceleration region, all intact ions with the same charge will ideally have the same kinetic energy, but different, mass-dependent, velocities. The ions are then allowed to drift in a field-free region towards a detector. The time difference between the start signal and the pulse generated when the ion hits the detector is the time of flight (t_{TOF}) and can be expressed as

$$t_{\mathrm{TOF}} = \frac{L}{v} = L\sqrt{\frac{m}{2qU_a}} \propto \sqrt{m/z},$$

where L is the length of the field-free region, v is the ion velocity after acceleration, m is the mass of the ion, q the charge of the ion, U_a the accelerating electric potential difference, and z the charge state. The faster or lighter the ion, the shorter the time of flight and

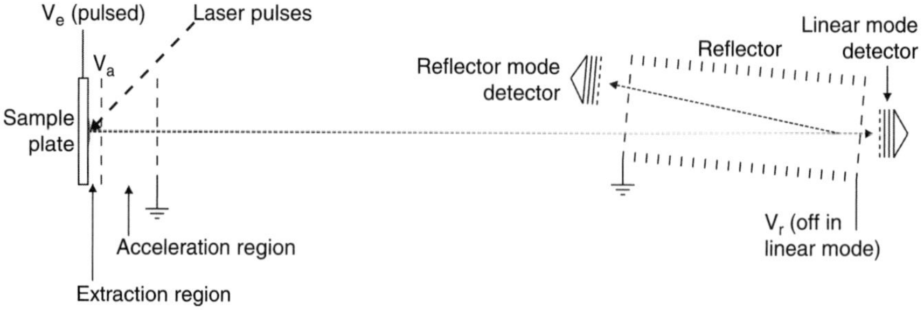

Figure 2.10. Schematic of a MALDI-TOF mass spectrometer with reflector and time-lag focusing. Reprinted from A. Westman-Brinkmalm and G. Brinkmalm (2002). In *Mass Spectrometry and Hyphenated Techniques in Neuropeptide Research*, J. Silberring and R. Ekman (eds.) New York: John Wiley & Sons, 47–105. With permission of John Wiley & Sons, Inc.

the time-of-flight spectrum obtained can be converted into a mass spectrum. There is no need to know the exact potentials and distances of the spectrometer, as the time/mass conversion is made by calibration with ions of known masses.

Continuous ion sources, such as ESI, can be connected to the TOF analyzers through orthogonal acceleration (oa-TOF) [205–209]. In oa-TOF, the ions generated by the source enter the TOF analyzer perpendicular to its main axis (Fig. 2.11). The acceleration potential is initially set to zero and the start pulse is generated instantaneously as the potential is raised and the ions are accelerated into the field-free flight tube.

2.2.1.2. Time Focusing Devices. The resolution of the TOF analyzer is limited by the initial velocity spread of the ions. However, there are powerful devices that can compensate for this velocity distribution, and the most widespread techniques at present are the electrostatic ion reflector (electrostatic mirror) and time-lag focusing (delayed extraction).

The reflector was introduced by Mamyrin et al. in 1973 [210]. When applying the reflector, the ions are not allowed to go all the way to the detector, but are instead reflected by a somewhat higher potential than the acceleration potential, after which they hit another detector, off axis (Figs. 2.10 and 2.12). The reflector increases the flight path, but more importantly, compensates for the initial velocity spread. This is accomplished when the slightly faster ions penetrate deeper into the electric field, thus getting a longer path in the spectrometer than the slower ions (Fig. 2.12). By optimizing the ratio of the acceleration and reflector voltages, ions of the same m/z but with different initial velocities can be pressed to arrive at the detector at (almost) the same time. The reflector has the advantage of being mass-independent, so the optimum setting applies to the full mass range. There are a few variants of the reflector. The

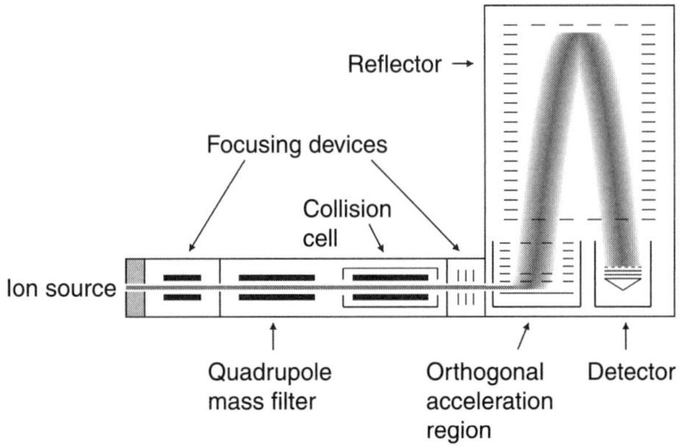

Figure 2.11. Schematic of a quadrupole time-of-flight (Qq-TOF) mass spectrometer. Reprinted from A. Westman-Brinkmalm and G. Brinkmalm (2002). In *Mass Spectrometry and Hyphenated Techniques in Neuropeptide Research*, J. Silberring and R. Ekman (eds.) New York: John Wiley & Sons, 47–105. With permission of John Wiley & Sons, Inc.

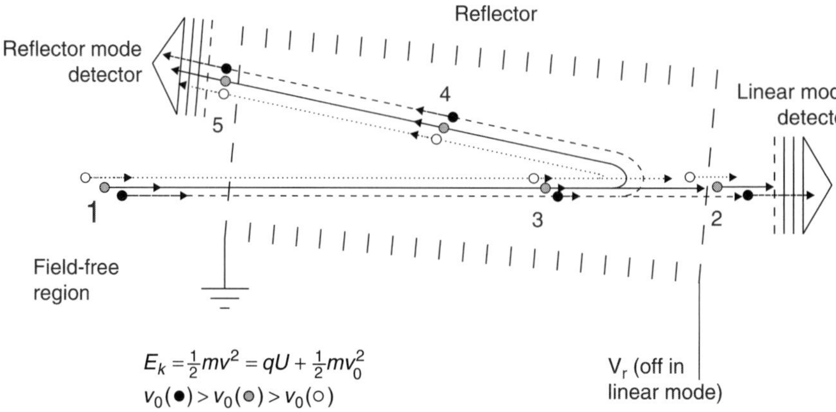

Figure 2.12. The working principle of a single-stage reflector. Desorbed ions will have different initial axial velocities, v_0. Consequently ions of the same m/z will at a certain time have a spread in space (1). With no reflector this spread will increase over time and result in peak broadening when registered by the linear mode detector (2). The electric field of the reflector will slow down the ions (3) and finally make them turn around (in noncoaxial designs the mirror is slightly tilted, typically $\sim 1^\circ$, compared to the spectrometer axis, and the turnaround angle will not be exactly 180° but rather $\sim 178^\circ$). Faster ions (higher v_0) will penetrate deeper into the reflector than slower ions (lower v_0), thus travelling a longer distance in the spectrometer. After reversing direction the ions will again gain speed (4) and if the mirror is properly set ions with the same m/z will arrive at the reflector mode detector (5) with much less time spread than in the linear mode case. Reprinted from A. Westman-Brinkmalm and G. Brinkmalm (2002). In *Mass Spectrometry and Hyphenated Techniques in Neuropeptide Research*, J. Silberring and R. Ekman (eds.) New York: John Wiley & Sons, 47–105. With permission of John Wiley & Sons, Inc.

two most common types are the linear reflector (shown in Fig. 2.12) and the two-stage reflector. However, there are also the quadratic field reflector and the curved field reflector [211–214]. Two or more reflectors can be arranged in series [215–217], thus increasing the flight path and increasing the mass resolving power. The drawback is loss in sensitivity, and analysis of intact proteins is often carried out in a linear mode.

Time-lag focusing, or delayed extraction (DE), was introduced by Wiley and McLaren in 1955 [218] and has, since the mid-1990s, been a standard feature of MALDI-TOF mass spectrometers [219–221]. The ion source is modified from a single-stage to a two-stage acceleration region. At the time of desorption the first region (extraction region) is kept field-free. After a short time, typically a few hundred nanoseconds, the electric field is switched on. Ions with lower initial velocity will travel a shorter distance than those with higher initial velocity. Hence, the previously slower ions will suddenly be at higher potential than the previously faster ones, which means that the originally slower ions will instead be the faster ones when exiting the acceleration region (Fig. 2.13). With properly set delay time and source voltages, the

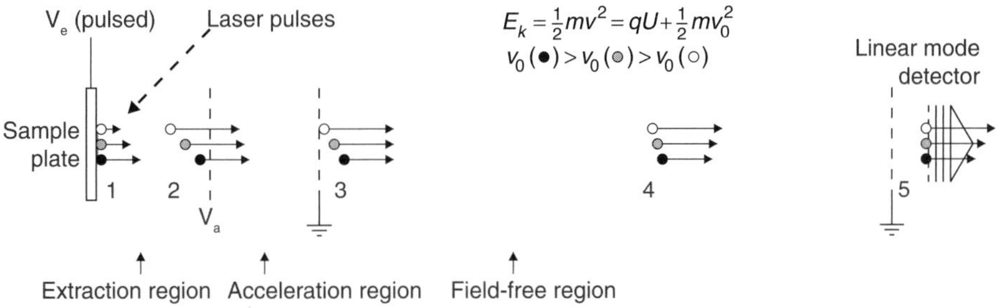

Figure 2.13. The principle of time-lag focusing. When the laser pulse desorbs the ions (1) the sample plate has the same potential as the first grid, so the ions do not experience any electric field but travel with their respective initial velocity, v_0. After a short delay time, τ, typically a few hundred nanoseconds, the sample plate potential is raised and the ions suddenly experience a force from the electric field (2). The slower ions (lower v_0) that are closer to the sample plate will be subject to a higher potential than the faster ions (higher v_0) further away and will therefore have gained more kinetic energy and speed when entering the field-free region (3). While traveling through the field-free region ions with the same m/z will gradually get closer to each other (4) and if the source voltages and delay time are properly set the ions will be detected (almost) simultaneously (5). Reprinted from A. Westman-Brinkmalm and G. Brinkmalm (2002). In *Mass Spectrometry and Hyphenated Techniques in Neuropeptide Research*, J. Silberring and R. Ekman (eds.) New York: John Wiley & Sons, 47–105. With permission of John Wiley & Sons, Inc.

spread in time of flight for ions of the same m/z is minimized and better resolution is achieved. Time-lag focusing can compensate both for initial velocity spread and for initial spatial dispersion, although optimal compensation conditions for the two types of dispersions cannot be achieved simultaneously. The former feature is utilized for axial acceleration instruments like standard MALDI-TOF mass spectrometers (Fig. 2.13), and the latter feature is utilized for orthogonal acceleration mass spectrometers (Fig. 2.11). Time-lag focusing is, unfortunately, mass dependent, so the optimum settings will vary depending on the mass region of interest. Contrary to the reflector, time-lag focusing does not decrease sensitivity but rather acts in an opposite way, since the ions formed enter the gas phase in a field-free environment. The desorbed compounds will then have some time to spread out before the accelerating field is applied and, therefore, there will be much fewer collisions causing fragmentation.

2.2.1.3. Performance Parameters. The maximum mass resolving power of a TOF analyzer is mass dependent, and is also a function of the length of the flight path of the instrument. Other factors that affect the resolving power are the detector surface, space focusing devices, voltage drifts in the power supplies, jitter in the timing electronics, ringings, and mechanical misalignments (most of these factors are limiting for all analyzer types). The best commercial MALDI-TOF instruments are, at present, capable of (employing time-lag focusing and reflector) a mass resolving power of more than 35,000 for bovine insulin (5734 Da). When the mass increases above the point when the isotopes no longer can be separated, the isotopic envelope will determine the width of the peak and, hence, there will be a sudden drop in mass resolving power.

In the linear mode, the attainable mass resoving power is about 3000 with time-lag focusing employed. When operating in this mode, usually the aim is to detect larger species, which will not be small enough to allow resolved isotopic distribution and, therefore, the practical resolving power is much lower (limited by the width of the isotopic envelope). Operation without time-lag focusing is seldom carried out, as there is no advantage to doing that except when, for example, studying fundamental issues of the desorption/ionization process.

The mass accuracy is highly dependent on the mode the instrument is operating in. In the reflector mode, with time-lag focusing, the best MALDI-TOF and oa-TOF instruments are capable of achieving <5 ppm with internal standards, provided that the isotopes are resolved. In many cases it is not possible to add internal calibrants, and then the error in mass accuracy is often increased to 50–100 ppm. Operation of an instrument in a linear mode will typically decrease the mass accuracy.

There is no theoretical upper, or lower, limit in the mass range for TOF analyzers. However, the detection efficiency of the normally employed detector, the microchannel plate (MCP), decreases with increasing mass (see Section 2.3.3.2). Other factors also limit the practical upper mass limit, such as broader peaks, due to wider isotopic distributions, more adduct formation, nonspecificity of the molecular species (e.g., glycoproteins or protein variants), and fragmentation (in-source or post-source decay). Due to the post-source fragmentation of larger molecular ions it is often difficult or even impossible to detect a signal in the reflector mode. This is not a problem for acquisition in a linear mode since the fragments will arrive at virtually the same time as the intact species, thus still contributing to the signal intensity.

In general, a TOF analyzer is much more sensitive than the sector or quadrupole analyzers operating in scanning mode (see Sections 2.2.2 and 2.2.3). For larger species, the sensitivity is highest when operating in a linear mode (because of shorter flight path and post-source fragmentation). For some smaller, more stable, compounds, the opposite situation is often the case. When the ions travel along the same path in an instrument, the time-focusing devices do not limit transmission but merely lower the time spread so that a peak with the same area (i.e., generated by the same number of ions) will now be more narrow, and thus higher and easier to observe. In the oa-TOF analyzers it is important to control the filling of the acceleration region with new ions with the TOF repetition rate analysis, so that loss of the analyte ions is minimized.

TOF analyzers require very fast detectors to provide high resolution. Moreover, the time-spread during the amplification has to be minimal. The detector type that satisfies these conditions is the MCP, but its dynamic range is often limited by detector saturation (see Section 2.3.3.2). In MALDI, the main problem arises from the very intense matrix signals. The low mass ions can be deflected by employing timed deflection plates, which reduces the saturation greatly. For the ion-counting detection systems, employing a time-to-digital converter (TDC), ions that have intensities higher than one per collection cycle will be discriminated against. The data acquisition software can, to a great extent, adjust for this.

The quantification capability is normally limited by the detector and/or the ion source. The MCP that is often utilized in TOF instruments cannot fully handle the ion currents that are produced in MALDI and are often saturated to some extent. With other ion sources, such as SIMS, the detection system is less strained so the detector is less limiting. Instead the ion source will limit the quality in quantification. Magnetic sectors and also qudrupoles are more often utilized when quantification is important.

The TOF analyzer is, in general, the fastest mass spectrometric analyzer. The repetition rate, that is, the number of start pulses per second, is fundamentally determined by the data collection time window; the longer time of flights, and the lower maximum repetition rate. In addition, other devices may limit the maximum repetition rate, for example, lasers, high voltage switches, and acquisition electronics. The newer MALDI-TOF instruments have lasers with repetition rates of 200 Hz. For oa-TOFs and TOF-SIMS instruments the repetition rates are typically several kilohertz. High speed is critical for applications such as chemical imaging and high throughput screening where many thousands of spectra are acquired.

In a reflector TOF instrument, PSD can be performed. The Qq-TOF is an oa-TOF with a quadrupole as a precursor selector. TOF-TOF is a type of instrument that has emerged in recent years and has enabled MALDI-TOF instruments that can deliver good MS/MS data. There are also examples of instruments with two consecutive reflector TOFs. TOF analyzers can also be coupled to ion traps and to sector instruments. See Chapter 3 for more thorough MS/MS descriptions.

Pulsed ion sources for axial injection such as MALDI and SIMS, and earlier plasma desorption are suitable for TOF analyzers. For continuous ion sources such as ESI the oa configuration is suitable.

TOF instruments come in a variety of sizes and complexity, typically correlated with cost, ranging from the bench-top linear instruments to the several-hundred-kilogram TOF-TOFs. Generally TOF instruments are somewhat larger than quadrupole instruments but smaller than sector and FTICR instruments.

2.2.2. Magnetic/Electric Sector

In the 1910s Thomson used magnetic and electric fields to separate ions of different mass and energy [222]. A few years later, Dempster employed a variable magnetic field to scan an m/z range [6]. High resolution, double-focusing instruments were developed in the 1930s by Mattauch and Herzog [223] and in the 1950s by Johnson and Nier [224]. Sector analyzers are easily adaptable to continuous ion sources, such as EI, dynamic SIMS, ICP, and ESI. Although more difficult to implement, there are several examples of MALDI-sector mass spectrometers [225–229]. Some years ago, double-focusing sector instruments were the flagships of MS, providing the best overall performance, except for high m/z range where the TOF analyzer performed better, but before MALDI an upper limit of $m/z = 10$ kTh was mostly quite sufficient. The top-line instruments were generally space-demanding, four-sector machines that could provide MS/MS data. More recently these instruments have largely been replaced by Qq-TOFs and FTICR instruments, depending on the demands. Qq-TOFs typically provide good enough MS/MS data and are much less demanding instruments, being less expensive and much smaller. High resolution and high mass accuracy applications can be performed by FTICRs (and lately the orbitrap) that are in many ways more flexible and also consume less space. However, the sector instruments still are "king" in high accuracy quantitative measurements, such as isotope ratio determination or analysis of toxic compounds and their congeners, such as dioxins. Since these applications typically involve smaller species and often atomic species, there is no need to have the really large machines any more, and thus modern sector instruments are typically two-sector devices that are relatively compact.

2.2.2.1. Principle. As in the TOF case, the ions leaving the source in a magnetic sector mass spectrometer are accelerated to a high velocity. In a reverse-geometry instrument, the ions then pass through a magnetic sector, in which the magnetic field is applied perpendicularly to the direction of the ion beam. The magnetic field will not change the velocity of the ions, but force them into a circular motion with a radius that depends on the magnetic field strength, and the mass-to-charge ratio and velocity of the ions (Fig. 2.14) so that

$$\frac{m}{q} = \frac{B^2 r^2}{2U_a},$$

where m is the ion mass, q the ion charge, $\vec{\mathbf{B}}$ the magnetic field strength (or more correctly the magnetic flux density), r the radius of the ion trajectory in the magnetic field, and U_a the potential difference over the acceleration region. Only ions that pass through a narrow slit will be detected. A magnetic sector will also act as a lens, and ions of the same m/z but with slightly different velocity will not be focused at the same point. Hence, the resolution will be limited by the initial velocity spread of the ions. To obtain better resolution, it is necessary to add an electric sector that focuses ions according to their kinetic energy. The electric sector also applies a force perpendicular to the ion beam direction and is, therefore, also shaped as an arc. If the electric sector is designed so that the dispersion of ions due to their velocity spread is exactly equal and opposite to that of the magnetic sector, the result of combining the two types of sectors is a zero net velocity dispersion, that is, double focusing (Fig. 2.14). In a forward-geometry sector instrument the electric sector is placed before the magnetic sector. Reverse-geometry means that the magnetic sector is placed before the electric sector. Reference 230 describes a number of geometries, including ion optics theory.

2.2.2.2. Operating Modes. The simplest mode of operating a magnetic sector mass spectrometer is by scanning the magnet, that is, to keep the acceleration potential and the electric sector potential constant and vary the magnetic field strength. Ions with different m/z escape through the slit in front of the detector at different magnetic field strengths. When operating in magnetic scan mode, the performance is decreased and

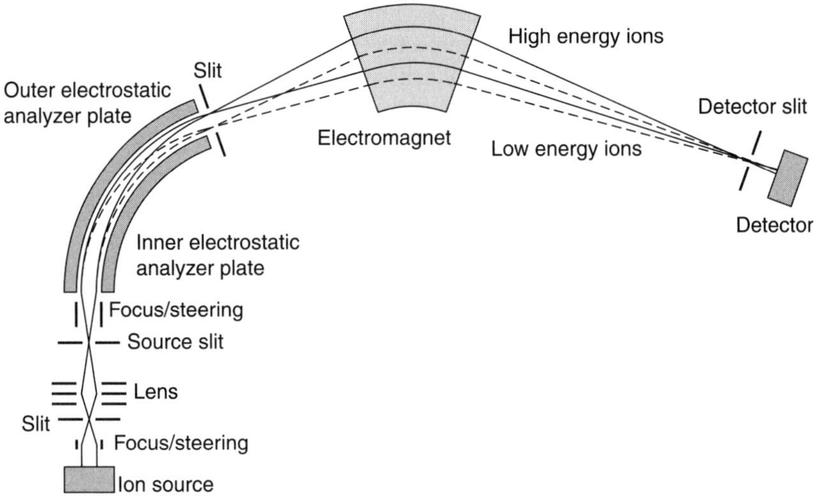

Figure 2.14. Schematic of a magnetic and electric analyzer with so-called forward geometry. Ions coming from the source are focused and steered into the electric analyzer. The electric analyzer does not disperse ions according to m/z but focuses ions of the same kinetic energy to the same position at the exit slit. Ions with different kinetic energy will be focused to different positions. The magnetic sector then disperses the ions according to m/z but also (like the electric sector) focuses ions of different kinetic energy to different positions. In principle both sectors create dispersions of the ions according to kinetic energy but in opposite directions. With a proper arrangement these dispersions are balanced so that ions of only the desired m/z but with a small spread in kinetic energy are focused to the same position at the detector slit. The desired resolution (and transmission) is controlled by the opening of the detector slit. Reprinted from A. Westman-Brinkmalm and G. Brinkmalm (2002). In *Mass Spectrometry and Hyphenated Techniques in Neuropeptide Research*, J. Silberring and R. Ekman (eds.) New York: John Wiley & Sons, 47–105. With permission of John Wiley & Sons, Inc.

also dependent on the scan rate due to the reluctance of the magnet. If the mass spectrometer is coupled to a chromatographic system, a relatively high scan rate is desired and the mass resolving power, as well as the mass accuracy, have to be compromised even further. The mass resolving power and mass accuracy obtained under such conditions is, however, usually good enough to provide information about the elemental composition of a molecule with a mass of several hundred daltons. Scanning decreases sensitivity since ion species are detected one at a time. A method to increase the sensitivity is to use a focal-plane or array detector (see Section 2.3.4) where a portion of the mass range can be detected simultaneously. The magnetic field can then be changed in steps rather than scanned, thus providing shorter acquisition time and, consequently, higher sensitivity.

Alternatively, the instrument can operate in the voltage scanning mode, that is, the magnetic field strength is kept constant and the electric field is varied. The electric sector potential is coupled to the accelerating potential. The advantage of this mode of operation is that the electric field is not subject to the magnetic hysteresis effects, so the relationship between m/z and the accelerating potential is linear. With this mode, a better mass accuracy is obtained. The disadvantage is that the sensitivity will now be roughly proportional to m/z.

Finally, the instrument can be operated in the peak-matching mode, which provides optimum mass resolving power and mass accuracy. Here the magnetic field strength is kept constant and the electric sector and acceleration voltages are scanned over a relatively small m/z range. This mode of operation is suitable when two ions that are very close in mass need to be separated or when the elemental composition of a molecule is to be determined at high resolution.

2.2.2.3. Performance Parameters. The mass resolving power and mass accuracy depend on the way the analyzer is operated. Under optimum conditions in the peak-matching mode, the more powerful instruments can achieve a mass resolving power well over 100,000 ($m/\Delta m$ at 10% valley definition, corresponding to approximately the double for the FWHM definition for a Gaussian peak shape). The resolution is determined by the slit widths. Higher mass resolving power is obtained by decreasing the slit widths, and thereby also reducing the number of ions reaching the detector. Hence, in a sector instrument sensitivity and resolution are mutually limiting parameters.

A mass accuracy better than 1 ppm is routinely obtained by the better instruments when they are operated in a peak matching mode. An interlaboratory study of mass accuracy of different instruments and operating modes can be found in Reference 231.

The upper m/z limit of a magnetic sector analyzer depends on the magnet. Both its maximum magnetic field strength (the larger commercial instruments have magnets with a maximum field strength of 1 to 2 T) and its arc radius determine the maximum m/z at a given acceleration potential. The higher the acceleration potential (and kinetic energy of the ions) the lower the mass range. The maximum transmission and sensitivity occur at the maximum working acceleration potential, so there is a trade-off between high mass range and high sensitivity. Commercial instruments offer an m/z range up to about 10 kTh.

The sensitivity depends on the mode of operation. In the full scan mode the sensitivity is limited, while in the selected ion monitoring mode the sensitivity is high. The desired resolution also affects sensitivity (see above).

Magnetic sector instruments generally have very high reproducibility, very good quantitative performance, and very high dynamic range (depending on detector and ion source). Relatively short ion source residence times and ion flight times in a magnetic sector mass spectrometer limit the interactions of ions with neutral molecules and other ions. This minimizes space-charge effects (ion-ion repulsion) and ion-molecule reactions that would adversely affect the reproducibility of mass spectra. Magnetic sector instruments are often used in conjunction with detectors that can handle a large span in ion current. Generally, they have the best quantitative performance of all mass spectrometer types.

The speed of the analyzer highly depends on the mode of operation. The scan speed affects both the resolution and the mass accuracy, so if high quality data is needed, a relatively slow scan rate must be considered. For applications that do not require scanning, this is, of course, a minor problem.

With linked scan MS/MS, another analyzer is not required, but this mode gives either limited precursor selectivity with unit product-ion resolution, or unit precursor selection with poor product-ion resolution. Four-sector instruments can give very high quality data, but are costly, very space demanding, and complex. Hybrids like sector-TOF or sector-QIT are good (and cheaper) alternatives. High-energy CID MS/MS spectra are very reproducible (see Chapter 3).

The compatibility is excellent with continuous ion sources such as ESI, dynamic SIMS, CF-FAB, ICP, EI, CI, etc. Sector instruments are not well-suited for pulsed ionization methods, although there are examples where MALDI sources have been utilized [225–229]. Sector instruments are usually larger and more expensive than other mass analyzers, such as TOFs, quadrupole filters, and traps.

2.2.3. Quadrupole Mass Filter

The quadrupole mass filter is a relatively small device that can be set to let through ions within a very limited m/z range only. The trajectories of the ions with either higher or lower m/z will bend and the ions will never escape the filter. A mass spectrum is acquired by scanning through the whole m/z range of interest and detecting how many ions pass the filter at each m/z.

2.2.3.1. Principle. The basic principles of the quadrupole mass filter were published in 1953 [232, 233]. The analyzer employs a combination of direct-current (DC) and radio frequency (RF) potentials. Four parallel rods are arranged symmetrically as is shown in Fig. 2.15. Opposite rods are connected electrically in pairs. The two pairs will, at any given time, have potentials of the same magnitude, but of opposite sign. Ions emerging from the source, typically accelerated over a potential of 5 to 20 V, enter the analyzer region between the rods and travel parallel to the rods. At given values of the DC and RF potentials and the RF frequency, only ions within a certain narrow m/z range will have stable trajectories through the quadrupole. The m/z range for ions allowed to pass through depends on the ratio between the DC and RF potentials. Ions that do not have a stable trajectory will collide with the rods, never reaching the detector.

The motion of an ion traveling in the quadrupole field is described by the Mathieu equation [234, 235]. Qualitatively, the heavy (more inert) ions mainly respond to the DC component of the field while the lighter (quicker) ions also respond to the alternating RF component. One pair of rods will act as a high pass filter and force the heavy ions to the middle between the electrodes. Low mass ions respond faster when, once every cycle, the net force from the DC and RF components is attractive for a short time. Then, if the mass of the ion is low enough, the ion will be accelerated towards one of the electrodes and hit it before the force changes direction again. The opposite pair of rods will instead act as a low pass filter. High mass ions experience an attractive force and will hit one of the electrodes. The low mass ions also experience an attractive force most of the time, but they will, once every cycle, respond to the repulsive force and be pushed back towards the middle between the electrodes. One can consider the analogy of a ball placed

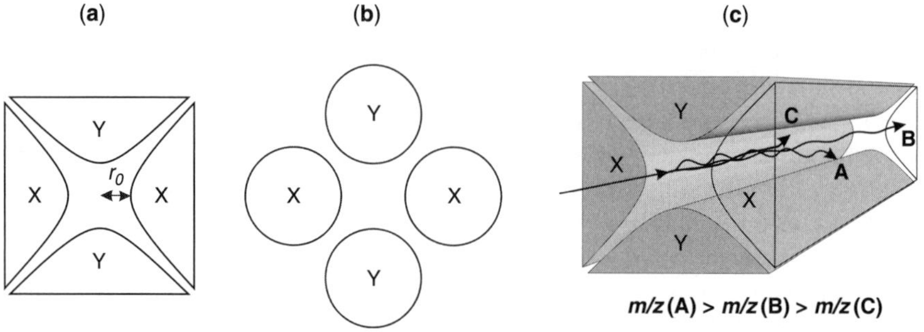

Figure 2.15. Schematic of a quadrupole analyzer. (a) A hyperbolic cross-section; (b) cross-section of cylindrical rods; (c) the operating principle of a quadrupole mass filter. The *x*-direction pair of rods acts like a high pass filter so ion C (with low m/z) is not allowed through, and the *y*-direction pair of rods acts like a low pass filter and takes care of ion A (with high m/z). Only ion B having an m/z in the stable range is allowed through the quadrupole mass filter for subsequent detection. Reprinted from A. Westman-Brinkmalm and G. Brinkmalm (2002). In *Mass Spectrometry and Hyphenated Techniques in Neuropeptide Research*, J. Silberring and R. Ekman (eds.) New York: John Wiley & Sons, 47–105. With permission of John Wiley & Sons, Inc.

on the top of a curved surface (e.g., the side of a cylinder). If the ball is situated exactly on the top and undisturbed, it will remain there. However, any perturbation will cause the ball to roll to one of the sides. If the surface is instead wiggled back and forth in an appropriate way, the ball will not fall off and the situation can be mastered. The alternating field between the rods acts in a similar way. The ions that have a stable trajectory through both the high pass and low pass filter escape through a narrow slit, and are detected. A mass spectrum is acquired by scanning the ratio of the DC and RF potentials and monitoring the abundance of the detected ions. When operated in the RF-only mode (i.e., no DC voltage applied), a wide m/z range is allowed to pass through the filter.

2.2.3.2. Performance Parameters. Scanning speeds up to about 6 kTh/s are routine and unit mass resolution (FWHM) up to 2 kTh is attainable with cylindrical quadrupole rods. With hyperbolic rods the transmission is significantly higher and a mass resolving power exceeding 25,000 (at 1 kTh) has been achieved. Note that for a certain resolution setting, the peak width does not vary with m/z, so the relative resolution will improve with increasing mass. The mass selectivity (i.e., resolution) of the quadrupole is dependent on the number of cycles an ion undergoes while in the analyzer. Thus, the resolution is negatively affected by an increased ion velocity and decreased RF frequency. Quadrupole mass filters are not, however, typically chosen to achieve high mass accuracy and resolution, but rather for maintaining good speed and sensitivity. Often the analyzer is set to operate at somewhat better than unit resolution since sensitivity drops when the resolution is substantially increased. This way the isotopes are resolved for singly charged ions, and, depending on application, high speed, high sensitivity, or a compromise of these factors can be selected.

Quadrupole analyzers have generally been considered to give poor mass accuracy. Recently, however, with better machined parts and better electronics, commercially available instruments can perform quite well. A mass accuracy ≤ 5 ppm can be obtained with internal calibrants.

The mass range depends on the settings of the DC and RF voltages and the RF frequency. Because the lower RF frequency means higher m/z range, there is a trade-off between resolution and m/z range. Typical m/z ranges are 25 to 2000 Th with unit mass resolution or better and up to 4000 Th with less than unit resolution and still reasonable sensitivity. Similarly to the sector instruments, a quadrupole analyzer operating in the scanning mode detects one ionic species at any given time, so most of the ions produced are not detected, thus decreasing the sensitivity. The sensitivity is vastly improved when scanning a narrow m/z range only or when operating in a single ion monitoring mode. Quadrupole-based instruments generally have good reproducibility, but the sensitivity is strongly mass dependent and drops for higher m/z.

The quadrupole mass filter has relatively good dynamic range. Space-charge problems are limited. Normally the ion source or detector limits the practical dynamic range. The quantification capability of quadrupole instruments is generally very good. The magnetic sectors are more stable in nonscanning mode but more costly, so quadrupoles are normally used except when the highest performance is necessary. Under normal operation, quadrupoles are intermediate devices concerning scanning speed. TOF analyzers are much faster (since they are nonscanning) while magnetic analyzers are slower.

With triple quadrupole instruments (QqQ), tandem mass spectrometry can be performed. The second quadrupole, which nowadays often is a hexapole or octapole, is not used for m/z selection or scanning, but serves as a collision cell containing a gas. With such a setup, the low-energy CID can be performed. This technique is efficient in producing fragment ions, but reproducibility between instruments is limited since the spectra obtained depend heavily on the nature of the collision gas selected, its pressure, the collision energy, as well as other parameters (see Chapter 3). A feature of the QqQ that most other instruments lack is the ability to perform precursor ion scans, that is, to select a specific fragment and look for possible precursors. A common type of hybrid instrument is the Qq-TOF, where a quadrupole can be used for precursor selection and the TOF analyzer for fragment separation.

Like sector analyzers, quadrupole analyzers are well suited for continuous ion sources such as ESI, but are not well-suited for pulsed ionization methods. Quadrupole mass spectrometers are generally substantially cheaper and smaller than sector instruments and Qq-TOFs. They are very often used in combination with GC and LC, and single or triple quadrupole mass filters are very common benchtop instruments for routine measurements.

2.2.4. Quadrupole Ion Trap

In the quadrupole ion trap (QIT), ions are trapped and stored in a potential well. A mass spectrum is acquired by ejecting the ions from the potential well in order of ascending m/z and detecting them. The trap can also be used to selectively store ions with a particular m/z, and perform further experiments, for example, fragmentation in several consecutive steps.

2.2.4.1. Principle. The cylindrical quadrupole ion trap is based on the same principle as the quadrupole mass filter, but the geometry is different (Fig. 2.16). The cylindrical QIT, or Paul trap, was developed almost simultaneously with the quadrupole mass filter [232, 233]. Recently, a variant of the theme has emerged, the linear quadrupole ion trap [236], which is a device built like a quadrupole mass filter with extra trapping end electrodes for the axial direction. Under stable conditions, ions moving around inside such traps will ideally continue to do that forever.

Ions generated in the source have to be guided into the ion trap and they typically enter through an opening in one of the end-cap electrodes (Fig. 2.16). In the linear trap, the "end caps" consist of short quadrupole mass filters that can be set to block passage of ions in the axial direction (Fig. 2.17). For simplicity, the following description applies to the cylindrical QIT, unless otherwise stated. The linear QIT works quite similar to the cylindrical device. However, it has some important advantages that will be pointed out at the end of this section. During injection, the amplitude of the main RF voltage may either be held constant or changed. With constant RF voltage amplitude, where ions enter an active trapping field, the ion kinetic energy is dissipated during repetitive collisions with damping gas atoms. A normal operational background pressure in the trap is 1 mtorr of helium. The increase in amplitude is synchronized with the ion injection. This method is particularly well suited for pulsed ion injection, such as MALDI where the initial velocity spread is rather large. In this case, the packet of ions to be injected will have m/z dispersion like in a TOF analyzer, and the RF voltage amplitude can be set to match the ion energy and m/z, thus increasing the trapping efficiency [237, 238].

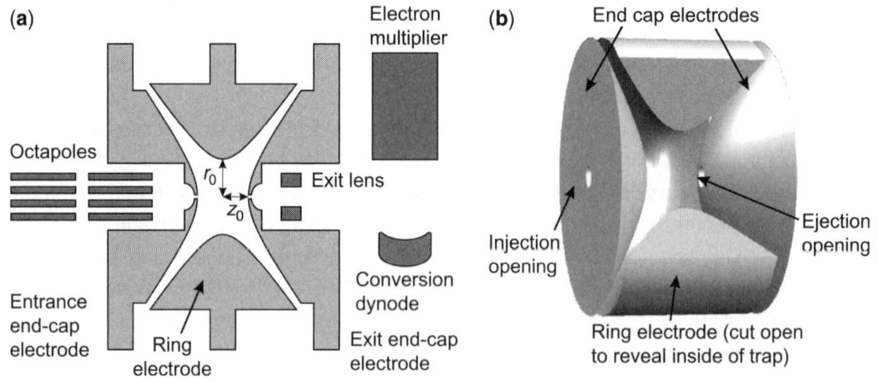

Figure 2.16. A cross-section schematic of a quadrupole ion trap mass spectrometer (a), with a three-dimensional perspective view of the quadrupole ion trap (b). Reprinted from A. Westman-Brinkmalm and G. Brinkmalm (2002). In *Mass Spectrometry and Hyphenated Techniques in Neuropeptide Research*, J. Silberring and R. Ekman (eds.) New York: John Wiley & Sons, 47–105. With permission of John Wiley & Sons, Inc.

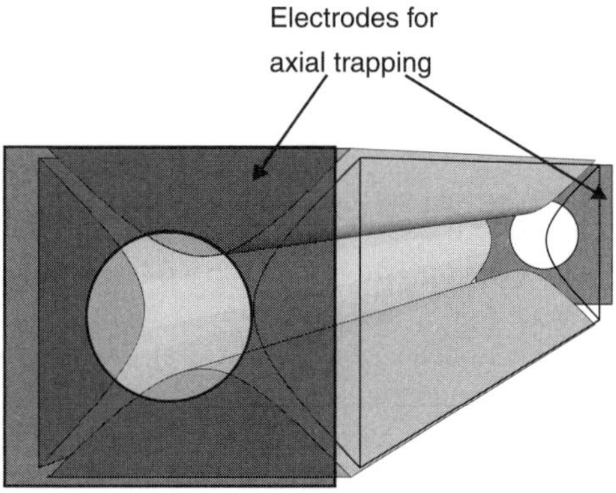

Figure 2.17. Schematic of a linear quadrupole ion trap (QIT). This type of analyzer consists in principle of a quadrupole analyzer with electrodes at the ends to block ion passage in the *z*-direction.

The helium gas in the trap not only helps in trapping the ions but also cools them (i.e., the kinetic energy of a trapped ion is dissipated through repeated collisions with the He gas), thus forcing the ions to the center of the trap where the quadrupole field is best defined. Both sensitivity and mass resolution are significantly enhanced by the presence of the He gas. Moreover, the same He can also be used to induce fragmentation when working in the MS^n mode (see below).

In most commercial cylindrical ion trap instruments the end-cap electrodes are held at ground potential and usually only a RF potential is applied to the ring electrode. When the RF amplitude is set to a low, so-called storage voltage, all ions above a certain m/z are trapped. This voltage is usually chosen so the lowest trapped m/z is greater than those of water, air, and solvent ions (i.e., above 100 to 150 Th), depending on the nature of the measured species.

One method of acquiring a mass spectrum is the mass selective instability scan. As the RF voltage increases, the ions with lowest m/z become unstable and are ejected through small holes in the end cap to hit a detector. As the RF voltage is further increased, heavier ions become successively unstable and are ejected, thus yielding a mass spectrum.

An important additional technique is resonant ejection. In this method, a supplementary RF voltage is applied to the end caps, with a 180° phase shift between the two caps. The trapped ions oscillate with different frequencies depending on their m/z. To eject ions of a certain m/z, a supplementary RF voltage with corresponding frequency (typically between a few and hundreds of kilohertz) is applied. The ions fall into resonance with the oscillating potential, their oscillation amplitudes in the axial direction increase, and finally they are ejected. When resonant ejection is combined with an RF scan, larger and larger ions are successively moved to the position in the stability diagram where they fall into resonance with the applied supplementary RF voltage and are ejected.

By applying the supplementary RF voltage during injection (without scanning the RF voltage on the ring electrode) it is possible to prevent certain ions from being trapped.

This can be useful when working with several species with large differences in abundance. Choosing not to trap the dominant species allows for higher concentration of the others, thus increasing sensitivity without decreasing performance due to space-charging.

The supplementary RF voltage may also be set to contain all frequencies but a narrow band, corresponding to a narrow m/z range, forcing all ions outside this range to leave the trap. This mode of operation is usually referred to as ion isolation, and is the first step when performing MS/MS (see below).

The most common way to detect the ions is to eject them from the trap and have them hit a detector situated outside the trap, as seen in Figs. 2.16 and 2.17. A standard detector is the conversion dynode together with an electron multiplier. Ions ejected from the trap are accelerated towards the detector and then amplified (see Section 2.3.3).

2.2.4.2. Performance Parameters. By reducing the scan speed to ~0.02 Th/s, a mass resolving power of $3 \cdot 10^7$ (FWHM) has been achieved in a single scan for $m/z =$ 414 Th [239]. More typical resolving powers for commercial instruments are up to about 30,000 (FWHM) for $m/z = 1500$ Th, although QITs are usually operated at a mass resolving power of ~2000. Often the instrument is set to operate at a little better than unit resolution, because of the decrease in scan speed and sensitivity that comes with the higher resolution. For example, when using the instrument online with a chromatographic system, speed is often crucial.

The quadrupole ion trap still suffers from mass accuracy problems (in comparison with the high resolution attainable) but <30 ppm has been obtained for peptides [240]. This reference nicely addresses performance features of the linear QIT in particular, but mentions other instrument types as well. As with the resolution, to obtain better mass accuracy, sensitivity and speed have to be reduced.

The mass range depends on the settings of the main RF voltage and frequency. Unfortunately, selecting a lower frequency in order to extend the m/z range means sacrificing resolution. Typical m/z ranges are 15 to 3000 Th but can be extended (up to 20 kTh for commercial instruments) with reduction in other performance parameters. These m/z ranges usually cover most analytes when employing an ESI source, but for MALDI an upper m/z limit of even 20 kTh is still constraining.

The sensitivity of the QIT is rather high. Peptide levels in the low attomole range can be detected when operated with an LC system. With CE coupled to an ESI source, 100 pM peptide concentrations have been detected [241] and 100 amol of tryptic bovine serum albumin peptides deposited on a target have been detected with both vacuum MALDI and AP-MALDI sources [201]. The sensitivity would be increased even further with higher trapping efficiency. Usually the major portion of the ions entering the trap are never trapped, but goes straight through or hits the end cap at the far end.

The maximum number of ions in a cylindrical QIT is limited to about 10^5 before space-charge effects seriously affect the performance, so the dynamic range is rather poor. The poor dynamic range can sometimes be compensated for by using automatic gain control. The linear QIT has a larger volume and can store more ions before space-charge affects the performance.

Due to the limited dynamic range the QIT is not a particularly good analyzer for quantification. For pure samples the problem is less pronounced, but if a background is present the background ions will constitute a substantial portion of the total number of ions entering the QIT and, hence, affect the quality of the quantification. Typical scan speeds are ~5 kTh/s (several commercial systems offer the possibility of different scan speeds). Higher resolution is attained by reducing the scan speed.

The ion trap is excellent for multistage mass spectrometry (analogous to FTICR experiments, see Section 2.2.6), utilizing low-energy collision induced dissociation (CID). One small drawback is the practical mass selection interval, which often cannot be set narrow enough to select a single isotope, especially not in the case of multiply charged ions. There is a practical cut-off for detection of low mass fragments. Often this limit is set to 25% of the precursor m/z.

Quadrupole ion traps were originally coupled to continuous ion sources but also work well with pulsed ion sources. The cylindrical QIT is a compact device with a diameter and length of ~5 cm and the linear QIT is the size of a quadrupole mass filter, that is, ~20 cm long. As stand-alone instruments they are nowadays of benchtop size and, together with quadrupole mass filters, are considered to be standard low-cost devices, which are commonly coupled to LC systems.

2.2.5. Orbitrap

The orbitrap is the most recently invented mass analyzer. Like with the QIT, ions are trapped and stored in a potential well. However, instead of ejecting the ions for external detection the frequency of the trapped oscillationg ions is measured. This method provides substantially better resolution and mass accuracy in normal operation.

2.2.5.1. Principle. The orbitrap is a pure electrostatic device operated using a logarithmic electrostatic field between its inner and outer electrodes, as well as a quadrupolar field between its end caps. The orbitrap is inspired by the Kingdon trap invented in 1923 [242], which consists of a thin wire central electrode, a coaxial outer electrode, and two end-cap electrodes. Between the inner and outer electrodes a DC voltage is applied giving a logarithmic potential. Ions injected (through a narrow slit in the outer electrode) perpendicular to the wire and having an appropriate velocity will circulate in an orbit around the wire. By applying a potential to the end caps, the ions will be confined also axially. In 1981 Knight reshaped the outer electrode [243] so that an additional axial quadrupolar term was added to the potential. Now the ions can be trapped in an orbit by the radial logarithmic field and axially by the qadrupole field, which forces the ions into a harmonic oscillation in the z-direction. However, there were distortion problems that were proposed to originate from the central wire influencing the quadrupolar field. Hence, cross-terms in r and z might be needed to accurately describe the potential.

The orbitrap (Fig. 2.18) was invented by Makarov in 1999 [244, 245]. It can be seen as a modified Knight/Kingdon trap because of its general construction. It can also be seen as a modified quadrupole trap that uses electrostatic fields instead of dynamic. Ions move in stable trajectories both around the central electrode and in harmonic

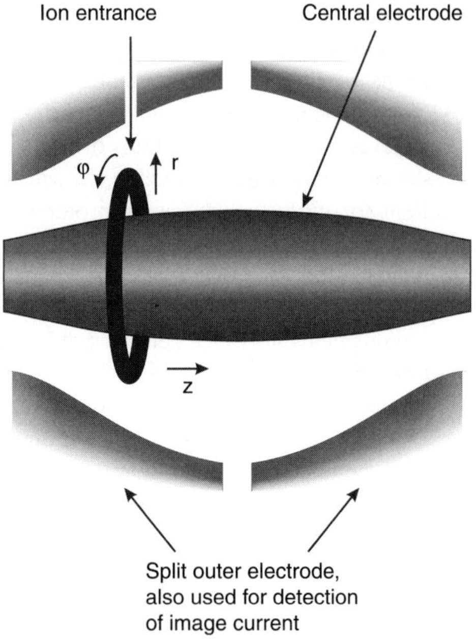

Figure 2.18. Schematic of an orbitrap analyzer. The *z*-direction oscillary motion of the ions induces an image current that is detected by the electrodes.

oscillations in the z-direction. The specially shaped electrodes give a potential that has no cross terms in r and z, meaning that the z-direction potential is purely quadratic. Thus, ion motion in the z-direction is independent and m/z is related to the ion oscillation frequency, f_z, along the z-axis as

$$f_z \propto \frac{1}{\sqrt{m/z}}.$$

See Reference 245 for a detailed description of the equations of motions.

As for the other trapping instruments, ions from the source have to be guided into the orbitrap. In the original orbitrap described in the literature, an ion accumulation quadrupole was employed that utilized collisional cooling of the ions and had an S-shaped exit path to eliminate residual gas flow into the orbitrap [246, 247]. However, the orbitrap itself does not utilize collisional cooling and therefore the final step of injecting the ions has special requirements. Ions injected into Paul traps, linear ion traps, and ion cyclotron resonance (ICR) cells in some instruments, are collisionally cooled in the trap, allowing for a quite generous velocity range and a rather long injection time. To obtain stable ion trajectories in the orbitrap, the injected ions must have sufficient tangential velocity, otherwise they will collide with the inner electrode. Such long injection times would correspond to hundreds of ion oscillations and would compromise ion stability. A fast injection is obtained by applying a high voltage on the exit lens from the accumulating trap and minimizing the distance to the orbitrap. Thus, ion injection can be minimized to only a few microseconds, and ions with the same m/z will enter the orbitrap in small millimeter-long packets. Each m/z packet has then a flight time spread of typically less than a picosecond, which is considerably shorter than half a period of axial oscillation. When these packets are injected into the orbitrap off its equator they begin coherent axial oscillations without the need for excitation. In the commercial hybrid linear ion trap-orbitrap instrument by Thermo Fischer Scientific, an additional C-shaped quadrupole trap is introduced between the linear (accumulation) trap and the orbitrap [248–250]. This trap is used to store and collisionally cool the ions before delivery to the orbitrap. The ions are pulsed into the central point of the C-trap arc, which coincides with the entrance to the orbitrap. With this arrangement, the practical m/z range of simultaneously injected ions is vastly improved.

Ion detection is carried out using image current detection with subsequent Fourier transform of the time-domain signal in the same way as for the Fourier transform ion cyclotron resonance (FTICR) analyzer (see Section 2.2.6). Because frequency can be measured very precisely, high m/z separation can be attained. Here, the axial frequency is measured, since it is independent to the first order on energy and spatial spread of the ions. Since the orbitrap, contrary to the other mass analyzers described, is a recent invention, not many variations of the instrument exist. Apart from Thermo Fischer Scientific's commercial instrument, there is the earlier setup described in References 245 to 247.

2.2.5.2. Performance Parameters. Typical resolving power for the commercial instrument is up to about 130,000 (FWHM) for m/z 400 Th. The mass resolving power is m/z dependent; it decreases with $\sqrt{m/z}$ (see the FTICR, described in Section 2.2.6, which decreases linearly with m/z).

The orbitrap mass accuracy is better than all quadrupoles and TOF instruments; just right after the FTICR and sector instruments, that is, around 2 ppm with internal calibration [248].

The mass range depends on the settings of the voltages. The m/z range offered by Thermo Fischer Scientific's orbitrap is 50 to 4000 Th [251; and supporting material at the internet site], which covers most analytes when employing an ESI source since larger species often are multiply charged, but is insufficient for many MALDI analyses.

Examples are given in References 249 and 250 of about 100 ions detected in a single scan. This is about the practical detection limit for image current detection due to thermal noise in the detection system. Bradykinin has been detected from a sample concentration of 3 nM [249] and detection of sub-femtomole levels on a column is readily obtained [251].

The maximum number of ions that can be contained in Thermo Fischer Scientific's orbitrap is higher than 10^6 without space-charge effects seriously affecting the performance, and the dynamic range is about 5000 with still good mass accuracy [249, 250]. At present, the C-trap limits the maximum number of ions that can be injected without performance loss. Being a trapping device the same issues as for the QIT apply. Although the orbitrap has a larger capacity, background ions may still influence the quantification capability.

The acquisition speed is, as for the FTICR, resolution dependent. With Thermo Fischer Scientific's orbitrap the desired mass resolving power can be selected. With the lowest setting (7500 FWHM) the acquisition time for one ion injection is ~0.3 s and with the highest setting (100,000 FWHM) it is ~1.9 s.

In principle, it would be possible to perform multistage mass spectrometry like in an ICR analyzer although with no gas CID would of course not be possible, but other dissociation methods could be employed. There might, however, be technical issues. At the time of writing, fragmentation is performed in the linear QIT preceeding the orbitrap in Thermo Fischer Scientific's instrument. Both pulsed and continuous ion sources can be employed. There are several ion sources that can be employed with Thermo Fischer Scientific's orbitrap.

The orbitrap itself is a compact device with a diameter and length less than 10 cm. Compared to an ICR instrument the ease of maintainance is superior since no liquid cooling is required. Also, complexity of the orbitrap is lower compared to most magnetic sector instruments. With no heavy magnet it is also relatively easy to move and transport, similar to a large TOF instrument. Ultrahigh vacuum ($\sim 10^{-10}$ torr) is required to obtain high resolution, since collisions lead to scattering, fragmentation, and transient decay.

2.2.6. Fourier Transform Ion Cyclotron Resonance

In the ion cyclotron resonance (ICR) analyzer, ions are trapped by a strong magnetic field. The magnetic field will cause the ions to move in a circular motion with a frequency that depends on their m/z. Ions to be detected are excited to make them move closer to the detection plates. Then a small current will be induced in the plate each time an ion passes by. Since the ions with different m/z have different ICR frequencies, each generated current frequency will correspond to a certain m/z value.

2.2.6.1. Principle. The principle of the ion cyclotron resonance was developed in the early 1930s by Lawrence and coworkers [252, 253]. The utilization of the ion cyclotron resonance (ICR) technique for mass spectrometry was introduced around 1950 by Sommer et al. [254, 255], and combination with the Fourier transform (FT) technique was developed by Comisarow and Marshall in 1974 [256]. Coupling of external sources to an FTICR analyzer was first done in 1985 [257, 258].

Like the quadrupole ion trap, the ICR mass analyzers are capable of storing ions in a cell. The geometry of the cell varies between different instruments—a cylindrical cell is shown in Fig. 2.19. It consists of three pairs of opposing plates used for trapping, excitation, and detection. The analyzer takes advantage of the same physical laws as the magnetic sector (one can think of the ICR analyzer as a sector with an arc of 360°). Ions with mass m and charge q moving in a spatially uniform magnetic field of strength $\vec{\mathbf{B}}$, directed perpendicular to the ions' direction of motion, will circulate in the cell with a cyclotron frequency f_c according to

$$f_c = \frac{qB}{2\pi \cdot m} \propto \frac{1}{m/z}.$$

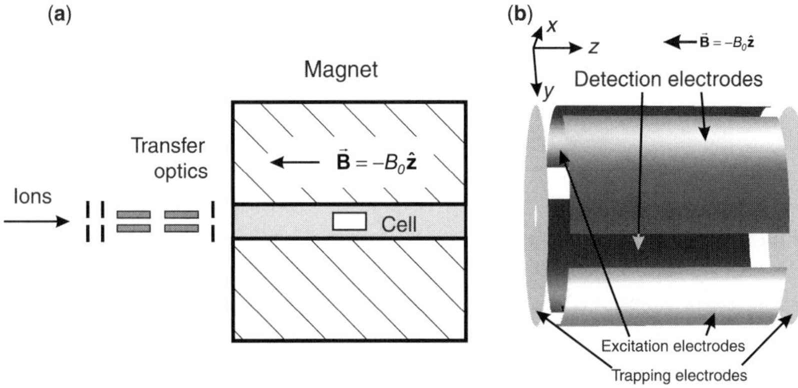

Figure 2.19. Schematic of a Fourier transform ion cyclotron resonance (FTICR) mass spectrometer (a) and a cylindrical cell (b). Reprinted from A. Westman-Brinkmalm and G. Brinkmalm (2002). In *Mass Spectrometry and Hyphenated Techniques in Neuropeptide Research*, J. Silberring and R. Ekman (eds.) New York: John Wiley & Sons, 47–105. With permission of John Wiley & Sons, Inc.

An interesting feature of this equation is that all ions of a certain m/z have the same cyclotron frequency, independent of their velocity. Hence, energy focusing is not essential for precise determination of m/z.

Ions generated in the source have to be guided into the cell, much in the same way as with the quadrupole ion trap. An additional challenge with the ICR analyzer is the very high magnetic field gradient that has to be overcome by the ions. If the radial (sideways) velocity component is large enough, compared to the axial velocity, the ions will, when experiencing an increasing magnetic field strength, turn around and never enter the cell. The two principal ways to deal with this magnetic mirror effect are either to accelerate the ions to sufficiently high velocity to pass the barrier or to hold the ions close to the magnet central axis, because the magnetic mirror force is zero on-axis. The problem with high ion velocity is that it is then necessary to slow them down for capturing in the cell.

Ions can be captured in the cell by allowing a short burst of gas into the cell. When the ions collide with the gas molecules, they will change direction and will thus be affected by the magnetic field and begin their cyclotron motion (an ion with no radial movement will ideally be unaffected by the magnetic field). This increases the trapping efficiency and also cools the ions in the same way as in the quadrupole ion trap, thus forcing them close to the center of the cell. Another method, which involves no gas, is trapping with a sidekick [259]. Here, electrodes convert the ions' kinetic energy component parallel to the magnetic field (i.e., in the z-direction) to kinetic energy perpendicular to the magnetic field. With no or almost no velocity component left in the z-direction, the ions can no longer escape the trap. Reference 260 describes several ways of trapping the ions.

The method for ion detection in an FTICR instrument is different from the majority of other mass spectrometers, where the ions hit a detector and are lost in the process.

To detect the ions trapped in the ICR cell, an RF voltage is applied to the two excitation electrodes (Fig. 2.19). Ions with the same ICR frequency as the RF voltage frequency are excited to a larger orbital radius. When the excitation is high enough so that the ions will pass close to the detection plates, a small current (or image charge) will be induced in the plate each time an ion passes by. Since the ions have different ICR frequencies depending on their m/z, each generated current frequency will correspond to a certain m/z.

Not only one type of ion, but ions of a certain m/z range can be excited to a larger orbital radius by applying an RF voltage to the excitation plates that includes all the corresponding frequencies. The detector electrodes will now pick up signals generated by all the ion packages that circle in the excited orbit, the so-called broadband detection. Each bundle will induce a signal every time it passes one of the electrodes and a complicated time domain signal is acquired. By Fourier transformation of the time spectrum, a frequency spectrum is obtained, which in turn can be converted to a mass spectrum. Even if recorded for a very short time, all frequencies corresponding to the excited ions will be present in the transformed data. However, the longer the acquisition time, the better the resolution.

Because long acquisition times are required for maximum resolution, it is essential that the ions can survive in the trap for extended periods of time. A main reason for ion loss is collisions with residual gas in the cell. Therefore, it is essential to keep the pressure as low as possible, preferably in the region of 10^{-9} torr or below. It is also important not to allow too many ions to enter the cell. When more than 10^6 to 10^7 ions are present in the cell, the coulomb repulsion can shift or broaden peaks in the mass spectra. By using an ion gate, the number of ions entering the cell can be limited in a controlled way.

To enhance the signal-to-noise ratio, several scans can be acquired and averaged. The nondestructive detection of the FTICR analyzer allows for remeasurement of the already detected ions (rather than to repeat the injection and trapping procedures).

2.2.6.2. Performance Parameters. For most mass spectrometric applications, the mass resolving power of the FTICR analyzer is unprecedented. A list of different world records is presented in Reference 260, among them being a mass resolving power of 8×10^6 (FWHM) for bovine ubiquitin (monoisotopic mass 8569.6 Da) and 5×10^6 (FWHM) for a depleted p16 tumor suppressor protein (monoisotopic mass 15,792.1 Da) [261]. In both cases the isotopic fine structure is revealed. The very high mass resolving power also allows for analysis of complex mixtures even in combination with ESI, which generates a multitude of charge states for each species.

A mass calibration for FTICR analyzers with superconducting magnets is very stable and is valid for many days for normal applications. Mass accuracy <1 ppm can be obtained over a fairly wide mass range. Unique elemental composition can be determined for masses over 800 Da [262]. Recently, 0.1 ppm mass accuracy, which required a mass resolving power $>300{,}000$, has been achieved for several thousand peaks by a 14.5 T instrument [263] and commercial instruments with mass accuracy <0.2 ppm are available. As with the orbitrap (see Section 2.2.5) the frequency is measured, which can be done very precisely, allowing for the very high resolution attainable with FTICR analyzers. For a magnetic sector the weak point is the determination of the variable magnetic field strength, but since that is kept constant in the FTICR it does not influence the measurement.

The m/z range is limited by the magnetic field strength of the magnet and the kinetic energy of the ions being trapped. For the commercial 9.4 T instruments, a typical m/z range is between ~30 and ~10,000. For example, the heaviest molecule detected at present is an ESI produced single DNA molecule of ~100 MDa with ~30,000 elementary charges [91]. The m/z of such a species is only ~3300 Th, well below the upper limit for normal operation.

The minimum amount of analyte required for analysis depends strongly (as always) on the nature of the substance and on the properties of the ion source. Less than 100 zeptomole of tryptic peptides from bovine serum albumin has been detected by coupling nanoflow LC to an ESI source [264] and 0.4 nM of equine cytochrome c by employing an RF-field focusing funnel to improve transmission [265].

Since a minimum of about 100 ions is needed to generate a detectable signal under normal circumstances (ion counting is inherently more sensitive than image current detection) and space-charge effects become influential with more than 10^6 to 10^7 ions, the dynamic range is relatively poor, about 10^4. The same applies to the FTICR as to the QIT and orbitrap. The signal depends on other species present in the trap at the same time, which limits quantification quality.

The FTICR analyzer is relatively slow. In a low resolution mode (<25,000 FWHM) scans can be performed in substantially less than a second. A high resolution scan is more time demanding, and more than 1 s is often required for mass resolving powers of 100,000 or more.

The FTICR analyzer has powerful capabilities for ion chemistry and MS/MS experiments. Several different fragmentation techniques can be utilized, such as CID, ECD, and IRMPD to mention a few; see Chapter 3 for descriptions. CID requires a collision gas in the FTICR cell and the gas has to be pumped away after fragmentation, which slows down the throughput. Since the detection is nondestructive, a precursor ion spectrum can first be recorded and then an ion species of interest can be isolated and subsequently fragmented, after which the fragment ion spectrum can be recorded, all in one single injection. Such a procedure is, however, somewhat cumbersome and sample demanding. Instead, the present trend for commercial instruments is to combine the FTICR with a preceding quadrupole ion trap, thus allowing for greater flexibility when taking advantage of both the superior resolution of the FTICR analyzer and the greater speed and general sensitivity of the quadrupole ion trap. Like with the QIT, both continuous (e.g., ESI) and pulsed (e.g., MALDI) ion sources can successfully be coupled to an ICR analyzer.

The dominant feature of an FTICR instrument is the magnet. Commercial instruments typically have superconducting magnets, and are presently available with different magnetic field strengths up to 15 T, but there are also examples of instruments with up to 25 T field strength. The present designs facilitate active shielding of the magnetic field, thus considerably decreasing the size and weight compared to the previous passive shielding designs. The heart of the instrument, the cell, has typical dimensions of

$\sim$10 cm, depending somewhat on the geometry. See Reference 266 for a description of a commercial 9.4 T instrument with examples of performance and applications, and Reference 267 for a review of protein identification with an FTICR analyzer.

2.2.7. Accelerator Mass Spectrometry

An accelerator mass spectrometer is not really just an instrument with another type of analyzer, but rather a system that utilizes magnetic/electric sectors to separate ion species. In fact, it can also be looked upon as an ion source. However, we felt it belonged best among the analyzers, because of the way it is utilized.

It is used to measure isotope ratios with very high sensitivity. Normally, a tandem accelerator is employed, but, for example, a cyclotron has also been used [268]. Compared to other mass spectrometric instruments accelerators are large installations and can require a whole building, or at least a substantial part of one, for itself. The last years, however, there has been a trend towards smaller devices with lower high voltage potentials [269]. Originally these devices were used for nuclear physics experiments, which required particles with kinetic energies in the megaelectronvolt range. Tandem accelerators have also been used as ion sources for plasma desorption mass spectrometry (see Section 2.1.20), where they could provide a variety of more defined ion beams, as compared to the more commonly used ^{252}Cf source. As early as the late 1930s, an accelerator was used as a sensitive mass spectrometer and the ^{3}He isotope was discovered [270]. Then the method was laid to rest until 1977 when several laboratories revived the technique [268, 271, 272]. Since then the method has found more applications and currently there are some tens of facilities in the world.

Accelerator mass spectrometry (AMS) is a very sensitive tool to count individual atoms of low abundance in a sample, usually radionuclides. The typical example is ^{14}C dating of archaeological items, where samples younger than about 10,000 years can be dated with a precision of $\sim$0.3% [273]. There are, however, many other applications where the abundance of other rare isotopes can be determined, for example, ice core dating with ^{10}Be; hydrology using ^{36}Cl; biomedicine using ^{14}C, ^{26}Al, ^{41}Ca, and ^{32}Si; nuclear weapons and the nuclear industry using ^{63}Ni and ^{41}Ca; cosmic ray interaction with the atmosphere using ^{10}Be and ^{26}Al; rock age using ^{10}Be, ^{26}Al, and ^{36}Cl; and global climate change using ^{14}C. For more thorough reviews see, for example, References 269 and 274 to 276.

2.2.7.1. Principle. A tandem accelerator mass spectrometer (Fig. 2.20) consists of a source where the sample is bombarded by positive cesium ions from the same type of ion gun as used in SIMS (see Section 2.1.18). Negative (secondary) ions coming off the sample are accelerated from a high negative potential (-30 to -200 kV) towards ground potential, thus obtaining a relatively high kinetic energy. Thereafter, the ions of interest are m/z selected by a magnet, like in a sector instrument. The ions that come out of the magnet travel towards the accelerator. The accelerator is a Van de Graaff apparatus consisting of a large metal tank that protects the so-called terminal that is set at a very high positive potential, 1 to 25 MV. This potential is achieved by continously transporting charges on a large rubber belt. The terminal and the tank walls

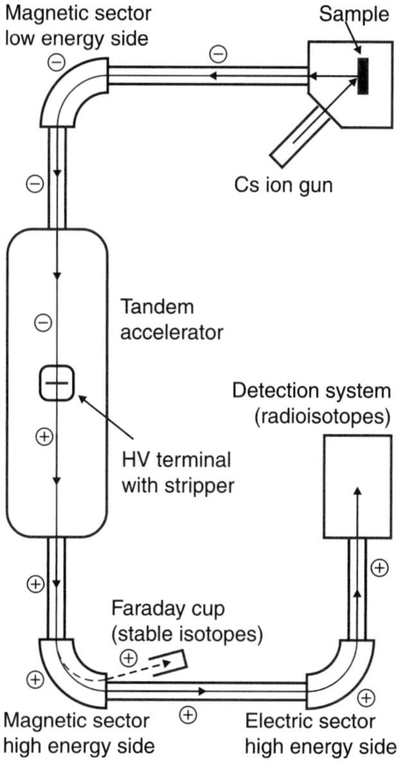

Figure 2.20. Schematic of a tandem accelerator set-up. A Cs gun irradiates the material to be analyzed. Negative secondary ions are accelerated and separated according to their m/z in the low-energy side magnetic sector. The selected negative ions are then accelerated towards the HV terminal where they pass through a stripper gas or foil and lose many of their electrons. The positive ions will have several different charge states. The vast majority of molecules entering the terminal do not survive but dissociate in the stripping process. The positive ions created are now repelled from the terminal and further accelerated towards the high energy side magnetic sector. In this sector the positive ions undergo another m/z separation. In this example ions belonging to a selected stable isotope are directed towards a Faraday cup and the intensity of that ion beam component is registered. The radioisotope component of the ion beam continues towards an electric sector that allows for further separation in order to separate interfering components with m/z close to that of the isotope of interest. Finally, the portion of the ion beam that passes through even this selection is registered by the detector, which may provide even further separation.

are separated by a chamber that is under vacuum or contains an insulating gas. In the terminal either thin carbon foils can be placed or gas can be let in (like in the collision cell of other instruments). The negative ions that now enter the accelerator are accelerated to a very high kinetic energy and, when they enter the terminal and pass through the foil or the gas, they will be stripped of several of their electrons, and thus become positively charged ions. Therefore, they will now be repelled by the terminal and be further accelerated towards the other end of the accelerator; hence the term tandem accelerator. The outcoming ions will then be subject to a second magnet and an electric sector to single out the ions of interest.

The measurement is always relative, that is, in ^{14}C measurements the amount of ^{14}C is always related to that of ^{13}C or ^{12}C. The more abundant isotope is detected by a Faraday detector while the less abundant isotope is detected by a specialized device (Fig. 2.20).

Why then, is such a complicated and expensive set up necessary? AMS combines mass spectrometric features with efficient discrimination of isobaric and molecular interferences. Therefore, it can detect and quantify atomic species of very low abundance. In the case of ^{14}C dating, before AMS was utilized, about 1 g of carbon was needed to date an archaeological item. One gram of fresh carbon contains about 6×10^{10} ^{14}C atoms, of which 14 decay per minute. To get 0.5% statistical precision using decay counting, a 48 h acquisition time is necessary. The same result can be obtained with AMS in about 10 min and with only 1 mg of carbon.

With a standard low energy mass spectrometer the signal from the very low abundant ^{14}C would drown in the intense ^{14}N signal, and signals from molecular species such as $^{13}C^{1}H$, $^{12}C^{1}H_2$, $^{12}C^{2}H$, and $^{7}Li_2$, as well as from background of ^{12}C and ^{13}C. By initially creating negative ions, the problem with ^{14}N is overcome because the $^{14}N^-$ ion is very unstable and will not pass through the first magnet. Because of the high energy impact in the stripping process the vast majority of molecular ions will fragment to atomic species in the terminal. Energy levels in the megaelectronvolt region are also critical for complete separation of isobaric species that still pass through the system. Accelerators with higher energy can better separate the isobars exiting the accelerator stage. Since the scattering cross section decreases with increasing ion energy, spatially better defined beams can be produced with higher energy. The less diffuse beams can thus be more easily separated by the magnetic and electric sectors.

The detection system depends on the actual application, that is, the isotopes that are analyzed. A standard detector is the solid state detector (see Section 2.3.5), which measures the energy of the incoming particle. However, to separate certain isobars, such as ^{41}Ca and ^{41}K, energy loss analysis is necessary [274]. This can be obtained by using a gas ionization detector. There are other ways to discriminate between isobars, for example, TOF systems, gas-filled magnets, and X-ray detectors [275]. The method that is utilized depends on the specific isobars. The very high kinetic energy of the ions impinging the detector also leads to a single ion counting detection efficiency of virtually 100%, which is not possible in standard mass spectrometers, which detect ions in the kiloelectronvolt range. Finally, the sample preparation is of great importance for the result. Utter care is mandatory in order to minimize sample loss, for example, in the purification process, and to avoid contamination with isobaric components.

2.2.7.2. Performance Parameters. Since the detector is often involved in the separation of isobars, normal mass analyzer resolution and mass accuracy do not really apply. The mass spectrometric resolution would be determined by the magnetic and electric sectors. Only atomic species are analyzed, so that sets the upper m/z

limit. At the other end of the range hydrogen can be analyzed. Apart from these restrictions the magnetic and electric sectors determine the upper m/z limit.

The sensitivity is very high and subattomole levels can be detected. In AMS different detectors are utilized for monitoring the different species analyzed. The more abundant isotopes like ^{12}C and ^{13}C are measured by a Faraday detector (see Section 2.3.2). Low abundant isotopes like ^{14}C are detected by an ion counting device. This way an extremely high sensitivity can be obtained. The best systems can provide a dynamic range of $>10^{15}$. Filling an average house from floor to ceiling with sugar grains, we can fill in roughly 10^{12} grains. A ^{14}C measurement of modern organic carbon with AMS is capable of finding one sugar grain with different mass in the house filled with sugar. Furthermore, a ^{14}C measurement in a 57,300-year-old sample is the equivalent of finding one sugar grain with different mass in 1000 houses filled from floor to ceiling with sugar [269]. AMS is only used as a quantification tool. The precision varies from $\sim$0.3% for ^{14}C to a few percent for other isotopes.

Analysis time is typically of the order of minutes to hours depending on the sample. Normally the time spent in actual AMS analysis is not the constraining factor, but rather sample purification prior to the spectrometric analysis. Accelerator mass spectrometers are space demanding facilities that typically occupy hundreds of square meters. Normally, dedicated personnel operate the device. Considerable effort is directed into refining the methods to allow operation by smaller, less costly facilities.

2.3. DETECTORS

The role of the detector is to convert the energy of incoming particles into a current signal that is registered by the electronic devices and transferred to the computer of the acquisition system of the mass spectrometer. When an in-coming particle strikes the detector the energy from the impact causes emission of secondary particles, for example, electrons or photons. The number of secondary particles created by an impact most often depends on the energy and/or the velocity of the incoming ion (see Section 2.3.6 on cryogenic detectors for velocity independent detection). Hence, if all ions are accelerated to the same kinetic energy, as in the TOF mass analyzer, then the detection sensitivity is lower for high mass (slow) ions then for low mass (fast) ions. To increase the detection sensitivity the ions are often post-accelerated before they strike the detector. A detector preferably should have high efficiency for converting the energy of the incoming ion to electrons or photons, a linear response, low noise, short recovery time, and minimal variations in transit time (narrow peak width). In this section some of the detectors currently (or recently) used in mass spectrometers are described. Hybrids of the different detectors are also frequently used. See Reference 277 for an excellent review of MS detectors.

2.3.1. Photoplate Detector

The photoplate detector is one of the oldest types of MS detectors (Fig. 2.21a). J.J. Thompson used photoplates to record mass spectra [4]. Photoplates were for a

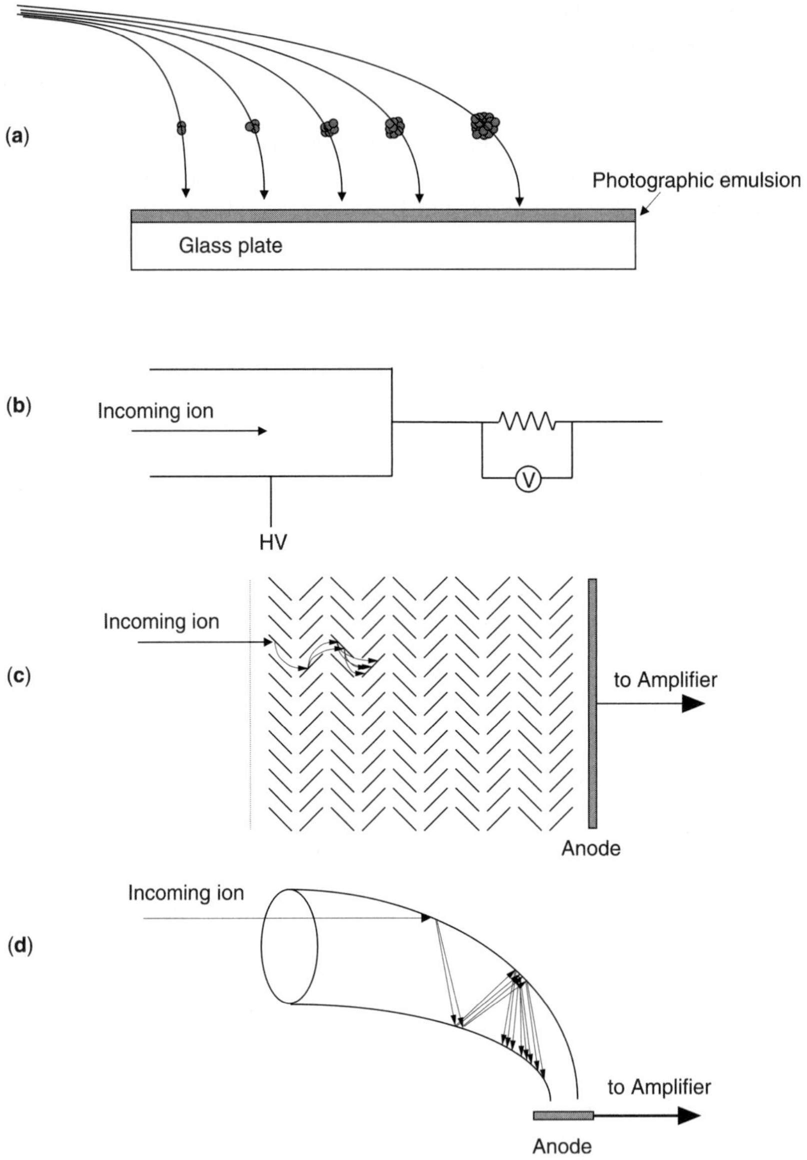

Figure 2.21. Schematic of (a) a photoplate detector; (b) a Faraday cup; (c) a discrete-dynode electron multiplier (EM) of venetian blind type; and (d) a continuous dynode EM. Parts (c) and (d) reprinted from A. Westman-Brinkmalm and G. Brinkmalm (2002). In *Mass Spectrometry and Hyphenated Techniques in Neuropeptide Research*, J. Silberring and R. Ekman (eds.) New York: John Wiley & Sons, 47–105. With permission of John Wiley & Sons, Inc.

long time one of the most common detectors for multicomponent detection. In more recent times the photoplate has mainly been used in conjunction with spark source (SS) MS (see Section 2.1.3). In SSMS ion-sensitive photoplates are used to simultaneously integrate ion beams separated by a double focusing sector analyzer (see Section 2.2.2). Even though the photoplate has advantages such as simultanous detection and possibilities of signal integration, compared to modern electrical devices the disadvantages are numerous, including poor sensitivity, short linear range, off-line image processing, off-line calibration, and shortage of commercial suppliers.

2.3.2. Faraday Detector

Unlike the photoplate, the Faraday detector (or Faraday cup) is still very much in use today. The main reasons for its lasting popularity are accuracy, reliability, and rugged construction. The simplest form of Faraday detector is a metal (conductive) cup that collects charged particles and is electrically connected to an instrument that measures the produced current (Fig. 2.21b). Faraday cups are not particularly sensitive and the signal produced must in most applications be significantly amplified. An important application for Faraday detectors is precise measurements of ratios of stable isotopes [278]. See, for example, Section 2.2.7 and Chapter 11 for examples of applications and methods in which Faraday detectors are utilized.

2.3.3. Electron Multipliers

An electron multiplier (EM) amplifies a weak current of incoming particles by using a series of secondary emission electrodes or dynodes to produce a considerably higher current at the anode [279]. When a particle impinges on the dynode, energy is transferred directly to the electrons in the dynode material and a number of secondary electrons are emitted. The dynode often consists of an alloy of an alkali or alkali earth metal with a more noble metal. Thus, a thin insulating film of oxidized alkaline metal is formed on a conducting support. A good dynode material should emit many secondary electrons per primary incoming particle, have linear gain for high currents, and have low thermionic emission, that is, low noise. EM detectors are today the most common MS detectors and are available in a variety of designs.

2.3.3.1. Discrete-Dynode Electron Multiplier. In a discrete-dynode EM the string of dynodes is connected via a chain of resistors. A high voltage, typically 1 to 3 kV, is applied to the first dynode and when a particle strikes the detector secondary electrons are accelerated from one dynode to the next towards the anode at ground potential. The velocity of the secondary electrons and hence the gain of the electron multiplier depends on the voltage applied to the dynode string.

Dynode strings can be constructed in many ways and the response time and range of linearity of the detector depend on the configuration. In the venetian blind configuration (Fig. 2.21c) the dynodes are wide strips of material placed at an angle of 45° with respect to the electron cascade axis. This system offers a large input area to the incident primary particles. The advantage is that the dynodes are easily placed in line and the dimensions are not critical. The disadvantage is that it is impossible to prevent a fraction of the primary electrons from passing straight through. This results in low gain and large variations in transit time. In the box and grid configuration, the linear focused configuration, and the circular focused configuration the electrons are reflected from one dynode to the next, which isolates the cathode and anode well from each other, minimizing the risk of feedback. Another advantage of these latter detector configurations is that the efficient spacing allows many dynodes, that is, high amplification.

2.3.3.2. Continuous-Dynode Electron Multiplier. Very compact electron multipliers that consist of one continuous dynode (Fig. 2.21d) are also often used [280]. The upper limit of the gain is here set by the onset of ion feedback. Ion feedback is caused by positive ions produced in the high charge density region at the output of the channel that drift back to the input. In the case of large single channels, ion feedback can be suppressed simply by bending or twisting the channels. Another example of a compact electron multiplier configuration with short transit times is one of the most frequently used detectors in mass spectrometers, the microchannel plate (MCP, also sometimes referred to as multichannel plate) [280]. An MCP consists of a parallel array of channel electron multipliers (Fig. 2.22). The inner surfaces of the channels are treated with a semiconductor material acting as a secondary electron emitter. The flat end surfaces of the channels are coated with a metallic alloy to allow a potential difference to be applied over the MCP. When a particle strikes the channel wall, secondary electrons are generated and accelerated down the channel toward the output end. The secondary electrons also hit the channel wall and an avalanche of electrons is generated. Each channel thus acts as a continuous dynode that generates up to 10^4 electrons per impact. Two MCPs are often stacked together to achieve higher gain.

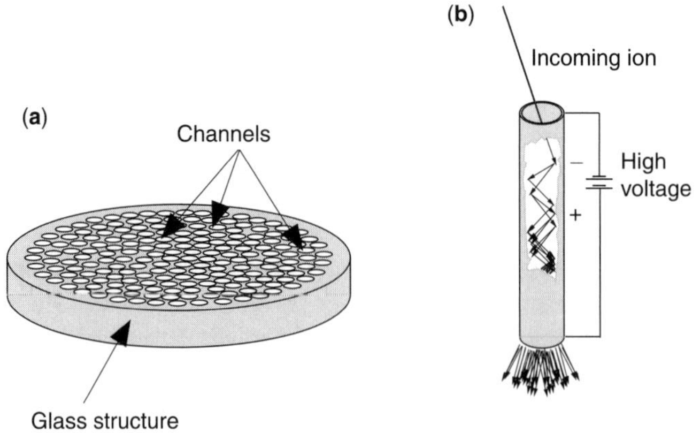

Figure 2.22. Schematic of an MCP (a) with a close-up of one channel (b). The size of the channels in the MCP schematic is greatly exaggerated compared to the size of the plate. Reprinted from A. Westman-Brinkmalm and G. Brinkmalm (2002). In *Mass Spectrometry and Hyphenated Techniques in Neuropeptide Research*, J. Silberring and R. Ekman (eds.) New York: John Wiley & Sons, 47–105. With permission of John Wiley & Sons, Inc.

A drawback of MCPs is that an electron cascade in one channel drains the neighboring channels for several microseconds, leading to nonlinearities for count rates above a few thousand incident particles per second. This saturation effect is a major disadvantage when utilizing MCPs in a MALDI-TOF mass spectrometer. The yield of matrix ions is normally very high and since the molecular mass of the matrix is low ($\sim$200 Da), their flight time is short, and they arrive at the detector before the analyte ions. Consequently, the gain of the detector will be low when the heavier (slower) analyte molecular ions arrive, thereby considerably reducing the sensitivity of the MCP detector. The saturation is more pronounced in the second MCP in a tandem MCP arrangement since each ion hitting the first MCP produces up to 10^4 electrons that strike the second MCP.

However, the gain of the MCP depends strongly on the potential applied and a modification involving pulsing of the voltage over the first MCP in a tandem MCP detector has been successfully employed to reduce detector saturation [281, 282]. When the matrix ions arrive to the first MCP the voltage, and consequently the amplification, is low. Thus, fewer electrons strike the second MCP and fewer channels are saturated. The potential applied to the first MCP is then increased and when the sample ions arrive to the MCP detector both MCPs have maximum amplification. The time between the firing of the laser and the rise of the potential, that is, increase in amplification, can easily be varied and different portions of the spectrum are thus attenuated or amplified. Another way of increasing the dynamic range of the detector is to exchange the second MCP with a conventional electron multiplier.

2.3.4. Focal Plane Detector

Focal plane detectors are used primarily to detect ions separated in space by, for example, magnetic sector analyzers (see Section 2.2.2). The objective of an ideal focal plane detector is to simultaneously record the location of every ion in the spectrum. In many ways the photoplate (see Section 2.3.1) is the original focal plane detector, but it has today been more or less replaced with designs that rely on EM detectors (see Section 2.3.3). A common arrangement is to allow the spatially disperse ion beams simultaneously to impinge on an MCP (see Section 2.3.3.2). The secondary electrons generated by the ion impacts then strike a one- or two-dimensional array of metal strips and the current from the individual electrodes is recorded. A tutorial on the fundamentals of focal plane detectors is found in Reference 283. Reference 284 provides a relatively recent review of MS detector-array technology.

2.3.5. Scintillation Detector

The scintillation detector makes use of the fact that certain materials, when struck by a particle, emit a small flash of light, so-called scintillation [279]. Generally the detector consists of a scintillating material that is optically coupled to an amplifying device such as a photomultiplier. When an ion passes through the scintillator, it excites the atoms and molecules, causing light to be emitted. The light is transmitted to the photomultiplier, where it is converted into a weak current of photoelectrons, which is further amplified by an electron-multiplier system. The light output of a scintillator is directly proportional to the excitation energy of the incoming particle. Since the photomultiplier also is a linear device, the amplitude of the final electrical signal will also be proportional to the energy of the incoming particle. Scintillation detectors have fast recovery time and therefore accept high count rates.

2.3.6. Cryogenic Detector

Cryogenic or energy-sensitive calorimetric detectors measure the heat generated by a particle impacting on a superconducting thin film. One example of such a cryogenic detector is the so-called superconducting tunnel junction (STJ) [285]. Cryogenic detectors measure low energy solid-state excitations (below 5 millielectronvolts) and must therefore be operated at temperatures typically below 2 K to avoid excessive thermal excitations. Compared to ionization-based detectors, which rely on electronvolt energies needed to produce secondary electrons or electronic excitations, cryogenic detectors are more sensitive to slow-moving (large) ions. Relatively large numbers of excitations are created for a given energy deposition, which allows the energy to be measured with smaller statistical error and, thus, greater precision. An advantage of cryogenic detectors is that they are able to distinguish the charge state of the ion. Another advantage is that, contrary to EM detectors such as the MCP, for example, the detection efficiency of cryogenic detectors is independent of the velocity of the ions. Cryogenic detectors have recently been coupled to TOF mass spectrometers and shown encouragingly good sensitivity for high-mass ions in the megadalton mass range [286]. The main disadvantages of cryogenic detectors are the extensive cooling necessary for operation and the practical problems when connecting the cooled detector with a spectrometer at room temperature. Reference 286 provides an indepth account of the analytical possibilities and practical difficulties with a MALDI-TOF instrument equipped with a cryogenic STJ detector.

2.3.7. Solid-State Detector

Solid-state detectors (SSD) consist of silicon or germanium that emits electrons in response to ionizing radiation. SSDs are mainly used to measure the energy of a particle as a compliment to an m/z determination. The addition of an SSD to determine the residual energy of the ion allows the unambiguous determination of mass, charge, and energy independently, instead of obtaining just ratios of energy to charge or mass to charge. Two areas of application are space science (see Chapter 11) and isotope ratio measurements [278].

2.3.8. Image Current Detection

Image current detection is (currently) the only nondestructive detection method in MS. The two mass analyzers that employ image current detection are the FTICR and the orbitrap. In the FTICR ions are trapped in a magnetic field and move in a circular motion with a frequency that depends on their m/z. Correspondingly, in the orbitrap ions move in harmonic oscillations in the z-direction with a frequency that is m/z dependent but independent of the energy and spatial spread of the ions. For detection ions are made to pass close to a pair of detection plates. Each time an ion passes a plate a small current (or image charge) will be induced. Thus, the frequency of each generated current will correspond to a certain m/z. An important advantage of nondestructive detection is that it allows for remeasurement of the already detected ions. More detailed descriptions of the respective methods of ion detection are found in the sections describing the FTICR (see Section 2.2.6) and the orbitrap (see Section 2.2.5) mass analyzers. See also Reference 287 for a comprehensive account of the principles of FTICR detection.

REFERENCES

1. E. Goldstein. Über eine noch nicht untersuchte Strahlungsform an der Kathode inducirter Entladungen. *Sitzungsber. Königl. Preuss. Akad. Wiss. Berlin*, **2**(1886): 691–699.
2. W. Wien. Untersuchungen über die electrische Entladung in verdünnten Gasen. *Ann. Physik*, **301**(1898): 440–452.
3. J. J. Thomson. On Rays of Positive Electricity. *Phil. Mag.*, **Ser. 6, 13**(1907): 561–575.
4. J. J. Thomson. Rays of Positive Electricity. *Proc. Roy. Soc. London*, **A89**(1913): 1–20.
5. E. Gehrcke and O. Reichenheim. *Verh. Deutsch. Phys. Ges.*, **8**(1906): 559.
6. A. J. Dempster. A New Method of Positive Ray Analysis. *Phys. Rev.*, **11**(1918): 316–325.
7. M. G. Inghram and W. A. Chupka. Surface Ionization Source Using Multiple Filaments. *Rev. Sci. Instrum.*, **24**(1953): 518–520.
8. K. G. Heumann. Isotope Dilution Mass Spectrometry (IDMS) of the Elements. *Mass Spectrom. Rev.*, **11**(1992): 41–67.
9. K. G. Heumann. Isotope-dilution ICP–MS for Trace Element Determination and Speciation: From a Reference Method to a Routine Method? *Anal. Bioanal. Chem.*, **378**(2004): 318–329.
10. T. Walczyk. Iron Isotope Ratio Measurements by Negative Thermal Ionisation Mass Spectrometry using FeF_4^- Molecular Ions. *Int. J. Mass Spectrom. Ion Proc.*, **161**(1997): 217–227.
11. A. Deyhle. Improvements of Boron Isotope Analysis by Positive Thermal Ionization Mass Spectrometry Using Static Multicollection of $Cs_2BO_2^+$ Ions. *Int. J. Mass Spectrom.*, **206**(2001): 79–89.
12. R. Doucelance and G. Manhes. Reevaluation of Precise Lead Isotope Measurements by Thermal Ionization Mass Spectrometry: Comparison with Determinations by Plasma Source Mass Spectrometry. *Chem. Geol.*, **176**(2001): 361–377.
13. J. L. Mann and W. R. Kelly. Measurement of Sulfur Isotope Composition ($\delta^{34}S$) by Multiple-Collector Thermal Ionization Mass Spectrometry Using a ^{33}S–^{36}S Double Spike. *Rapid Commun. Mass Spectrom.*, **19**(2005): 3429–3441.
14. M. D. Schmitz, S. A. Bowring, and T. R. Ireland. Evaluation of Duluth Complex Anorthositic Series (AS3) Zircon as a U-Pb Geochronological Standard: New High-Precision Isotope Dilution Thermal Ionization Mass Spectrometry Results, *Geochim. Cosmochim. Acta*, **67**(2003): 3665–3672.
15. A. J. Dempster. Isotopic Structure of Iridium. *Nature*, **136**(1935): 909.
16. A. J. Dempster. Ion Sources for Mass Spectroscopy. *Rev. Sci. Instrum.*, **7**(1936): 46–49.

17. J. G. Gorman, E. J. Jones, and J. A. Hipple. Analysis of Solids with Mass Spectrometer. *Anal. Chem.*, **23**(1951): 438–440.
18. N. B. Hannay. A Mass Spectrograph for the Analysis of Solids. *Rev. Sci. Instrum.*, **25**(1954): 644–648.
19. N. B. Hannay and A. J. Ahearn. Mass Spectrographic Analysis of Solids. *Anal. Chem.*, **26**(1954): 1056–1058.
20. N. B. Hannay. Mass Spectrographic Analysis of Solids: High Sensitivity for Bulk and Surface Impurities is Provided by a New Analytical Method. *Science*, **134**(1961): 1220–1225.
21. J. S. Becker and H.-J. Dietze. Inorganic Mass Spectrometric Methods for Trace, Ultratrace, Isotope, and Surface Analysis. *Int. J. Mass Spectrom.*, **197**(2000): 1–35.
22. W. W. Harrison and C. W. Magee. Hollow Cathode Ion Source for Solids Mass Spectrometry. *Anal. Chem.*, **46**(1974): 461–464.
23. W. W. Harrison, C. M. Barschick, J. A. Klingler, P. H. Ratliff, and Y. Mei. Glow Discharge Techniques in Analytical Chemistry. *Anal. Chem.*, **62**(1990): 943A–949A.
24. F. L. King, J. Teng, and R. E. Steiner. Glow Discharge Mass Spectrometry: Trace Element Determinations in Solid Samples. *J. Mass Spectrom.*, **30**(1995): 1061–1075.
25. V. Hoffmann, M. Kasik, P. K. Robinson, and C. Venzago. Glow Discharge Mass Spectrometry. *Anal. Bioanal. Chem.*, **381**(2005): 173–188.
26. R. S. Houk, V. A. Fassel, G. D. Flesch, H. J. Svec, A. L. Gray, and C. E. Taylor. Inductively Coupled Argon Plasma as an Ion Source for Mass Spectrometric Determination of Trace Elements. *Anal. Chem.*, **52**(1980): 2283–2289.
27. A. L. Gray. Mass-Spectrometric Analysis of Solutions Using an Atmospheric Pressure Ion Source. *Analyst*, **100**(1975): 289–299.
28. R. S. Houk. Mass Spectrometry of Inductively Coupled Plasmas. *Anal. Chem.*, **58**(1986): 97A–105A.
29. A. Montaser, S. K. Chan, and D. W. Koppenaal. Inductively Coupled Helium Plasma as an Ion Source for Mass Spectrometry. *Anal. Chem.*, **59**(1987): 1240–1242.
30. J. A. C. Broekaert. Mass Spectrometry with Plasma Sources at Atmospheric Pressure: State-of-the-Art and Some Developmental Trends. *Fresenius. J. Anal. Chem.*, **368**(2000): 15–22.
31. A. J. Dempster. Positive Ray Analysis of Lithium and Magnesium. *Phys. Rev.*, **18**(1921): 415–422.
32. W. Bleakney. A New Method of Positive Ray Analysis and Its Application to the Measurement of Ionization Potentials in Mercury Vapor. *Phys. Rev.*, **34**(1929): 157–160.
33. A. O. Nier. A Mass Spectrometer for Routine Isotope Abundance Measurements. *Rev. Sci. Instrum.*, **11**(1940): 212–216.
34. A. O. Nier. A Mass Spectrometer for Isotope and Gas Analysis. *Rev. Sci. Instrum.*, **18**(1947): 398–411.
35. H. Deutsch, K. Becker, S. Matt, and T. D. Mark. Theoretical Determination of Absolute Electron-Impact Ionization Cross Sections of Molecules. *Int. J. Mass Spectrom.*, **197**(2000): 37–69.
36. M. S. B. Munson and F. H. Field. Chemical Ionization Mass Spectrometry. I. General Introduction. *J. Am. Chem. Soc.*, **88**(1966): 2621–2630.
37. F. H. Field. The Early Days of Chemical Ionization: A Reminiscence. *J. Am. Soc. Mass Spectrom.*, **1**(1990): 277–283.

38. B. Munson. Development of Chemical Ionization Mass Spectrometry. *Int. J. Mass Spectrom.*, **200**(2000): 243–251.

39. D. F. Hunt, G. C. Stafford, F. W. Crow, and J. W. Russell. Pulsed Positive Negative Ion Chemical Ionization Mass Spectrometry. *Anal. Chem.*, **48**(1976): 2098–2104.

40. A. G. Harrison. *Chemical Ionization Mass Spectrometry*. CRC Press, Roca Baton, FL, 1992.

41. R. W. Giese. Electron-Capture Mass Spectrometry: Recent Advances. *J. Chromatogr.*, **A892**(2000): 329–346.

42. E. C. Horning, M. G. Horning, D. I. Carroll, I. Dzidic, and R. N. Stillwell. New Picogram Detection System Based on a Mass Spectrometer with an External Ionization Source at Atmospheric Pressure. *Anal. Chem.*, **45**(1973): 936–943.

43. D. I. Carroll, I. Dzidic, R. N. Stillwell, M. G. Horning, and E. C. Horning. Subpicogram Detection System for Gas Phase Analysis Based upon Atmospheric Pressure Ionization (API) Mass Spectrometry. *Anal. Chem.*, **46**(1974): 706–710.

44. E. C. Horning, D. I. Carroll, I. Dzidic, K. D. Haegele, M. G. Horning, and R. N. Stillwell. Liquid Chromatograph-Mass Spectrometer-Computer Analytical Systems: A Continuous-Flow System Based on Atmospheric Pressure Ionization Mass Spectrometry. *J. Chromatogr.*, **A99**(1974): 13–21.

45. D. I. Carroll, I. Dzidic, R. N. Stillwell, K. D. Haegele, and E. C. Horning. Atmospheric Pressure Ionization Mass Spectrometry. Corona Discharge Ion Source for Use in a Liquid Chromatograph-Mass Spectrometer-Computer Analytical System. *Anal. Chem.*, **47**(1975): 2369–2373.

46. H. Hertz. Ueber einen Einfluss des ultravioletten Lichtes auf die electrische Entladung. *Ann. Physik Chemie*, **267**(1887): 983–1000.

47. A. Einstein. Über einen die Erzeugung und Verwandlung des Lichtes betreffenden heuristischen Gesichtspunkt. *Ann. Physik*, **322**(1905): 132–148.

48. R. W. Ditchburn and F. L. Arnot. The Ionisation of Potassium Vapour. *Proc. Roy. Soc. London*, **A123**(1929): 516–536.

49. H. Hurzeler, M. G. Inghram, and J. D. Morrison. Photon Impact Studies of Molecules Using a Mass Spectrometer. *J. Chem. Phys.*, **28**(1958): 76–82.

50. J. N. Driscoll and J. B. Clarici. Ein neuer Photoionisationsdetektor für die Gas-Chromatographie. *Chromatographia*, **9**(1976): 567–570.

51. D. C. Locke, B. S. Dhingra, and A. D. Baker. Liquid-Phase Photoionization Detector for Liquid Chromatography. *Anal. Chem.*, **54**(1982): 447–450.

52. M. A. Baim, R. L. Eatherton, and H. H. Hill, Jr. Ion Mobility Detector for Gas Chromatography with a Direct Photoionization Source. *Anal. Chem.*, **55**(1983): 1761–1766.

53. T. A. Cool, K. Nakajima, C. A. Taatjes, A. McIlroy, P. R. Westmoreland, M. E. Law, and A. Morel. Studies of a Fuel-rich Propane Flame with Photoionization Mass Spectrometry. *Proc. Combust. Inst.*, **30**(2005): 1681–1688.

54. N. K. Berezhetskaya, G. A. Varanov, G. A. Delone, N. B. Delone, and G. K. Piskova. *Sov. Phys. JETP*, **31**(1970): 403.

55. R. V. Ambartzumian and V. S. Letokhov. Selective Two-Step (STS) Photoionization of Atoms and Photodissociation of Molecules by Laser Radiation. *Appl. Opt.*, **11**(1972): 354–358.

56. V. S. Antonov, I. N. Knyazev, V. S. Letokhov, V. M. Matiuk, V. G. Movshev, and V. K. Potapov. Stepwise Laser Photoionization of Molecules in a Mass Spectrometer: A New Method for Probing and Detection of Polyatomic Molecules. *Opt. Lett.*, **3**(1978): 37.
57. U. Boesl, H. J. Neusser, and E. W. Schlag. Two-Photon Ionization of Polyatomic Molecules in a Mass Spectrometer. *Z. Naturforsch.*, **A33**(1978): 1546–1548.
58. L. Zandee, R. B. Bernstein, and D. A. Lichtin. Vibronic/Mass Spectroscopy via Multiphoton Ionization of a Molecular Beam: The I_2 Molecule. *J. Chem. Phys.*, **69**(1978): 3427–3429.
59. V. Antonov and V. Letokhov. Laser Multiphoton and Multistep Photoionization of Molecules and Mass Spectrometry. *Appl. Phys.*, **24**(1981): 89–106.
60. U. Boesl. Multiphoton Excitation and Mass-Selective Ion Detection for Neutral and Ion Spectroscopy. *J. Phys. Chem.*, **95**(1991): 2949–2962.
61. U. Boesl, J. Grotemeyer, K. Müller-Dethlefs, H. J. Neusser, H. L. Selzle, and E. W. Schlag. Multiphoton and Soft X-Ray Ionization Mass Spectrometry. *Int. J. Mass Spectrom. Ion Proc.*, **118/119**(1992): 191–220.
62. J. E. Braun and H. J. Neusser. Threshold Photoionization in Time-of-Flight Mass Spectrometry. *Mass Spectrom. Rev.*, **21**(2002): 16–36.
63. K. Wendt, K. Blaum, B. A. Bushaw, C. Grüning, R. Horn, G. Huber, J. V. Kratz, P. Kunz, P. Müller, W. Nörtershäuser, M. Nunnemann, G. Passler, A. Schmitt, N. Trautmann, and A. Waldek. Recent Developments in and Applications of Resonance Ionization Mass Spectrometry. *Fresenius. J. Anal. Chem.*, **364**(1999): 471–477.
64. K. W. D. Ledingham and R. P. Singhal. High Intensity Laser Mass Spectrometry: A Review. *Int. J. Mass Spectrom. Ion Proc.*, **163**(1997): 149–168.
65. I. A. Revel'skii, Y. S. Yashin, V. N. Voznesenskii, V. K. Kurochkin, and R. G. Kostyanovskii. Mass Spectrometry with Photoionization of *n*-Alkanes, Alcohols, Ketones, Esters, and Amines at Atmospheric Pressure. *Izv. Akad.Nauk SSSR Ser. Khim.*, **9**(1986): 1987–1992.
66. I. A. Revel'skii, Y. S. Yashin, V. N. Voznesenskii, V. K. Kurochkin, and R. G. Kostyanovskii. Mass Spectrometry with Photoionization of *n*-Alkanes, Alcohols, Ketones, Esters, and Amines at Atmospheric Pressure. *Russ. Chem. Bull.*, **35**(1986): 1806–1810.
67. D. R. Robb, T. R. Covey, and A. P. Bruins. Atmospheric Pressure Photoionization: An Ionization Method for Liquid Chromatography-Mass Spectrometry. *Anal. Chem.*, **72**(2000): 3653–3659.
68. K. A. Hanold, S. M. Fischer, P. H. Cormia, C. E. Miller, and J. A. Syage. Atmospheric Pressure Photoionization. I. General Properties for LC/MS. *Anal. Chem.*, **76**(2004): 2842–2851.
69. T. J. Kauppila, T. Kuuranne, E. C. Meurer, M. N. Eberlin, T. Kotiaho, and R. Kostiainen. Atmospheric Pressure Photoionization Mass Spectrometry: Ionization Mechanism and the Effect of Solvent on the Ionization of Naphthalenes. *Anal. Chem.*, **74**(2002): 5470–5479.
70. A. Raffaelli and A. Saba. Atmospheric Pressure Photoionization Mass Spectrometry. *Mass Spectrom. Rev.*, **22**(2003): 318–331.
71. E. W. Müller. Das Feldionenmikroskop. *Z. Phys.*, **A131**(1951): 136–142.
72. M. G. Inghram and R. Gomer. Mass Spectrometric Analysis of Ions from the Field Microscope. *J. Chem. Phys.*, **22**(1954): 1279–1280.
73. H. D. Beckey, E. Hilt, A. Maas, M. D. Migahed, and E. Ochterbeck. A Method for Strong Activation of Field Ion Emitters. *Int. J. Mass Spectrom. Ion Phys.*, **3**(1969): 161–165.
74. H. R. Schulten and H. D. Beckey. Field Desorption Mass Spectrometry with High Temperature Activated Emitters. *Org. Mass Spectrom.*, **6**(1972): 885–895.

75. J. N. Damico and R. P. Barron. Application of Field Ionization to Gas-Liquid Chromatography-Mass Spectrometry (GLC-MS) Studies. *Anal. Chem.*, **43**(1971): 17–21.

76. H. D. Beckey. Field Desorption Mass Spectrometry: A Technique for the Study of Thermally Unstable Substances of Low Volatility. *Int. J. Mass Spectrom. Ion Phys.*, **2**(1969): 500–503.

77. H. U. Winkler and H. D. Beckey. Field Desorption Mass Spectrometry of Peptides. *Biochem. Biophys. Res. Commun.*, **46**(1972): 391–398.

78. H.-R. Schulten and H. M. Schiebel. Principle and Technique of Field-Desorption Mass Spectrometry Analysis of Corrins and Vitamin B_{12}. *Naturwiss.*, **65**(1978): 223–230.

79. C. R. Blakley and M. L. Vestal. Thermospray Interface for Liquid Chromatography/Mass Spectrometry. *Anal. Chem.*, **55**(1983): 750–754.

80. P. Arpino. Combined Liquid Chromatography Mass Spectrometry. II. Techniques and Mechanisms of Thermospray. *Mass Spectrom. Rev.*, **9**(1990): 631–669.

81. P. Arpino. Combined Liquid Chromatography Mass Spectrometry. III. Applications of Thermospray. *Mass Spectrom. Rev.*, **11**(1992): 3–40.

82. M. Dole, L. L. Mack, R. L. Hines, R. C. Mobley, L. D. Ferguson, and M. B. Alice. Molecular Beams of Macroions. *J. Chem. Phys.*, **49**(1968): 2240–2249.

83. M. Yamashita and J. B. Fenn. Electrospray Ion Source: Another Variation on the Free-Jet Theme. *J. Phys. Chem.*, **88**(1984): 4451–4459.

84. M. L. Aleksandrov, L. N. Gall, V. N. Krasnov, V. I. Nikolaev, V. A. Pavlenko, and V. A. Shkurov. Ion Extraction from Solutions at Atmospheric Pressures: A Mass Spectrometric Method of Analysis of Bioorganic Compounds. *Dokl. Akad. Nauk SSSR*, **277**(1984): 379–383.

85. J. B. Fenn, M. Mann, C. K. Meng, S. F. Wong, and C. M. Whitehouse. Electrospray Ionization for Mass Spectrometry of Large Biomolecules. *Science*, **246**(1989): 64–71.

86. J. F. Mora, G. J. Van Berkel, C. G. Enke, R. B. Cole, M. Martinez-Sanchez, and J. B. Fenn. Electrochemical Processes in Electrospray Ionization Mass Spectrometry. *J. Mass Spectrom.*, **35**(2000): 939–952.

87. J. V. Iribarne and B. A. Thomson. On the Evaporation of Small Ions from Charged Droplets. *J. Chem. Phys.*, **64**(1976): 2287–2294.

88. M. Wilm and M. Mann. Analytical Properties of the Nanoelectrospray Ion Source. *Anal. Chem.*, **68**(1996): 1–8.

89. M. Wilm, A. Shevchenko, T. Houthaeve, S. Breit, L. Schweigerer, T. Fotsis, and M. Mann. Femtomole Sequencing of Proteins from Polyacrylamide Gels by Nano-Electrospray Mass Spectrometry. *Nature*, **379**(1996): 466–469.

90. M. A. Tito, K. Tars, K. Valegard, J. Hajdu, and C. V. Robinson. Electrospray Time-of-Flight Mass Spectrometry of the Intact MS2 Virus Capsid. *J. Am. Chem. Soc.*, **122**(2000): 3550–3551.

91. R. D. Chen, X. H. Cheng, D. W. Mitchell, S. A. Hofstadler, Q. Y. Wu, A. L. Rockwood, M. G. Sherman, and R. D. Smith. Trapping, Detection, and Mass Determination of Coliphage T4 DNA Ions of 10^8 Da by Electrospray Ionization Fourier Transform Ion Cyclotron Resonance Mass Spectrometry. *Anal. Chem.*, **67**(1995): 1159–1163.

92. L. Konermann and D. J. Douglas. Unfolding of Proteins Monitored by Electrospray Ionization Mass Spectrometry: A Comparison of Positive and Negative Ion Modes. *J. Am. Soc. Mass Spectrom.*, **9**(1998): 1248–1254.

93. A. T. Iavarone, J. C. Jurchen, and E. R. Williams. Effects of Solvent on the Maximum Charge State and Charge State Distribution of Protein Ions Produced by Electrospray Ionization. *J. Am. Soc. Mass Spectrom.*, **11**(2000): 976–985.

94. Y.-T. Li, Y.-L. Hsieh, J. Henion, M. W. Senko, F. McLafferty, and B. Ganem. Mass Spectrometric Studies on Noncovalent Dimers of Leucine Zipper Peptides. *J. Am. Chem. Soc.*, **115**(1993): 8409–8415.

95. S. A. Carr, M. E. Hemling, M. F. Bean, and G. D. Roberts. Integration of Mass Spectrometry in Analytical Biotechnology. *Anal. Chem.*, **63**(1991): 2802–2824.

96. M. Karas, U. Bahr, and T. Dulcks. Nano-Electrospray Ionization Mass Spectrometry: Addressing Analytical Problems beyond Routine. *Fresenius. J. Anal. Chem.*, **366**(2000): 669–676.

97. Y. Wang, M. Schubert, A. Ingendohm and J. Franzen. Analysis of Non-Covalent Protein Complexes up to 290 kDa Using Electrospray Ionization and Ion Trap Mass Spectrometry. *Rapid Commun. Mass Spectrom.*, **14**(2000): 12–17.

98. B. N. Pramanik, P. L. Bartner, U. A. Mirza, Y. H. Liu, and A. K. Ganguly. Electrospray Ionization Mass Spectrometry for the Study of Non-Covalent Complexes: An Emerging Technology. *J. Mass Spectrom.*, **33**(1998): 911–920.

99. S. A. Hofstadler and K. A. Sannes-Lowery. Applications of ESI-MS in Drug Discovery: Interrogation of Noncovalent Complexes. *Nat. Rev. Drug Discov.*, **5**(2006): 585–595.

100. R. D. Smith, J. A. Loo, C. G. Edmonds, C. J. Barinaga, and H. R. Udseth. New Developments in Biochemical Mass Spectrometry: Electrospray Ionization. *Anal. Chem.*, **62**(1990): 882–899.

101. M. Carrascal, K. Schneider, R. E. Calaf, S. van Leeuwen, D. Canosa, E. Gelpi, and J. Abian. Quantitative Electrospray LC-MS and LC-MS/MS in Biomedicine. *J. Pharm. Biomed. Anal.*, **17**(1998): 1129–1138.

102. Z. Takáts, J. M. Wiseman, B. Gologan, and R. G. Cooks. Mass Spectrometry Sampling Under Ambient Conditions with Desorption Electrospray Ionization. *Science*, **306**(2004): 471–473.

103. J. M. Wiseman, D. R. Ifa, Q. Song, and R. G. Cooks. Tissue Imaging at Atmospheric Pressure Using Desorption Electrospray Ionization (DESI) Mass Spectrometry. *Angewandte Chemie International Edition*, **45**(2006): 7188–7192.

104. Z. Takáts, J. M. Wiseman, and R. G. Cooks. Ambient Mass Spectrometry Using Desorption Electrospray Ionization (DESI): Instrumentation, Mechanisms and Applications in Forensics, Chemistry, and Biology. *J. Mass Spectrom.*, **40**(2005): 1261–1275.

105. R. G. Cooks, Z. Ouyang, Z. Takats, and J. M. Wiseman. Ambient Mass Spectrometry. *Science*, **311**(2006): 1566–1570.

106. Z. Takáts, I. Cotte-Rodríguez, N. Talaty, H. Chen, and R. G. Cooks. Direct, Trace Level Detection of Explosives on Ambient Surfaces by Desorption Electrospray Ionization Mass Spectrometry. *Chem. Commun.*, no. 15 (2005): 1950–1952.

107. C. N. McEwen, R. G. McKay, and B. S. Larsen. Analysis of Solids, Liquids, and Biological Tissues Using Solids Probe Introduction at Atmospheric Pressure on Commercial LC/MS Instruments. *Anal. Chem.*, **77**(2005): 7826–7831.

108. J. Shiea, M.-Z. Huang, H.-J. HSu, C.-Y. Lee, C.-H. Yuan, I. Beech, and J. Sunner. Electrospray-Assisted Laser Desorption/Ionization Mass Spectrometry for Direct Ambient Analysis of Solids. *Rapid Commun. Mass Spectrom.*, **19**(2005): 3701–3704.

109. R. B. Cody, J. A. Laramée, and H. D. Durst. Versatile New Ion Source for the Analysis of Materials in Open Air under Ambient Conditions. *Anal. Chem.*, **77**(2005): 2297–2302.

110. J. P. Williams, V. J. Patel, R. Holland, and J. H. Scrivens. The Use of Recently Described Ionisation Techniques for the Rapid Analysis of Some Common Drugs and Samples of Biological Origin. *Rapid Commun. Mass Spectrom.*, **20**(2006): 1447–1456.

111. J. J. Thomson. On Rays of Positive Electricity. *Phil. Mag.*, **Ser. 6, 20**(1910): 752–767.

112. R. F. K. Herzog and F. P. Viehböck. Ion Source for Mass Spectrography. *Phys. Rev.*, **76**(1949): 855–856.

113. R. E. Honig. Sputtering of Surfaces by Positive Ion Beams of Low Energy. *J. Appl. Phys.*, **29**(1958): 549–555.

114. R. E. Honig. The Development of Secondary Ion Mass Spectrometry (SIMS): A Retrospective. *Int. J. Mass Spectrom. Ion Proc.*, **66**(1985): 31–54.

115. R. E. Honig. Stone-Age Mass Spectrometry: The Beginnings of "SIMS" at RCA Laboratories, Princeton. *Int. J. Mass Spectrom. Ion Proc.*, **143**(1995): 1–10.

116. R. Castaing and G. J. Slodzian. Optique corpusculaire—premiers essais de microanalyse par emission ionique secondaire. *J. Microscopie*, **1**(1962): 395–399.

117. H. Liebl. Ion Microprobe Mass Analyzer. *J. Appl. Phys.*, **38**(1967): 5277–5283.

118. A. Benninghoven, D. Jaspers, and W. Sichtermann. Secondary-Ion Emission of Amino Acids. *Appl. Phys.*, **11**(1976): 35–39.

119. P. Sigmund and C. Claussen. Sputtering from Elastic-Collision Spikes in Heavy-Ion-Bombarded Metals. *J. Appl. Phys.*, **52**(1981): 990–993.

120. L. A. McDonnell and R. M. A. Heeren. Imaging Mass Spectrometry. *Mass Spectrom. Rev.*, **26**(2007): 606–643.

121. D. McPhail. Applications of Secondary Ion Mass Spectrometry (SIMS) in Materials Science. *J. Material Sci.*, **41**(2006): 873–903.

122. R. N. S. Sodhi. Time-of-Flight Secondary Ion Mass Spectrometry (TOF-SIMS): Versatility in Chemical and Imaging Surface Analysis. *Analyst*, **129**(2004): 483–487.

123. S. Hofmann. Sputter-dDepth Profiling for Thin-Film Analysis. *Phil. Trans. Royal Soc.*, **A362**(2004): 55–75.

124. E. A. Jones, J. S. Fletcher, C. E. Thompson, D. A. Jackson, N. P. Lockyer, and J. C. Vickerman. ToF-SIMS Analysis of Bio-Systems: Are Polyatomic Primary Ions the Solution? *Appl. Surf. Sci.*, **252**(2006): 6844–6854.

125. H. Oechsner and W. Gerhard. A Method for Surface Analysis by Sputtered Neutrals. *Phys. Lett.*, **40**(1972): 211–212.

126. H. Oechsner. Secondary Neutral Mass Spectrometry (SNMS): Recent Methodical Progress and Applications to Fundamental Studies in Particle/Surface Interaction. *Int. J. Mass Spectrom. Ion Proc.*, **143**(1995): 271–282.

127. W. Husinsky and G. Betz. Fundamental Aspects of SNMS for Thin Film Characterization: Experimental Studies and Computer Simulations. *Thin Solid Films*, **272**(1996): 289–309.

128. M. Barber, R. S. Bordoli, R. D. Sedgewick, and A. N. Tyler. Fast Atom Bombardment of Solids (F. A. B.): A New Ion Source for Mass Spectrometry. *J. Chem. Soc. Chem. Commun.*, no. 7 (1981): 325–327.

129. Š. Beranova-Giorgianni and D. M. Desiderio. Fast Atom Bombardment Mass Spectrometry of Synthetic Peptides. In: Methods in Enzymology: Solid-Phase Peptide Synthesis, ed. G. B. Fields. *Methods in Enzymology* **289**. Academic Press, San Diego, 1997, 478–499.

130. A. Benninghoven and W. K. Sichtermann. Detection, Identification, and Structural Investigation of Biologically Important Compounds by Secondary Ion Mass Spectrometry. *Anal. Chem.*, **50**(1978): 1180–1184.

131. M. Barber, R. S. Bordoli, G. J. Elliott, R. D. Sedgwick, and A. N. Tyler. Fast Atom Bombardment Mass Spectrometry. *Anal. Chem.*, **54**(1982): 645A–657A.

132. C. E. Heine, J. F. Holland, and J. T. Watson. Influence of the Ratio of Matrix to Analyte on the Fast Atom Bombardment Mass Spectrometric Response of Peptides Sampled from Aqueous Glycerol. *Anal. Chem.*, **61**(1989): 2674–2682.

133. H. Oka, Y. Ikai, T. Ohno, N. Kawamura, J. Hayakawa, K. Harada, and M. Suzuki. Identification of Unlawful Food Dyes by Thin-Layer Chromatography-Fast Atom Bombardment Mass Spectrometry. *J. Chromatogr.*, **A674**(1994): 301–307.

134. H. Y. Lin, G. Gonyea, S. Killeen, and S. K. Chowdhury. Negative Fast Atom Bombardment Ionization of Aromatic Sulfonic Acids: Unusual Sample-Matrix Interaction. *Rapid Commun. Mass Spectrom.*, **14**(2000): 520–522.

135. B. J. Sweetman and I. A. Blair. 3-Nitrobenzyl Alcohol Has Wide Applicability as a Matrix for FAB-MS. *Biomed. Environmental Mass Spectrom.*, **17**(1988): 337–340.

136. L. M. Mallis, H. M. Wang, D. Loganathan, and R. J. Linhardt. Sequence Analysis of Highly Sulfated, Heparin-Derived Oligosaccharides Using Fast Atom Bombardment Mass Spectrometry. *Anal. Chem.*, **61**(1989): 1453–1458.

137. C. Dass. The Role of a Liquid Matrix in Controlling FAB-Induced Fragmentation. *J. Mass Spectrom.*, **31**(1996): 77–82.

138. M. A. Moseley, L. J. Deterding, K. B. Tomer, and J. W. Jorgenson. Nanoscale Packed-Capillary Liquid Chromatography Coupled with Mass Spectrometry Using a Coaxial Continuous-Flow Fast Atom Bombardment Interface. *Anal. Chem.*, **63**(1991): 1467–1473.

139. R. M. Caprioli, T. Fan, and J. S. Cottrell. Continuous-Flow Sample Probe for Fast Atom Bombardment Mass Spectrometry. *Anal. Chem.*, **58**(1986): 2949–2954.

140. R. M. Caprioli and W. T. Moore. Continuous-Flow Fast Atom Bombardment Mass Spectrometry. In Methods in Enzymology: Mass Spectrometry, ed. J. A. McCloskey. *Methods in Enzymology* **193**, Academic Press, San Diego, 1990, 214–237.

141. D. F. Torgerson, R. P. Skowronski, and R. D. Macfarlane. New Approach to the Mass Spectrometry of Non-Volatile Compounds. *Biochem. Biophys. Res. Commun.*, **60**(1974): 616–621.

142. P. Dück, W. Treu, W. Galster, H. Frohlich, and H. Voit. Heavy Ion Induced Desorption of Organic Compounds. *Nucl. Instr. Meth.*, **168**(1980): 601–605.

143. P. Håkansson and B. U. R. Sundqvist. Electronic Sputtering of Biomolecules and its Application in Mass Spectrometry. *Vacuum*, **39**(1989): 397–399.

144. S. Della-Negra, J. Depauw, H. Joret, Y. Le Beyec, I. Bitensky, G. Bolbach, R. Galera, and K. Wien. Heavy Particle Induced Ion Emission from Langmuir-Blodgett Films: Dependence on the Charge State and the Angle of Incidence. *Nucl. Instr. Meth. Phys. Res.*, **B52**(1990): 121–128.

145. G. Brinkmalm, P. Håkansson, J. Kjellberg, P. Demirev, B. U. R. Sundqvist, and W. Ens. A Plasma Desorption Time-of-Flight Mass Spectrometer with a Single-Stage Ion Mirror: Improved Resolution and Calibration Procedure. *Int. J. Mass Spectrom. Ion Proc.*, **114**(1992): 183–207.

146. G. Jonsson, A. Hedin, P. Håkansson, B. U. R. Sundqvist, H. Bennich, and P. Roepstorff. Compensation for Non-Normal Ejection of Large Molecular Ions in Plasma-Desorption Mass Spectrometry. *Rapid Commun. Mass Spectrom.*, **3**(1989): 190–191.

147. B. Sundqvist and R. D. Macfarlane. ^{252}Cf-Plasma Desorption Mass Spectrometry. *Mass Spectrom. Rev.*, **4**(1985): 421–460.

148. K. Wien. Fast Heavy Ion Induced Desorption. *Rad. Eff. Defect. Solids*, **109**(1989): 137–167.

149. K. Wien. Fast Heavy Ion Induced Desorption of Insulators. *Nucl. Instr. Meth. Phys. Res.*, **B65**(1992): 149–166.

150. R. D. Macfarlane. Mass Spectrometry of Biomolecules: From PDMS to MALDI. *Brazilian J. Phys.*, **29**(1999): 415–420.

151. R. E. Honig and J. R. Woolston. Laser-Induced Emission of Electrons, Ions, and Neutral Atoms from Solid Surfaces. *Appl. Phys. Lett.*, **2**(1963): 138–139.

152. F. Hillenkamp, E. Unsöld, R. Kaufmann, and R. Nitsche. A High-Sensitivity Laser Microprobe Mass Analyzer. *Appl. Phys.*, **8**(1975): 341–348.

153. M. A. Posthumus, P. G. Kistemaker, H. L. C. Meuzelaar, and M. C. Ten Noever de Brauw. Laser Desorption-Mass Spectrometry of Polar Nonvolatile Bio-Organic Molecules. *Anal. Chem.*, **50**(1978): 985–991

154. R. J. Conzemius and J. M. Capellen. A Review of the Applications to Solids of the Laser Ion Source in Mass Spectrometry. *Int. J. Mass Spectrom. Ion Phys.*, **34**(1980): 197–271.

155. T. K. Dutta and S. Harayama. Time-of-Flight Mass Spectrometric Analysis of High-Molecular-Weight Alkanes in Crude Oil by Silver Nitrate Chemical Ionization after Laser Desorption. *Anal. Chem.*, **73**(2001): 864–869.

156. P. R. Buseck. Geological Fullerenes: Review and Analysis. *Earth Planetary Sci. Lett.*, **203**(2002): 781–792.

157. M. Kalberer, D. Paulsen, M. Sax, M. Steinbacher, J. Dommen, A. S. H. Prevot, R. Fisseha, E. Weingartner, V. Frankevich, R. Zenobi, and U. Baltensperger. Identification of Polymers as Major Components of Atmospheric Organic Aerosols. *Science*, **303**(2004): 1659–1662.

158. H. Gnaser, M. R. Savina, W. F. Calaway, C. E. Tripa, I. V. Veryovkin, and M. J. Pellin. Photocatalytic Degradation of Methylene Blue on Nanocrystalline TiO_2: Surface Mass Spectrometry of Reaction Intermediates. *Int. J. Mass Spectrom.*, **245**(2005): 61–67.

159. J. Wei, J. M. Buriak, and G. Siuzdak. Desorption-Ionization Mass Spectrometry on Porous Silicon. *Nature*, **399**(1999): 243–246.

160. W. G. Lewis, Z. Shen, M. G. Finn, and G. Siuzdak. Desorption/Ionization On Silicon (DIOS) Mass Spectrometry: Background and Applications. *Int. J. Mass Spectrom.*, **226**(2003): 107–116.

161. M. Karas, D. Bachmann, U. Bahr, and F. Hillenkamp. Matrix-Assisted Ultraviolet Laser Desorption of Non-Volatile Compounds. *Int. J. Mass Spectrom. Ion Proc.*, **78**(1987): 53–68.

162. M. Karas, U. Bahr, and F. Hillenkamp. UV Laser Desorption/Ionization Mass Spectrometry of Proteins in the 100,000 Dalton Range. *Int. J. Mass Spectrom. Ion Proc.*, **92**(1989): 231–242.

163. M. Karas and F. Hillenkamp. Laser Desorption Ionization of Proteins with Molecular Masses Exceeding 10,000 Daltons. *Anal. Chem.*, **60**(1988): 2299–2301.

164. K. Tanaka, H. Waki, Y. Ido, S. Akita, Y. Yoshida, and T. Yoshida. Protein and Polymer Analyses up to m/z 100,000 by Laser Ionization Time-of-Flight Mass Spectrometry. *Rapid Commun. Mass Spectrom.*, **2**(1988): 151–153.

165. R. C. Beavis and B. T. Chait. Cinnamic Acid Derivatives as Matrices for Ultraviolet Laser Desorption Mass Spectrometry of Proteins. *Rapid Commun. Mass Spectrom.*, **3**(1989): 432–435.

166. P. Juhasz, C. E. Costello, and K. Biemann. Matrix-Assisted Laser Desorption Mass Spectrometry with 2-(4-Hydroxyphenylazo)benzoic Acid Matrix. *J. Am. Soc. Mass Spectrom.*, **4**(1993): 399–409.

167. S. Berkenkamp, M. Karas, and F. Hillenkamp. Ice as a Matrix for IR-Matrix-Assisted Laser Desorption/Ionization: Mass Spectra from a Protein Single Crystal. *Proc. Natl. Acad. Sci. USA*, **93**(1996): 7003–7007.

168. S. Ring and Y. Rudich. A Comparative Study of a Liquid and a Solid Matrix in Matrix-Assisted Laser Desorption/Ionization Time-of-Flight Mass Spectrometry and Collision Cross Section Measurements. *Rapid Commun. Mass Spectrom.*, **14**(2000): 515–519.

169. J. Bai, X. Liang, Y. H. Liu, Y. Zhu, and D. M. Lubman. Characterization of Two New Matrices for Matrix-Assisted Laser Desorption/Ionization Mass Spectrometry. *Rapid Commun. Mass Spectrom.*, **10**(1996): 839–844.

170. D. Schleuder, F. Hillenkamp, and K. Strupat. IR-MALDI-Mass Analysis of Electroblotted Proteins Directly from the Membrane: Comparison of Different Membranes, Application to On-membrane Digestion, and Protein Identification by Database Searching. *Anal. Chem.*, **71**(1999): 3238–3247.

171. B. A. Budnik, K. B. Jensen, T. J. Jorgensen, A. Haase, and R. A. Zubarev. Benefits of 2.94 Micron Infrared Matrix-Assisted Laser Desorption/Ionization for Analysis of Labile Molecules by Fourier Transform Mass Spectrometry. *Rapid Commun. Mass Spectrom.*, **14**(2000): 578–584.

172. I. D. Figueroa, O. Torres, and D. H. Russell. Effects of the Water Content in the Sample Preparation for MALDI on the Mass Spectra. *Anal. Chem.*, **70**(1998): 4527–4533.

173. S. L. Cohen and B. T. Chait Influence of Matrix Solution Conditions on the MALDI-MS Analysis of Peptides and Proteins. *Anal. Chem.*, **68**(1996): 31–37.

174. Y. Dai, R. M. Whittal, and L. Li. Confocal Fluorecence Microscopic Imaging for Investigating the Analyte Distribution in MALDI Matrices. *Anal. Chem.*, **68**(1996): 2494–2500.

175. J. Axelsson, A.-M. Hoberg, C. Waterson, P. Myatt, G. L. Shield, J. Varney, D. M. Haddleton, and P. J. Derrick. Improved Reproducibility and Increased Signal Intensity in Matrix-Assisted Laser Desorption/Ionization as a Result of Electrospray Sample Preparation. *Rapid Commun. Mass Spectrom.*, **11**(1997): 209–213.

176. A. Westman, C. L. Nilsson, and R. Ekman. Matrix-Assisted Laser Desorption/Ionization Time-of-Flight Mass Spectrometry Analysis of Proteins in Human Cerebrospinal Fluid. *Rapid Commun. Mass Spectrom.*, **12**(1998): 1092–1098.

177. M. Mann and M. Wilm. Error-Tolerant Identification of Peptides in Sequence Databases by Peptide Sequence Tags. *Anal. Chem.*, **66**(1994): 4390–4399.

178. J. Zheng, N. Li, M. Ridyard, H. Dai, S. M. Robbins, and L. Li. Simple and Robust Two-Layer Matrix/Sample Preparation Method for MALDI MS/MS Analysis of Peptides. *J. Proteome Res.*, **4**(2005): 1709–1716.

179. J. Gobom, M. Schuerenberg, M. Mueller, D. Theiss, H. Lehrach, and E. Nordhoff. α-Cyano-4-hydroxycinnamic Acid Affinity Sample Preparation. A Protocol for MALDI-MS Peptide Analysis in Proteomics. *Anal. Chem.*, **73**(2001): 434–438.

180. S. D. Hanton and D. M. Parees. Extending the Solvent-Free MALDI Sample Preparation Method. *J. Am. Soc. Mass Spectrom.*, **16**(2005): 90–93.

181. E. Nordhoff, M. Schurenberg, G. Thiele, C. Lubbert, K.-D. Kloeppel, D. Theiss, H. Lehrach, and J. Gobom. Sample Preparation Protocols for MALDI-MS of Peptides and Oligonucleotides Using Prestructured Sample Supports. *Int. J. Mass Spectrom.*, **226**(2003): 163–180.

182. A. M. Haag, J. Chaiban, K. H. Johnston, and R. B. Cole. Monitoring of Immune Response by Blood Serum Profiling Using Matrix-Assisted Laser Desorption/Ionization Time-of-Flight Mass Spectrometry. *J. Mass Spectrom.*, **36**(2001): 15–20.

183. J. Yu, E. R. Butelman, J. H. Woods, B. T. Chait, and M. J. Kreek. In Vitro Biotransformation of Dynorphin A (1-17) is Similar in Human and Rhesus Monkey Blood as Studied by Matrix-Assisted Laser Desorption/Ionization Mass Spectrometry. *J. Pharmacol. Exp. Ther.*, **279**(1996): 507–514.

184. J. Gobom, K. O. Kraeuter, R. Persson, H. Steen, P. Roepstorff, and R. Ekman. Detection and Quantification of Neurotensin in Human Brain Tissue by Matrix-Assisted Laser Desorption/Ionization Time-of-Flight Mass Spectrometry. *Anal. Chem.*, **72**(2000): 3320–3326.

185. T. Krishnamurthy and P. L. Ross. Rapid Identification of Bacteria by Direct Matrix-aAssisted Laser Desorption/Ionization Mass Spectrometric Analysis of Whole Cells. *Rapid Commun. Mass Spectrom.*, **10**(1996): 1992–1996.

186. D. S. Cornett, M. L. Reyzer, P. Chaurand, and R. M. Caprioli. MALDI Imaging Mass Spectrometry: Molecular Snapshots of Biochemical Systems. *Nat. Meth.*, **4**(2007): 828–833.

187. S. Ekström, D. Ericsson, P. Önnerfjord, M. Bengtsson, J. Nilsson, G. Marko-Varga, and T. Laurell. Signal Amplification Using "Spot-on-a-Chip" Technology for the Identification of Proteins via MALDI-TOF MS. *Anal. Chem.*, **73**(2001): 214–219.

188. R. W. Nelson, D. Dogruel, and P. Williams. Mass Determination of Human Immunoglobulin IgM Using Matrix-Assisted Laser Desorption/Ionization Time-of-Flight Mass Spectrometry. *Rapid Commun. Mass Spectrom.*, **8**(1994): 627–631.

189. R. S. Annan and S. A. Carr. Phosphopeptide Analysis by Matrix-Assisted Laser Desorption Time-of-Flight Mass Spectrometry. *Anal. Chem.*, **68**(1996): 3413–3421.

190. O. A. Mirgorodskaya, Y. P. Kozmin, M. I. Titov, R. Körner, C. P. Sönksen, and P. Roepstorff. Quantitation of Peptides and Proteins by Matrix-Assisted Laser Desorption/Ionization Mass Spectrometry Using ^{18}O-Labeled Internal Standards. *Rapid Commun. Mass Spectrom.*, **14**(2000): 1226–1232.

191. J. J. Lennon and K. A. Walsh. Locating and Identifying Posttranslational Modifications by In-Source Decay During MALDI-TOF Mass Spectrometry. *Protein Sci.*, **8**(1999): 2487–2493.

192. T. B. Farmer and R. M. Caprioli. Determination of Protein-Protein Interactions by Matrix-Aassisted Laser Desorption/Ionization Mass Spectrometry. *J. Mass Spectrom.*, **33**(1998): 697–704.

193. R. P. Pauly, F. Rosche, M. Wermann, C. H. McIntosh, R. A. Pederson, and H. U. Demuth. Investigation of Glucose-Dependent Insulinotropic Polypeptide-(1-42) and Glucagon-like Peptide-1-(7-36) Degradation *In Vitro* by Dipeptidyl Peptidase IV Using Matrix-Assisted

Laser Desorption/Ionization-Time of Flight Mass Spectrometry. A Novel Kinetic Approach. *J. Biol. Chem.*, **271**(1996): 23222–23229.

194. S. P. Gygi, B. Rist, S. A. Gerber, F. Turecek, M. H. Gelb, and R. Aebersold. Quantitative Analysis of Complex Protein Mixtures Using Isotope-Coded Affinity Tags. *Nat. Biotechnol.*, **17**(1999): 994–999.
195. F. Hillenkamp and M. Karas. Matrix-Assisted Laser Desorption/Ionisation, an Experience. *Int. J. Mass Spectrom.*, **200**(2000): 71–77.
196. T. W. Hutchens and T.-T. Yip. New Desorption Strategies for the Mass Spectrometric Analysis of Macromolecules. *Rapid Commun. Mass Spectrom.*, **7**(1993): 576–580.
197. H. J. Issaq, T. D. Veenstra, T. P. Conrads, and D. Felschow. The SELDI-TOF MS Approach to Proteomics: Protein Profiling and Biomarker Identification. *Biochem. Biophys. Res. Commun.*, **292**(2002): 587–592.
198. V. V. Laiko, M. A. Baldwin, and A. L. Burlingame. Atmospheric Pressure Matrix-Assisted Laser Desorption/Ionization Mass Spectrometry. *Anal. Chem.*, **72**(2000): 652–657.
199. S. C. Moyer, L. A. Marzilli, A. S. Woods, V. V. Laiko, V. M. Doroshenko, and R. J. Cotter. Atmospheric Pressure Matrix-Assisted Laser Desorption/Ionization (AP MALDI) on a Quadrupole Ion Trap Mass Spectrometer. *Int. J. Mass Spectrom.*, **226**(2003): 133–150.
200. P. V. Tan, V. V. Laiko, and V. M. Doroshenko. Atmospheric Pressure MALDI with Pulsed Dynamic Focusing for High-Efficiency Transmission of Ions into a Mass Spectrometer. *Anal. Chem.*, **76**(2004): 2462–2469.
201. B. B. Schneider, C. Lock, and T. R. Covey. AP and Vacuum MALDI on a QqLIT Instrument. *J. Am. Soc. Mass Spectrom.*, **16**(2005): 176–182.
202. J. S. Sampson, A. M. Hawkridge, and D. C. Muddiman. Generation and Detection of Multiply-Charged Peptides and Proteins by Matrix-Assisted Laser Desorption Electrospray Ionization (MALDESI) Fourier Transform Ion Cyclotron Resonance Mass Spectrometry. *J. Am. Soc. Mass Spectrom.*, **17**(2006): 1712–1716.
203. K. Blaum. High-Accuracy Mass Spectrometry with Stored Ions. *Physics Reports*, **425**(2006): 1–78.
204. W. E. Stephens. A Pulsed Mass Spectrometer with Time Dispersion. *Phys. Rev.*, **69**(1946): 691.
205. J. H. J. Dawson and M. Guilhaus. Orthogonal-Acceleration Time-of-Flight Mass Spectrometer. *Rapid Commun. Mass Spectrom.*, **3**(1989): 155–159.
206. A. F. Dodonov, I. V. Chernushevich, and V. V. Laiko. Atmospheric Pressure Ionization Time-of-Flight Mass Spectrometer (Abstr). *12th International Mass Spectrometry Conference, 26–30 August, 1991, Amsterdam, the Netherlands*, 153.
207. C. H. Sin, E. D. Lee, and M. L. Lee. Atmospheric Pressure Ionization Time-of-Flight Mass Spectrometry with a Supersonic Ion Beam. *Anal. Chem.*, **63**(1991): 2897–2900.
208. J. G. Boyle and C. M. Whitehouse. Time-of-flight Mass Spectrometry with an Electrospray Ion Beam. *Anal. Chem.*, **64**(1992): 2084–2089.
209. A. N. Verentchikov, W. Ens, and K. G. Standing. Reflecting Time-of-Flight Mass Spectrometer with an Electrospray Ion Source and Orthogonal Extraction. *Anal. Chem.*, **66**(1994): 126–133.
210. B. A. Mamyrin, V. I. Karataev, D. V. Shmikk, and V. A. Zagulin. The Mass-Reflectron, a New Non-Magnetic Time-of-Flight Mass Spectrometer with High Resolution. *Sov. Phys. JETP*, **37**(1973): 45–48.

211. B. A. Mamyrin. Time-of-Flight Mass Spectrometry (Concepts, Achievements, and Prospects). *Int. J. Mass Spectrom.*, **206**(2001): 251–266.

212. R. J. Cotter, W. Griffith, and C. Jelinek. Tandem Time-of-Flight (TOF/TOF) Mass Spectrometry and the Curved-Field Reflectron. *J. Chromatogr.*, **B855**(2007): 2–13.

213. J. Flensburg, D. Haid, J. Blomberg, J. Bielawski, and D. Ivansson. Applications and Performance of a MALDI-ToF Mass Spectrometer with Quadratic Field Reflectron Technology. *J. Biochem. Biophys. Meth.*, **60**(2004): 319–334.

214. A. E. Giannakopulos, B. Thomas, A. W. Colburn, D. J. Reynolds, E. N. Raptakis, A. A. Makarov, and P. J. Derrick. Tandem Time-of-Flight Mass Spectrometer (TOF-TOF) with a Quadratic-Field Ion Mirror. *Rev. Sci. Instrum.*, **73**(2002): 2115–2123.

215. T. Cornish and R. J. Cotter. A Compact Time-of-Flight Mass Spectrometer for the Structural Analysis of Biological Molecules Using Laser Desorption. *Rapid Commun. Mass Spectrom.*, **6**(1992): 242–248.

216. T. J. Cornish and R. J. Cotter. A Curved-Field Reflectron for Improved Energy Focusing of Product Ions in Time-of-Flight Mass Spectrometry. *Rapid Commun. Mass Spectrom.*, **7**(1993): 1037–1040.

217. C. K. G. Piyadasa, P. Håkansson, and T. R. Ariyaratne. A High Resolving Power Multiple Reflection Matrix-Assisted Laser Desorption/Ionization Time-of-Flight Mass Spectrometer. *Rapid Commun. Mass Spectrom.*, **13**(1999): 620–624.

218. W. C. Wiley and I. H. McLaren. Time-of-Flight Mass Spectrometer with Improved Resolution. *Rev. Sci. Instrum.*, **26**(1955): 1150–1157.

219. R. S. Brown and J. J. Lennon. Mass Resolution Improvement by Incorporation of Pulsed Ion Extraction in a Matrix-Assisted Laser Desorption/Ionization Linear Time-of-Flight Mass Spectrometer. *Anal. Chem.*, **67**(1995): 1998–2003.

220. M. L. Vestal, P. Juhasz, and S. A. Martin. Delayed Extraction Matrix-Assisted Laser Desorption Time-of-Flight Mass Spectrometry. *Rapid Commun. Mass Spectrom.*, **9**(1995): 1044–1050.

221. R. M. Whittal, L. M. Russon, S. R. Weinberger, and L. Li. Functional Wave Time-Lag Focusing Matrix-Assisted Laser Desorption/Ionization in a Linear Time-of-Flight Mass Spectrometer: Improved Mass Accuracy. *Anal. Chem.*, **69**(1997): 2147–2153.

222. J. J. Thomson. *Rays of Positive Electricity and Their Application to Chemical Analysis.* Longmans, Green and Co., London, 1913.

223. J. Mattauch and R. Herzog. Double-Focusing Mass Spectrograph and the Masses of N^{15} and O^{18}. *Phys. Rev.*, **50**(1936): 617–623.

224. E. G. Johnson and A. O. Nier. Angular Aberrations in Sector Shaped Electromagnetic Lenses for Focusing Beams of Charged Particles. *Phys. Rev.*, **91**(1953): 10–17.

225. R. S. Annan, H. J. Kochling, J. A. Hill, and K. Biemann. Matrix-Assisted Laser Desorption Using a Fast-Atom Bombardment Ion Source and a Magnetic Mass Spectrometer. *Rapid Commun. Mass Spectrom.*, **6**(1992): 298–302.

226. R. S. Bordoli, K. Howes, R. G. Vickers, R. H. Bateman, and D. J. Harvey. Matrix-Assisted Laser Desorption Mass Spectrometry on a Magnetic Sector Instrument Fitted with an Array Detector. *Rapid Commun. Mass Spectrom.*, **8**(1994): 585–589.

227. J. D. Lennon III, D. Shinn, R. W. Vachet, and G. L. Glish. Strategy for Pulsed Ionization Methods on a Sector Mass Spectrometer. *Anal. Chem.*, **68**(1996): 845–849.

228. K. F. Medzihradszky, G. W. Adams, A. L. Burlingame, R. H. Bateman, and M. R. Green. Peptide Sequence Determination by Matrix-Assisted Laser Desorption Ionization Employing a Tandem Double Focusing Magnetic-Orthogonal Acceleration Time-of-Flight Mass Spectrometer. *J. Am. Soc. Mass Spectrom.*, **7**(1996): 1–10.

229. V. S. K. Kolli and R. Orlando. A New Strategy for MALDI on Magnetic Sector Mass Spectrometers with Point Detectors. *Anal. Chem.*, **69**(1997): 327–332.

230. T. W. Burgoyne and G. M. Hieftje. An Introduction to Ion Optics for the Mass Spectrograph. *Mass Spectrom. Rev.*, **15**(1996): 241–259.

231. A. W. T. Bristow and K. S. Webb. Intercomparison Study on Accurate Mass Measurement of Small Molecules in Mass Spectrometry. *J. Am. Soc. Mass Spectrom.*, **14**(2003): 1086–1098.

232. W. Paul and H. Steinwedel. Ein neues Massenspektrometer ohne Magnetfeld. *Z. Naturforsch.*, **A8**(1953): 448–450.

233. W. Paul and H. Steinwedel. U. S. Patent 2939952, 1960.

234. M. E. Mathieu. Mémoire sur le mouvement vibratoire d'une membrane de forme elliptique. *J. Math. Pur. Appl.*, **13**(1868): 137–203.

235. D. W. McLachlan. *Theory and Applications of Mathieu Functions*. Clarendon Press, Oxford, 1947.

236. J. C. Schwartz, M. W. Senko, and J. E. P. Syka. A Two-Dimensional Quadrupole Ion Trap Mass Spectrometer. *J. Am. Soc. Mass Spectrom.*, **13**(2002): 659–669.

237. V. M. Doroshenko and R. J. Cotter. A New Method of Trapping Ions Produced by Matrix-Assisted Laser Desorption Ionization in a Quadrupole Ion Trap. *Rapid Commun. Mass Spectrom.*, **7**(1993): 822–827.

238. G. C. Eiden, M. E. Cisper, M. L. Alexander, P. H. Hemberger, and N. S. Nogar. A Method of Increasing the Sensitivity for Laser Desorption in an Ion Trap Mass Spectrometer. *J. Am. Soc. Mass Spectrom.*, **4**(1993): 706–709.

239. R. E. March. Quadrupole Ion Trap Mass Spectrometry: Theory, Simulation, Recent Developments and Applications. *Rapid Commun. Mass Spectrom.*, **12**(1998): 1543–1554.

240. M. V. Gorshkov and R. A. Zubarev. On the Accuracy of Polypeptide Masses Measured in a Linear Ion Trap. *Rapid Commun. Mass Spectrom.*, **19**(2005): 3755–3758.

241. M. Pelzing and C. Neusüß. Separation Techniques Hyphenated to Electrospray-Tandem Mass Spectrometry in Proteomics: Capillary Electrophoresis versus Nanoliquid Chromatography. *Electrophoresis*, **26**(2005): 2717–2728.

242. K. H. Kingdon. A Method for the Neutralization of Electron Space Charge by Positive Ionization at Very Low Gas Pressures. *Phys. Rev.*, **21**(1923): 408–418.

243. R. D. Knight. Storage of Ions from Laser-Produced Plasmas. *Appl. Phys. Lett.*, **38**(1981): 221–223.

244. A. A. Makarov. U. S. Patent 5886346, 1999.

245. A. Makarov. Electrostatic Axially Harmonic Orbital Trapping: A High-Performance Technique of Mass Analysis. *Anal. Chem.*, **72**(2000): 1156–1162.

246. M. Hardman and A. Makarov. Interfacing the Orbitrap Mass Analyzer to an Electrospray Ion Source. *Anal. Chem.*, **75**(2003): 1699–1705.

247. Q. Hu, R. J. Noll, H. Li, A. Makarov, M. Hardman, and R. G. Cooks. The Orbitrap: A New Mass Spectrometer. *J. Mass Spectrom.*, **40**(2005): 430–443.

248. J. R. Yates, D. Cociorva, L. Liao, and V. Zabrouskov. Performance of a Linear Ion Trap-Orbitrap Hybrid for Peptide Analysis. *Anal. Chem.*, **78**(2006): 493–500.

249. A. Makarov, E. Denisov, A. Kholomeev, W. Balschun, O. Lange, K. Strupat, and S. Horning. Performance Evaluation of a Hybrid Linear Ion Trap/Orbitrap Mass Spectrometer. *Anal. Chem.*, **78**(2006): 2113–2120.

250. A. Makarov, E. Denisov, O. Lange, and S. Horning. Dynamic Range of Mass Accuracy in LTQ Orbitrap Hybrid Mass Spectrometer. *J. Am. Soc. Mass Spectrom.*, **17**(2006): 977–982.

251. M. Scigelova and A. Makarov. Orbitrap Mass Analyzer: Overview and Applications in Proteomics. *Proteomics*, **6**(2006): 16–21.

252. E. O. Lawrence and N. E. Edlefsen. On the Production of High Speed Protons. *Science*, **72**(1930): 376–377.

253. E. O. Lawrence and M. S. Livingston. The Production of High Speed Light Ions Without the Use of High Voltages. *Phys. Rev.*, **40**(1932): 19–35.

254. H. Sommer, H. A. Thomas, and J. A. Hipple. A Precise Method of Determining the Faraday by Magnetic Resonance. *Phys. Rev.*, **76**(1949): 1877–1878.

255. H. Sommer, H. A. Thomas, and J. A. Hipple. The Measurement of e/M by Cyclotron Resonance. *Phys. Rev.*, **82**(1951): 697.

256. M. B. Comisarow and A. G. Marshall. Fourier Transform Ion Cyclotron Resonance Spectroscopy. *Chem. Phys. Lett.*, **25**(1974): 282–283.

257. R. T. McIver, Jr, R. L. Hunter, and M. T. Bowers. Coupling of a Quadrupole Mass Spectrometer and a Fourier Transform Mass Spectrometer. *Int. J. Mass Spectrom. Ion Proc.*, **64**(1985): 67–77.

258. P. Kofel, M. Alleman, H. P. Kellerhals, and K.-P. Wanczek. External Generation of Ions in ICR Spectrometry. *Int. J. Mass Spectrom. Ion Proc.*, **65**(1985): 97–103.

259. P. Caravatti. U. S. Patent 4924089, 1990.

260. A. G. Marshall. Milestones in Fourier Transform Ion Cyclotron Resonance Mass Spectrometry Technique Development. *Int. J. Mass Spectrom.*, **200**(2000): 331–356.

261. S. D.-H. Shi, C. L. Hendrickson, and A. G. Marshall. Counting Individual Sulfur Atoms in a Protein by Ultrahighresolution Fourier Transform Ion Cyclotron Resonance Mass Spectrometry: Experimental Resolution of Isotopic Fine Structure in Proteins. *Proc. Natl. Acad. Sci. U.S.A.*, **95**(1998): 11532–11537.

262. R. P. Rodgers, E. N. Blumer, C. L. Hendrickson, and A. G. Marshall. Stable Isotope Incorporation Triples the Upper Mass Limit for Determination of Elemental Composition by Accurate Mass Measurement. *J. Am. Soc. Mass Spectrom.*, **11**(2000): 835–840.

263. A. G. Marshall, C. L. Hendrickson, M. R. Emmett, R. P. Rodgers, G. T. Blakney, and C. L. Nilsson. Fourier Transform Ion Cyclotron Resonance: State of the Art (Abstr.). *17th International Mass Spectrometry Conference, 27 August–September 1, 2006, Prague, Czech Republic, p. MoOr-21.*

264. Y. Shen, N. Tolić, C. Masselon, L. Paša-Tolić, D. G. Camp, K. K. Hixson, R. Zhao, G. A. Anderson, and R. D. Smith. Ultrasensitive Proteomics Using High-Efficiency On-Line Micro-SPE-NanoLC-NanoESI MS and MS/MS. *Anal. Chem.*, **76**(2004): 144–154.

265. M. E. Belov, M. V. Gorshkov, H. R. Udseth, G. A. Anderson, and R. D. Smith. Zeptomole-Sensitivity Electrospray Ionization-Fourier Transform Ion Cyclotron Resonance Mass Spectrometry of Proteins. *Anal. Chem.*, **72**(2000): 2271–2279.

266. M. Palmblad, K. Håkansson, P. Håkansson, X. Feng, H. J. Cooper, A. E. Giannakopulos, P. S. Green, and P. J. Derrick. A 9.4 T Fourier Transform Ion Cyclotron Resonance Mass Spectrometer: Description and Performance. *Eur. J. Mass Spectrom.*, **6**(2000): 267–275.

267. B. Bogdanov and R. D. Smith. Proteomics by FTICR Mass Spectrometry: Top Down and Bottom Up. *Mass Spectrom. Rev.*, **24**(2005): 168–200.

268. R. A. Muller. Radioisotope Dating with a Cyclotron. *Science*, **196**(1977): 489–494.

269. W. Kutschera. Progress in Isotope Analysis at Ultra-Trace Level by AMS. *Int. J. Mass Spectrom.*, **242**(2005): 145–160.

270. L. W. Alvarez and R. Cornog. He^3 in Helium. *Phys. Rev.*, **56**(1939): 379.

271. D. E. Nelson, R. G. Korteling, and W. R. Stott. Carbon-14: Direct Detection at Natural Concentrations. *Science*, **198**(1977): 507–508.

272. C. L. Bennett, R. P. Beukens, M. R. Clover, H. E. Gove, R. B. Liebert, A. E. Litherland, K. H. Purser, and W. E. Sondheim. Radiocarbon Dating Using Electrostatic Accelerators: Negative Ions Provide the Key. *Science*, **198**(1977): 508–510.

273. L. K. Fifield. Advances in Accelerator Mass Spectrometry. *Nucl. Instr. Meth. Phys. Res.*, **B172**(2000): 134–143.

274. J. S. Vogel, K. W. Turteltaub, R. Finkel, and D. E. Nelson. Accelerator Mass Spectrometry. *Anal. Chem.*, **67**(1995): 353–359.

275. L. K. Fifield. Accelerator Mass Spectrometry and Its Applications. *Rep. Prog. Phys.*, **62**(1999): 1223–1274.

276. M. Suter. 25 Years of AMS: A Review of Recent Developments. *Nucl. Instr. Meth. Phys. Res.*, **B223-B224**(2004): 139–148.

277. D. W. Koppenaal, C. J. Barinaga, M. B. Denton, R. P. Sperline, G. M. Hieftje, G. D. Schilling, F. J. Andrade, and J. H. Barnes. IV. MS Detectors. *Anal. Chem.*, **77**(2005): 418A–427A.

278. J. S. Becker and H. J. Dietze. Precise and Accurate Isotope Ratio Measurements by ICP-MS. *Fresenius J. Anal. Chem.*, **368**(2000): 23–30.

279. W. R. Leo. *Techniques for Nuclear and Particle Physics Experiments*. Berlin: Springer-Verlag, 1987.

280. J. Wiza. Microchannel Plate Detectors. *Nucl. Instr. Meth.*, **162**(1979): 587–601.

281. O. Vorm and P. Roepstorff. Detector Bias Gating for Improved Detector Response and Calibration in Matrix-Assisted Laser Desorption/Ionization Time-of-Flight Mass Spectrometry. *J. Mass Spectrom.*, **31**(1996): 351–356.

282. A. Westman, G. Brinkmalm, and D. F. Barofsky. MALDI Induced Saturation Effects in Chevron Microchannel Plate Detectors. *Int. J. Mass Spectrom. Ion Proc.*, **169/170**(1997): 79–87.

283. K. Birkinshaw. Fundamentals of Focal Plane Detectors. *J. Mass Spectrom.*, **32**(1997): 795–806.

284. J. H. Barnes IV and G. M. Hieftje. Recent Advances in Detector-Array Technology for Mass Spectrometry. *Int. J. Mass Spectrom.*, **238**(2004): 33–46.

285. G. Westmacott, F. Zhong, M. Frank, S. Friedrich, S. E. Labov, and W. H. Benner. Investigating Ion-Surface Collisions with a Niobium Superconducting Tunnel Junction Detector in a Time-of-Flight Mass Spectrometer. *Rapid Commun. Mass Spectrom.*, **14**(2000): 600–607.

286. R. J. Wenzel, U. Matter, L. Schultheis, and R. Zenobi. Analysis of Megadalton Ions Using Cryodetection MALDI Time-of-Flight Mass Spectrometry. *Anal. Chem.*, **77**(2005): 4329–4337.

287. A. G. Marshall and C. L. Hendrickson. Fourier Transform Ion Cyclotron Resonance Detection: Principles and Experimental Configurations. *Int. J. Mass Spectrom.*, **215**(2002): 59–75.

3

TANDEM MASS SPECTROMETRY

Ann Westman-Brinkmalm and Gunnar Brinkmalm

Tandem mass spectrometry (MS/MS) is a technique where structural information on sample molecules is obtained by using multiple stages of mass selection and mass separation. A prerequisite is that the sample molecules can be transferred into gas phase and ionized intact and that they can be induced to fall apart in some predictable and controllable fashion after the first mass selection step. Multistage MS/MS, or MS^n, can be performed by first selecting and isolating a precursor ion (MS^2), fragmenting it, isolating a primary fragment ion (MS^3), fragmenting it, isolating a secondary fragment (MS^4), and so on as long as you can obtain meaningful information or the fragment ion signal is detectable. A variety of imaginative modes of tandem MS is described in the literature, but the four most common are precursor ion mode, product ion mode (also refered to as multireaction monitoring, MRM), neutral loss mode, and single reaction monitoring (Fig. 3.1 and Table 3.1). The product ion mode is the most explorative mode and is typically utilized to get structural information (and often thereby putative identity) of as many different sample molecules as possible. In precursor ion mode and neutral loss mode the objective is to detect a certain class of molecules with a common functional group, such as phosphopeptides or different classes of lipids. Selected reaction monitoring (SRM) is universally utilized in combination with online chromatographic separation to quantify a specific compound in a complex (often biological) matrix. Frequently three degrees of separation: elution time, precursor m/z, and fragment m/z are necessary to

Mass Spectrometry. Edited by Ekman, Silberring, Westman-Brinkmalm, and Kraj

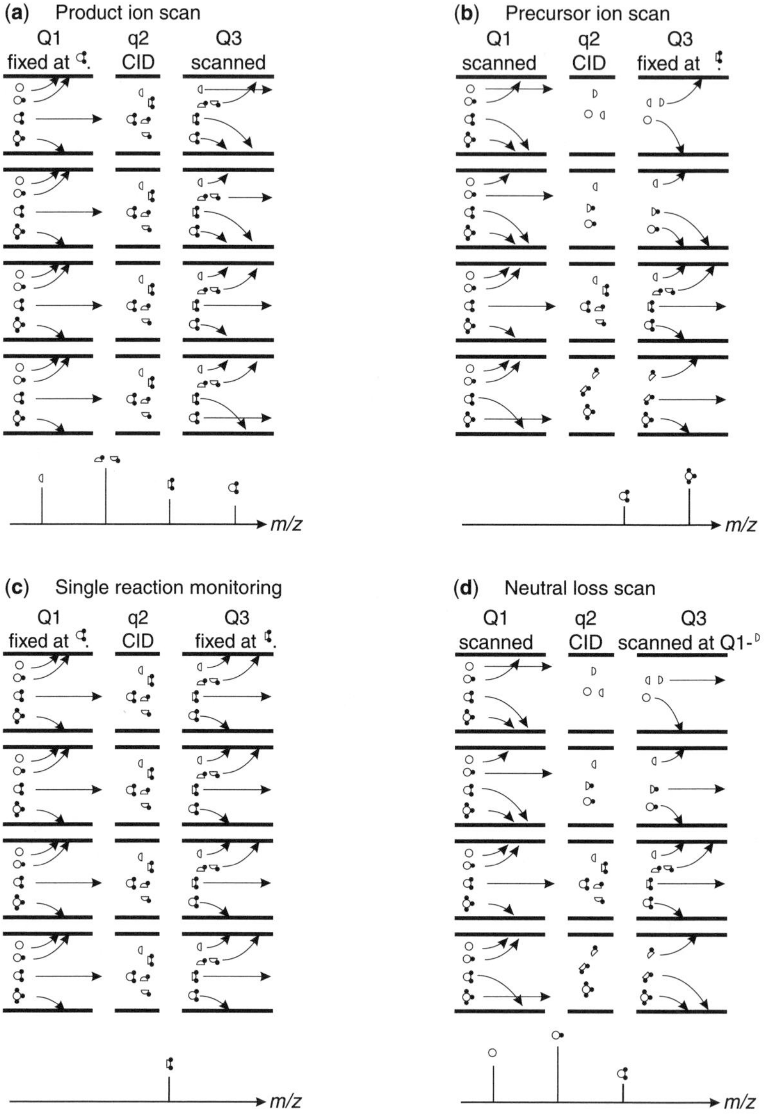

Figure 3.1. The principles of (a) product ion scanning, (b) precursor ion scanning, (c) selected reaction monitoring, and (d) neutral loss scanning. A triple-quadrupole analyzer consists of three main regions, two analyzer quadrupoles (Q1 and Q3) and an intermediate quadrupole (q2) that contains the collision cell. The voltages of Q1 and Q3 can be scanned independently to generate mass spectra. When operating in MS/MS mode, CID is achieved by letting gas into the collision cell. q2 is operated in RF-only mode and acts as a focusing device to keep fragments (and unfragmented precursors) on axis. Parts reprinted from A. Westman-Brinkmalm and G. Brinkmalam (2002). In Mass *Spectrometry and Hyphenated Techniques in Neuropeptide Research*, J. Silberring and R. Ekman (eds.) New York: John Wiley & Sons, 47–105. With permission of John Wiley & Sons, Inc.

TABLE 3.1. The Standard Tandem MS Modes

Mode	Purpose	Experiment
Product ion	To get structural information on ions produced in the ion source	MS1 selects one precursor ion, MS2 acquires a full mass spectrum of the fragment ions produced
Precursor ion	To find compounds that produce a common fragment	MS1 is scanning all precursor ions, MS2 selects one fragment ion
Neutral loss	To find compounds that lose a common neutral species	MS1 and MS2 are both scanning at a fixed m/z difference
Selected reaction monitoring	To monitor a selected reaction	MS1 selects one precursor ion, MS2 selects one fragment ion

obtain an unambiguous and quantifiable signal. In this chapter general descriptions of some common tandem MS analyzer combinations (Section 3.1) and ion activation methods (Section 3.2) are provided. A more detailed description of sequencing of peptides and proteins with tandem MS is provided in Section 2.2 of Chapter 2.

3.1. TANDEM MS ANALYZER COMBINATIONS

Tandem MS has been more or less successfully performed with a wide variety of analyzer combinations. What analyzers to combine for a certain application is determined by many different factors, such as sensitivity, selectivity, and speed, but also size, cost, and availability. The two major categories of tandem MS methods are tandem-in-space and tandem-in-time, but there are also hybrids where tandem-in-time analyzers are coupled in space or with tandem-in-space analyzers. Moreover, the ongoing development of faster electronics and high-voltage circuits as well as better software tools and hardware control devices constantly open up new possibilities of making innovative combinations of analyzers for specific applications. In this chapter a few examples of commercially available tandem MS instruments are presented. A brief summary of their weaknesses, strengths, and main areas of applications is given.

3.1.1. Tandem-in-Space

A tandem-in-space mass spectrometer consists of an ion source, a precursor ion activation device, and at least two nontrapping mass analyzers. The first mass analyzer is used to select precursor ions within a narrow m/z range. Isolated precursor ions are allowed to enter the ion activation device, for example, a gas-filled collision cell, where they dissociate. Created fragments continue on to the second mass analyzer for analysis. The second mass analyzer can either acquire a full mass fragment spectrum or be set to monitor a selected, narrow, m/z range. In principle the second mass analyzer could be followed by more ion activation devices and mass analyzers for MS^n experiments. However, due to rapidly decreasing transmission and increasing experimental complexity and instrument size, MS^3 experiments and beyond are seldom performed in tandem-in-space instruments. In tandem-in-space MS the ion activation has to be fast enough to produce fragment ions before the precursor ions enter the second mass analyzer. This means that tandem MS instruments with mass analyzers where precursor ions are accelerated to high kinetic energies, for example, time-of-flight (TOF) and sector analyzers, demand very fast ion activation events, such as in high-energy collision induced dissociation (CID; see Section 3.2.3.1). Triple-quadrupoles, on the other hand, where the precursor ions are comparatively slow, can accommodate slow ion activation techniques, such as low-energy CID (see Section 3.2.3.2). Very slow activation techniques, such as electron capture dissociation (ECD; see Section 3.2.6) and infrared multiphoton dissociation (IRMPD; see Section 3.2.4.2), on the other hand, have so far only been implemented in trapping instruments. Tandem-in-space instruments have a short duty cycle, which makes them highly suitable for coupling with online chromatography and they have found widespread use in high throughput applications.

3.1.1.1. Triple-Quadrupole. A widely used tandem-in-space instrument, especially in bioanalytical assays, is the triple-quadrupole (QqQ, see Section 2.2.3). The QqQ in combination with online LC separation is often used for quantitative analysis of low-abundance molecules in complex mixtures [1], see also Chapter 8 for an application example. In the SRM mode the first quadrupole of the QqQ is set to transmit a particular m/z. However, when the sample is very complex, for example, body fluids such as serum or urine, the resolution of the chromatographic separation is often not high enough and the compound of interest may co-elute with a number of molecules *within* the selected m/z range. These (approximately) isobaric interferences will all continue to the second quadrupole. In most QqQ instruments the second quadrupole is in reality an octopole or a hexapole, but it is nevertheless normally denoted quadrupole for historical reasons. This second "quadrupole" contains a gas-filled collision cell and here fragmentation occurs through low-energy CID (see Section 3.2.3.2). The third quadrupole is also set to transmit a particular m/z, namely a known fragment of the compound that is the object of the investigation. This two-step mass filtering process provides high specificity even for high sample complexity and comparatively fast chromatographic separations. The very rapid duty cycle (10 to 50 ms) of the QqQ also makes it particularly suitable for multicomponent monitoring and high throughput analysis.

Another major advantage of the rapid duty cycle is that many data points can be obtained over one chromatographic peak, thereby facilitating a reliable quantitation. In a recent comparison it was shown that the QqQ (SRM mode) reaches, at least, tenfold higher sensitivity for the majority of the compounds compared to the QIT (MRM mode) and Qq-TOF (MRM mode) [2]. Strong differences were also observed in the linear response range between the three mass analyzers. Also here the QqQ, which provided a linear response over three orders of magnitude, was roughly ten times better than its competitors. Even though the QqQ is often used in the SRM mode the instrument also performs very well in the precursor ion, product ion, and neutral loss modes. Then again, since the quadrupole filter is a scanning device the QqQ inherently has a lower sensitivity in the full mass scan than the Qq-TOF and the QIT.

3.1.1.2. Tandem Time-of-Flight. For more than a decade the speed, sensitivity, and reliability of MALDI-TOF-MS has made it the favored method for high throughput protein identification and characterization in proteomics. The only major downside has been the difficulty to obtain direct structural information through tandem MS. One method to obtain fragment data, which in principle can be performed in any MALDI-TOF instrument equipped with a reflector (RTOF), is termed MALDI-post-source decay (PSD). In MALDI-PSD intact ions are produced and accelerated in the MALDI ion source. If the ions have gained excess internal energy during the desorption/ionization or through collisions in the ion source they will decay in the field-free flight tube by post-source or metastable decay (see Section 3.2.2). Fragment ions have approximately the same velocity as the precursor ion, but their kinetic energies are mass dependent. Before the ions enter the reflector they pass a timed ion selector or an ion gate. The ion gate allows unaffected passage for only a brief period of time ideally selecting only ions originating from one precursor ion (in reality the selected m/z range is approximately 10 Th). Fragment ions with different m/z will now penetrate to different depths of the reflector, thus reaching the detector at different flight times (Fig. 3.2a). The conversion to a mass scale for the fragment spectrum is more complicated than for a normal spectrum where all ions have the same kinetic energy and travel the same distance. Moreover, the resolution of each fragment peak is strongly mass (energy) dependent. This can be remedied by collecting several fragment spectra, typically about ten, at different reflector potentials. After collection, the parts of the spectra that have best resolution are stitched together. However, stitching is time consuming, reduces sensitivity, and complicates calibration procedures. By inserting a collision cell in the field-free region of the TOF analyzer high energy CID can also be performed. The only difference from PSD is the type of fragments that are generated. For an extensive review of tandem MS with MALDI-PSD, see Reference 3.

Another method to obtain MS/MS data with a single MALDI-RTOF is to use a nonlinear field reflector instead of a conventional linear field reflector (see also Section 2.2.1). In theory a quadratic field reflector can focus fragment ions within a wide mass (energy) range simultaneously and no voltage stepping or stitching is necessary [4]. However, in practice a true quadratic field reflector is difficult to realize. Several RTOF instruments with close to quadratic reflecting fields have been designed [5–7]. Here the difficulty is to find a satisfactory compromise between the advantage of second-order energy focusing and the disadvantage of a divergent ion beam that is causing low transmission and hence low sensitivity.

Presently the most common tandem time-of-flight (TOF-TOF) configuration consists of a MALDI ion source, a short linear TOF analyzer, an ion gate, a collision cell for (moderately) high-energy CID, and a linear reflector TOF analyzer [8, 9] (Figs. 3.2b and c). The key feature in these configurations is that the fragments are created with a comparatively moderate kinetic energy (below 10 keV) and afterwards further accelerated (to at least 16 keV) when they enter the second TOF analyzer. When the fragment ions penetrate into the reflector they have a relatively narrow energy spread and can be sufficiently focused by the linear reflector. A comprehensive review of TOF-TOF in general and the curved-field reflectron in particular is given in

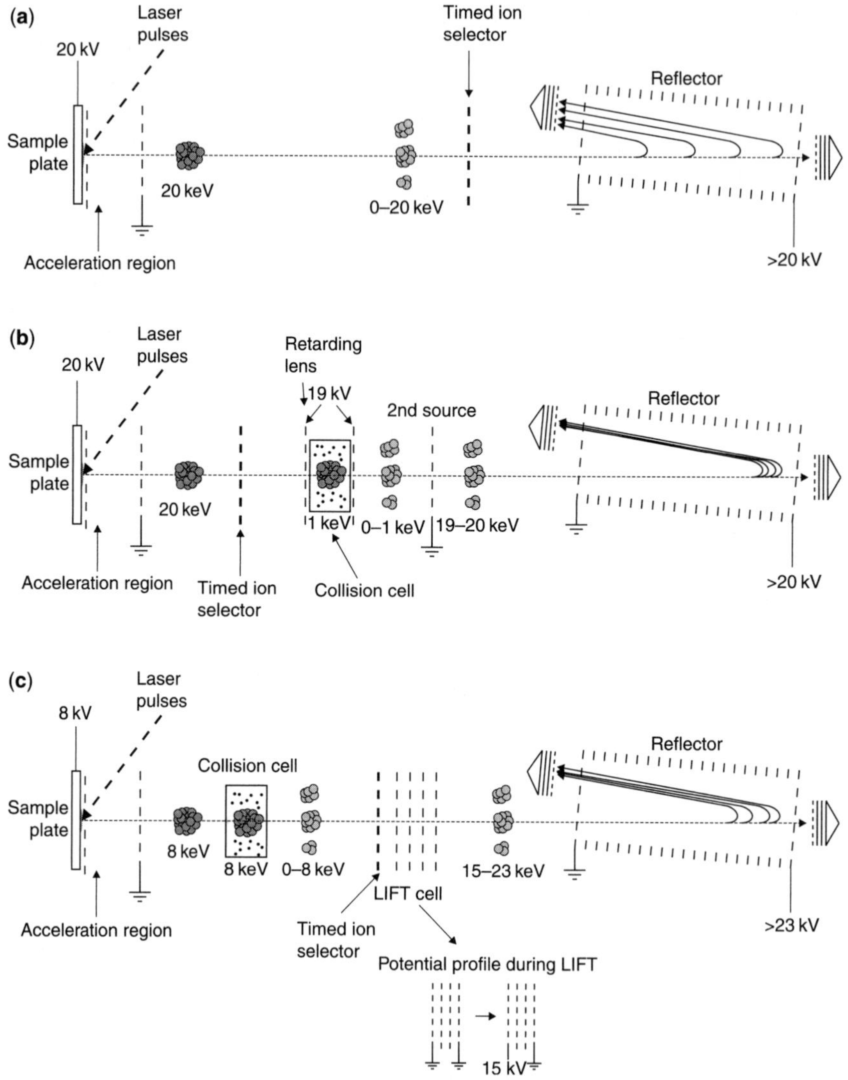

Figure 3.2. Schematics of MALDI-TOF tandem MS configurations. (a) Standard MALDI-TOF mass spectrometer equipped with a linear reflector. (b) MALDI-TOF-TOF configuration were the selected precursor ions are decelerated before they enter the collision cell. Product ions (and surviving intact ions) are subsequently accelerated in the second "source" and enter the linear reflector with a relatively narrow energy spread. (c) MALDI-TOF-TOF configuration were the ions are accelerated by an 8 kV potential gap in the source. After the collision cell product ions (and surviving intact ions) pass through a timed ion selector and the selected ions enter the LIFT cell. The potential in the LIFT cell is rapidly raised to 15 kV a short moment after the selected ions have entered it and these ions are subsequently accelerated. In this configuration the product ions enter the linear reflector with an energy spread, which is greater than for the configuration in (b), but can still be relatively well compensated for by the linear reflector.

Reference 5. With the advent of commercially available MALDI-TOF-TOF instruments the combination of off-line one- or two-dimensional LC-MALDI-MS/MS has become a popular alternative or rather a complement to LC-ESI-MS/MS in the proteomics community.

3.1.1.3. Quadrupole-Time-of-Flight. Hybrid mass spectrometers are instruments equipped with two or more different types of mass analyzers coupled together. A particularly fruitful hybrid tandem-in-time instrument is the quadrupole mass filter-TOF (Qq-TOF). The Qq-TOF can be regarded as a QqQ with the third quadrupole substituted by an orthogonal TOF equipped with a reflector (see Fig. 2.11). In MS mode the quadrupole mass filter only acts as a passive transmission element and the mass spectrum is acquired by the TOF. In MS/MS mode the quadrupole mass filter is set to transmit only the precursor ion of interest (m/z window typically 1 to 3 Th). Fragmentation of the precursor ions through low-energy CID occurs in the collision cell and the fragment mass spectrum is acquired by the TOF. Advantages with the TOF compared with a third quadrupole filter include higher resolution, better mass accuracy, and the opportunity to register all ions simultaneously without scanning. However, for quantification of targeted compounds the sensitivity and linear dynamic range of the Qq-TOF (MRM mode) is inferior to the QqQ (SRM mode) [2, 10]. On the other hand, the higher resolution of the TOF can provide a better selectivity, which can be beneficial in some analytical cases. Another advantage of the Qq-TOF is that since all fragments ions are detected simultanously (without compromising the sensitivity) it is possible to do post analysis of products.

3.1.2. Tandem-in-Time

In tandem-in-time MS ions produced in the ion source are trapped, isolated, fragmented, and m/z separated in the same physical device. This is only possible in trapping devices, such as the QIT and the FTICR, where ions can reside in the trap for a long time (see Sections 2.2.4 and 2.2.6). Long trapping times make it possible to manipulate the ions several times, thus making tandem-in-time mass spectrometers ideal for MS^n measurements. Moreover, since the precursor ions can be trapped for considerable amounts of time it is possible to use very slow activation methods, such as ECD and IRMPD. However, the relatively long duty cycle of tandem-in-time instruments limits the use in high throughput/high complexity applications. Nonetheless, several tandem-in-time mass spectrometers fast enough to enable online coupling to chromatography are commercially available.

3.1.2.1. Quadrupole Ion Trap. The quadrupole ion trap (QIT) is very sensitive and an excellent mass analyzer for performing MS^n experiments. A tandem-in-time experiment begins with ejection of all ions except those within a selected, narrow m/z range from the trap. Next, the selected ions are fragmented most often by low-energy CID. All fragments within the m/z acceptance range of the trap will remain in the trap, and an MS-MS spectrum can be obtained. To perform an MS^3 experiment the procedure is repeated, but instead of aquiring a mass spectrum, one of the observed fragment ions is isolated and excited to obtain an MS^3 spectrum, and so on. In practice the need for higher orders than MS^5 is rare. It is important to be aware that in the QIT the low mass cut-off for fragments is significantly higher than for the full mass scan. The reason is that in the QIT there is a trade-off between the depth of the trapping potential and the width of the m/z range. Hence, in order to still contain the precursor ions despite their increased kinetic energy the m/z range has to be compromised. Often the low mass cut-off is set to 25% of the precursor m/z. Advantages of the QIT are primarily its high sensitivity and MS^n capability, but its small size and relatively low price are also important reasons for its popularity. Despite a relatively long duty cycle (typically 100 ms or more) QIT-MS has also been successfully coupled to online separation techniques such as LC and CE. Like the Qq-TOF the QIT possesses the ability to obtain sensitive full scan CID fragment spectra, thus providing the advantage of performing post-acquisition data processing. However, the QIT is not particularly well suited for quantitative analysis of complex samples. One important reason is that the number of ions that can be stored simultaneously in a QIT is limited. This is not a problem when analyzing samples with few components because most of the time only one type of ion is trapped in the QIT in the same trapping event. On the contrary, with complex biological extracts many different ions are stored in the QIT at the same time and the detected analyte signal can be severely reduced by interference of high abundance "background" ions. The relative abundances of analyte and "background" ions are strongly sample- and method-dependent and make it difficult to find a robust and reliable quantitation method.

3.1.2.2. Fourier Transform Ion Cyclotron Resonance. Like the QIT the Fourier transform ion cyclotron resonance (FTICR) analyzer is excellent for MS^n measurements (see Section 2.2.6), perhaps even more so, since the ions remain in the cell after detection. In principle one injection of ions is enough for a whole MS^n sequence, including acquisition of a mass spectrum of each step.

The procedure is very similar to that of the quadrupole ion trap. First, all ions except those with the desired m/z are ejected from the cell by exciting them to a radius larger than the cell radius. Second, the selected ion species are excited by, for example, low-energy CID, ECD, or IRMPD. As with the quadrupole trap all fragments within the m/z acceptance range will remain trapped and can, after further dipolar excitation, be detected, thus yielding an MS-MS spectrum. The procedure is repeated giving an MS^3 spectrum, and so on. Like with the ion trap there is no theoretical upper limit for the number of fragmentations that can be induced but the need for higher orders than MS^5 is rare. As was stated in the section on QIT, the ability to record mass spectra from multiple fragmentation is a powerful tool for structural characterization. An advantage of the FTICR (besides the exceptional mass accuracy and resolving power) over the QIT is the ability to record low mass CID fragments (well below the 25% of the precursor m/z cut-off of the QIT). The QIT, on the other hand, has significantly better sensitivity than the FTICR and is also considerably faster. The space-demanding superconducting magnet that needs refilling of cryogenic gases is also a disadvantage of the FTICR instrument.

3.1.3. Other Tandem MS Configurations

The classical high performance tandem MS instrument is the four-sector. Two main types of four-sector instruments exist, the magnetic-electric-electric-magnetic or electric-magnetic-electric-magnetic, both with a collision cell located between the second and third sectors [11]. A four-sector tandem MS is capable of providing very high quality data, but is costly, very space demanding, and complex. Therefore, the four-sector tandem MS instruments have gradually (when possible) been replaced with smaller or more convenient instruments. Sector instruments have also been combined with, for example, quadrupole mass filters, TOFs, and QITs. There are also methods to perform MS-MS with limited performance using a standard double focusing instrument without an extra analyzer, for example, linked scan or mass-analyzed ion kinetic energy spectrometry (MIKES). Two relatively recent additions to the tandem MS instrument family are the QIT-FTICR and the QIT-orbitrap hybrids. Both these instruments combine the high sensitivity, high relative speed, and easy handling of the linear QIT with the high mass accuracy of the FTICR/orbitrap and have become widely popular in a short time. Another high performance hybrid instrument is the Qq-FTICR. Moreover, the quadrupole filter has also been used as a front end in combination with the QIT.

3.2. ION ACTIVATION METHODS

No tandem MS experiment can be successful if the precursor ions fail to fragment (at the right time and place). The ion activation step is crucial to the experiment and ultimately defines what types of products result. Hence, the ion activation method that is appropriate for a specific application depends on the MS instrument configuration as well as on the analyzed compounds and the structural information that is wanted. Various, more or less complementary, ion activation methods have been developed during the history of tandem MS. Below we give brief descriptions of several of these approaches. A more detailed description of peptide fragmentation rules and nomenclature is provided in Chapter 2. An excellent review of ion activation methods for tandem mass spectrometry is written by Sleno and Volmer, see Reference 12, and for a more detailed review on slow heating methods in tandem MS, see Reference 13.

3.2.1. In-Source Decay

In-source decay occurs when the precursor ions gain sufficient internal energy, either while being created or through early collision events, that they decay before exiting the source. The likelihood for in-source decay depends strongly on the type of ion source and the operating parameters chosen. One example of in-source decay is nozzle-skimmer dissociation (NSD), where fragment ions are generated in an electrospray ionization (ESI) source by accelerating the precursor ions as they pass through the atmospheric pressure/vacuum interface [14]. A modification of this method has recently been used to obtain top-down sequence information on purified 200 kDa proteins [15]. However, in-source decay does not allow for separation of precursor ions with different m/z before dissociation and is thereby not truly a tandem MS method.

3.2.2. Post-Source Decay

The main difference between postsource decay (PSD, also referred to as metastable ion dissociation) and in-source decay is that metastable ions are stable enough (or the time they spend in the source is short enough) for them to leave the source intact. Hence, they can be m/z separated before they fragment and a proper tandem MS measurement can be performed. An example where metastable fragmentation is used is vacuum MALDI-PSD in combination with a reflectron TOF or TOF-TOF analyzer (see Section 3.1.1.2). Post-source decay of vacuum MALDI produced precursor ions is probably a combination of contributions from primary laser induced processes and secondary processes, for example, high-energy collisions with the matrix plume during extraction [16]. Several factors influence the fragment spectrum, including the laser intensity and the effective temperature of the matrix.

3.2.3. Collision Induced/Activated Dissociation

In collision induced/activated dissociation (CID/CAD) precursor ions collide with gas atoms or molecules, such as nitrogen, argon, or helium, and fragment. In the collision a part of the kinetic energy is converted into vibrational/rotational energy of the parent ion. If the internal energy gained is high enough the precursor ion will fragment fast enough for the fragment ions to be observed in the mass spectrometer. Depending on the type of mass analyzer, either high-energy CID (kiloelectronvolt collision energy) or low-energy CID (<100 eV) is performed. Low-energy CID is common in quadrupole and ion trap instruments and high-energy CID is typical for sector and TOF instruments.

3.2.3.1. High-Energy CID. High-energy CID is performed by acceleration of the precursor ions to kiloelectronvolt energies and subsequent collision with gas atoms or molecules in a collision cell. One collision is enough to cause electronic excitation of the precursor ion and subsequently virtually all structurally possible fragmentations occur that have some probability of occurring. Fragment-ion mass spectra are often complex, with abundant low mass and internal fragment ions. The complexity of the MS-MS spectra can make the elucidation more challenging, but also increase confidence in the identification of the precursor ion. For example, high-energy CID often produces side-chain cleavage of peptides that makes it possible to distinguish between the amino acids leucine and isoleucine, which have the same elemental composition. High-energy CID is considered as very reproducible. This is due to the high kinetic energy of the accelerated precursor ion. Compared to the kiloelectronvolt energy of the precursor ion the kinetic energy of the gas atom is negligible in the collision. This means that changes in collision conditions, such as type of target gas, gas pressure, and temperature, do not largely influence the appearance of the fragment-ion spectrum.

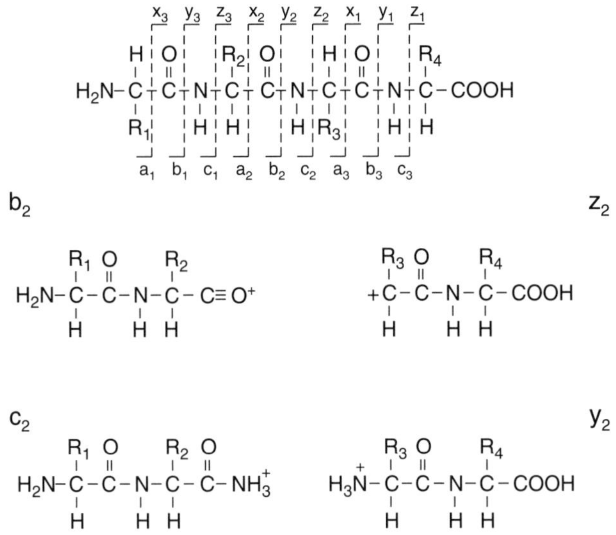

Figure 3.3. Nomenclature of peptide ions resulting from backbone fragmentation and chemical structure of b, c, z, and y product ions.

3.2.3.2. Low-Energy CID. In low-energy CID, only a small amount of energy is added to the internal energy of the precursor ion in each collision and energy from hundreds of collisions will have to be accumulated in order to induce fragmentation. Here, the time between the collisions is long enough for the internal energy to redistribute on all degrees of freedom thereby causing low-energy fragmentation pathways to dominate. In peptides low-energy CID often causes the amide bond of the backbone to dissociate, thereby producing b- and y-type fragment ions (Fig. 3.3).

In linear quadrupoles the CID activation time is equal to the time the accelerated precursor ion (<100 eV) takes to pass through the collision cell. The primary fragments produced will continue to undergo collisions that cause them to fragment as well, and so on, thus producing richer and more complex fragment-ion spectra.

In a QIT the precursor ion is selectively accelerated via resonance excitation and collided with the background gas. Since the ion acceleration is m/z dependent, no additional fragmentation occurs of the primary fragments. Hence, if the precursor ion contains groups that dissociate at a markedly lower internal energy than other groups (e.g., peptides containing phosphorylated serine or threonine residues) the resulting MS-MS spectra contain only a few intense fragment ions and thereby insufficient information for unambiguous precursor identification. An additional limitation is that it is not possible to simultaneously accelerate the precursor ion and trap the lightest fragment ions (see Section 3.1.2.1). In practice fragment ions with m/z below about one third of the m/z of the precursor ion are absent from the MS-MS spectra of a QIT.

In the ion cyclotron resonance (ICR) cell CID is performed by resonant excitation of the precursor ion and subsequent collisions with the background gas. One serious drawback with on-resonance excitation in the ICR cell is that the risk for ion losses due to ejection increases with increasing activation time. Several alternative ICR-CID methods with enhanced MS-MS sensitivity have been introduced, for example, sustained off resonance (SORI)-CID, very low energy (VLE)-CID, and multiple excitation collisional activation (MECA) [17].

3.2.4. Photodissociation

Photodissociation (PD) occurs when a precursor ion is excited by one or more photons and is subsequently fragmented. Dissociation can be induced by photons of a wide range of energies.

3.2.4.1. UV/Visible Photodissociation. In the case of UV/visible PD the photon energies are high enough to enable direct fragmentation. Previously, UV photodissociation was utilized mainly to study dissociation kinetics of gas-phase ions, but recently it was demonstrated that UV PD could be used for tandem MS of MALDI generated protonated peptides [18].

3.2.4.2. Infrared Multiphoton Dissociation. In infrared multiphoton dissociation (IRMPD) the precursor ions are irradiated with IR photons. IRMPD can be performed either by using high-power pulsed lasers or low-power continuous-wave (cw) lasers. The latter approach is a slow heating method with many similarities to low-energy CID and is the more common of the two. Typically cw-IRMPD is performed in QITs or ICR cells. The trapped ions are slowly heated by absorption of IR photons from a low-power cw CO_2 laser for tens to hundreds of milliseconds until dissociation starts. As in low-energy CID, the time between the excitation events is long enough for the energy to be distributed throughout the ion, principally leading to low-energy pathway fragmentation. A major difference between slow heating by CID and IRMPD is that all ions (independently of m/z) are activated at the same time. An important consequence of the m/z independence is that as long as the irradiation continues the primary fragments will dissociate and produce secondary fragments. Secondary fragmentation could be advantageous since it provides additional sequence information, but extensive fragmentation of primary fragments results in loss of information as well. It is therefore vital to tune laser energy and irradiance time in order to find the optimal balance between primary and secondary fragmentation. In QITs IRMPD has an advantage over CID since the kinetic energy of the precursor ion does not have to be increased and hence it is not necessary to make a compromise between sensitivity and m/z range (see Section 3.1.2.1). In the ICR cell it is an advantage that the background pressure does not need to be increased for IRMPD.

3.2.5. Blackbody Infrared Radiative Dissociation

In blackbody infrared radiative dissociation (BIRD) ions are activated by absorbtion of IR photons emitted from the walls of a heated ICR cell [19]. The ICR cell is so far the only mass analyzer that meets both essential requirements for successful

BIRD. The first requirement is that since BIRD is a very slow-heating method the observation time has to be long (seconds or minutes). Second, the pressure has to be low ($<10^{-6}$ torr) so that the risk of dissociation due to collisions with the background gas is negligible. There are close similarities between BIRD and another slow-heating method, IRMPD.

3.2.6. Electron Capture Dissociation

Electron capture dissociation (ECD) occurs when a precursor ion captures a low energy (<1 eV) electron, forms an excited radical species, and then rapidly dissociates. The dissociation in ECD is faster than the energy redistribution process (nonergodic) thus inducing more random fragmentation than observed with slow heating methods. For peptides ECD often creates c- and z-type fragment ions through cleavage of the C-α-N bond (Fig. 3.3). Another characteristic of peptide ECD is that several bonds normally stable to slow heating activation, such as disulfide bonds, dissociate with high probability. However, perhaps the most attractive trait of ECD is that labile modifications such as O/N-glycosylation, phosphorylation, and other labile posttranslational modifications are regularly retained [20]. Hence, ECD has a promising potential for being able to provide rapid and sensitive characterization of labile modifications in the future. ECD has almost exclusively been demonstrated in FTICR analyzers (see Section 2.2.6). The reason is that the electron capture probability is highest when the precursor ions are surrounded by a dense cloud of near thermal electrons. Consequently, both long (several milliseconds) irradiation times and efficient trapping of near-thermal electrons are prerequisites for efficient ECD. Several attempts have been made to implement ECD in mass analyzers that trap ions with time varying rather than static fields, for example, Qq-TOF, QIT, or linear QIT, but to date none have been implemented in a commercial instrument. The reason is that near thermal electrons are readily heated by the RF field and escape the trap within microseconds. For a comprehensive review of the ECD technique see Reference 20.

3.2.7. Electron Transfer Dissociation

Electron transfer dissociation (ETD) is an ECD-like method with most of the same characteristics [21]. Like ECD, ETD yields abundant peptide backbone c- and z-type ions while often retaining such labile groups as peptide O/N-glycosylation and phosphorylation [22]. Unlike ECD, ETD can be performed in the presence of an RF field. Here, anions created in a chemical ionization (CI) source (see Section 2.1.7) are used as electron donors but the fragmentation pathways are essentially the same as for ECD. Commercial linear QIT instruments have recently become available with the ETD option.

3.2.8. Surface-Induced Dissociation

Surface-induced dissociation (SID) is an extremely fast activation technique where precursor ions are m/z-selectively accelerated and collided with a solid surface. The amount of energy transferred can be varied from milli- to megaelectronvolts by changing the impact energy of the precursor ion. Fine tuning the collision energy can be valuable when identifying the dissociation energies needed for a specific fragmentation pathway to occur. The interaction time (of the order of picoseconds) is short compared to the dissociation time for polyatomic ions and SID probably involves collisional activation at the surface followed by delayed gas-phase dissociation of the scattered precursor ion. Collisions in the so called "hyperthermal" regime (energies from 1 to 100 eV) where the collision energies are of the same order of magnitude as the chemical bond energies tend to provide most specific information on the chemical nature of the precursor ion. SID has been implemented into several different mass analyzers, for example, FTICR, QIT, TOF-TOF, and QqQ instruments, but still awaits incorporation into commercial instruments. For a detailed discussion of fundamental principles of SID see References 12 and 17.

REFERENCES

1. G. Hopfgartner and E. Bourgogne. Quantitative High-Throughput Analysis of Drugs in Biological Matrices by Mass Spectrometry. *Mass Spectrom. Rev.*, **22**(2003): 195–214.
2. C. Soler, K. J. James, and Y. Pico. Capabilities of Different Liquid Chromatography Tandem Mass Spectrometry Systems in Determining Pesticide Residues in Food: Application to Estimate Their Daily Intake. *J. Chromatogr.*, **A1157**(2007): 73–84.
3. B. Spengler. Post-Source Decay Analysis in Matrix-Assisted Laser Desorption/Ionization Mass Spectrometry of Biomolecules. *J. Mass Spectrom.*, **32**(1997): 1019–1036.
4. B. A. Mamyrin. Time-of-Flight Mass Spectrometry (Concepts, Achievements, and Prospects). *Int. J. Mass Spectrom.*, **206**(2001): 251–266.
5. R. J. Cotter, W. Griffith, and C. Jelinek. Tandem Time-of-Flight (TOF/TOF) Mass Spectrometry and the Curved-Field Reflectron. *J. Chromatogr.*, **B855**(2007): 2–13.
6. J. Flensburg, D. Haid, J. Blomberg, J. Bielawski, and D. Ivansson. Applications and Performance of a MALDI-ToF Mass Spectrometer with Quadratic Field Reflectron Technology. *J. Biochem. Biophys. Meth.*, **60**(2004): 319–334.
7. A. E. Giannakopulos, B. Thomas, A. W. Colburn, D. J. Reynolds, E. N. Raptakis, A. A. Makarov, and P. J. Derrick. Tandem Time-of-Flight Mass Spectrometer (TOF-TOF) with a Quadratic-Field Ion Mirror. *Rev. Sci. Instrum.*, **73**(2002): 2115–2123.
8. K. F. Medzihradszky, J. M. Campbell, M. A. Baldwin, A. M. Falick, P. Juhasz, M. L. Vestal, and A. L. Burlingame. The Characteristics of Peptide Collision-Induced Dissociation Using a High-Performance MALDI-TOF/TOF Tandem Mass Spectrometer. *Anal. Chem.*, **72**(2000): 552–558.
9. D. Suckau, A. Resemann, M. Schuerenberg, P. Hufnagel, J. Franzen, and A. Holle. A Novel MALDI LIFT-TOF/TOF Mass Spectrometer for Proteomics. *Anal. Bioanal. Chem.*, **376**(2003): 952–965.
10. C. Soler, B. Hamilton, A. Furey, K. J. James, J. Manes, and Y. Pico. Comparison of Four Mass Analyzers for Determining Carbosulfan and Its Metabolites in Citrus by Liquid Chromatography/Mass Spectrometry. *Rapid Commun. Mass Spectrom.*, **20**(2006): 2151–2164.

11. M. L. Gross. Tandem Mass Spectrometry: Multisector Magnetic Instruments. In Methods in Enzymology: Mass Spectrometry, ed. J. A. McCloskey. *Methods in Enzymology* **193**, Academic Press, San Diego, 1990, 131–153.
12. L. Sleno and D. A. Volmer. Ion Activation Methods for Tandem Mass Spectrometry. *J. Mass Spectrom.*, **39**(2004): 1091–1112.
13. S. A. McLuckey and D. E. Goeringer. Slow Heating Methods in Tandem Mass Spectrometry. *J. Mass Spectrom.*, **32**(1997): 461–474.
14. J. A. Loo, C. G. Edmonds, and R. D. Smith. Primary Sequence Information from Intact Proteins by Electrospray Ionization Tandem Mass Spectrometry. *Science*, **248**(1990): 201–204.
15. X. Han, M. Jin, K. Breuker, and F. W. McLafferty. Extending Top-Down Mass Spectrometry to Proteins with Masses Greater Than 200 Kilodaltons. *Science*, **314**(2006): 109–112.
16. V. Gabelica, E. Schulz, and M. Karas. Internal Energy Build-Up in Matrix-Assisted Laser Desorption/Ionization. *J. Mass Spectrom.*, **39**(2004): 579–593.
17. J. Laskin and J. H. Futrell. Activation of Large Ions in FT-ICR Mass Spectrometry. *Mass Spectrom. Rev.*, **24**(2005): 135–167.
18. J. Y. Oh, J. H. Moon, Y. H. Lee, S. W. Hyung, S. W. Lee, and M. S. Kim. Photodissociation Tandem Mass Spectrometry at 266 nm of an Aliphatic Peptide Derivatized with Phenyl Isothiocyanate and 4-Sulfophenyl Isothiocyanate. *Rapid Commun. Mass Spectrom.*, **19**(2005): 1283–1288.
19. R. C. Dunbar. BIRD (Blackbody Infrared Radiative Dissociation): Evolution, Principles, and Applications. *Mass Spectrom. Rev.*, **23**(2004): 127–158.
20. R. A. Zubarev. Electron-Capture Dissociation Tandem Mass Spectrometry. *Curr. Op. Biotechnol.*, **15**(2004): 12–16.
21. J. E. Syka, J. J. Coon, M. J. Schroeder, J. Shabanowitz, and D. F. Hunt. Peptide and Protein Sequence Analysis by Electron Transfer Dissociation Mass Spectrometry. *Proc. Natl. Acad. Sci. U.S.A.*, **101**(2004): 9528–9533.
22. L. M. Mikesh, B. Ueberheide, A. Chi, J. J. Coon, J. E. Syka, J. Shabanowitz, and D. F. Hunt. The Utility of ETD Mass Spectrometry in Proteomic Analysis. *Biochim. Biophys. Acta*, **1764**(2006): 1811–1822.

4

SEPARATION METHODS

Ann Westman-Brinkmalm, Jerzy Silberring,
and Gunnar Brinkmalm

Many different separation methods have been used in conjunction with mass spectrometry. There are many advantages to combining MS and separation methods, including increased sensitivity, dynamic range, and selectivity. Various separation methods can be used to prefractionate or clean up samples off-line before they are introduced into the mass spectrometer, for example, analytical two-dimensional gel electrophoresis (2-DGE) combined with MALDI-TOF-MS. Other separation methods are well suited for online coupling with certain mass spectrometers, for example, nanoflow liquid chromatography in combination with ESI-QIT-MS. Most of the separation methods mentioned in this chapter are active areas of research and development of their own. However, the main goal of this chapter is to summarize the basic principles of some of the most common separation methods that have been used in conjunction with mass spectrometry. After the description of each separation method a reference to how it can be used together with MS is given. For an excellent review of the historical development of coupling of gas and liquid chromatography with mass spectrometry see Reference 1.

Mass Spectrometry. Edited by Ekman, Silberring, Westman-Brinkmalm, and Kraj

4.1. CHROMATOGRAPHY

Chromatography ("color writing") was developed by the Russian botanist Michael Tswett about a hundred years ago [2–4]. A chromatographic system consists of a mobile phase that carries the sample through a stationary phase. Separation of a complex mixture occurs when the sample molecules have different affinities for the stationary media. As a result, the time for a particular molecule to travel through the chromatographic medium (the retention time) will depend on its physicochemical properties. The two main types of chromatography are liquid chromatography, where the mobile phase is a liquid, and gas chromatography, where gas is used as the mobile phase. The mobile phase may also be operated under supercritical conditions. This type of chromatography is termed supercritical fluid chromatography.

4.1.1. Gas Chromatography

The mobile phase in gas chromatography (GC) is a gas, such as nitrogen, helium, or hydrogen, and the stationary phase is, in most cases, a solid support coated with liquid (gas-liquid chromatography, GLC) [5]. In general GC is used for the separation of low molecular weight, volatile materials that can be evaporated and remain stable at a temperature below the upper temperature limit of the chromatograph (i.e., $\sim 350^{\circ}C$). There are, however, many involatile substances that can be converted into volatile derivatives (derivatization) further separated by GC. A major general advantage of GC over LC is its simplicity, despite a time-consuming derivatization procedure. Hence, the GC separation process is faster and the columns are more efficient, and the separation times considerably shorter. There are two basic types of columns in GC; the packed column and the capillary column. Packed columns are usually made of stainless steel or glass and have larger inner diameter (2 to 4 mm) and shorter length (1 to 4 m) than capillary columns. When high sample loading capacity is important, as with gases, packed columns are preferred. However, packed columns have lower resolution than capillary columns and are more difficult to couple directly to mass spectrometers because of the high gas flows. Today packed columns have more and more been replaced with capillary columns. Capillary columns use lower gas flows, have higher resolving power, and lower sample loading capacity. They are made of fused silica. The internal diameter varies between 0.1 mm and 0.5 mm and the columns range in length from 10 m to 100 m. Capillary columns can be coated internally with a liquid (wall coated open tubular [WCOT] columns) or a solid (porous layer open tubular [PLOT] columns) stationary phase. The mobile phase (the carrier gas) has to be inert to protect both sample and columns. Helium is the most commonly used carrier gas. Hydrogen provides better separation, but helium is nonflammable and works with a greater number of detectors. In GC practically no interactions take place between the sample components and the gas (the mobile phase). Thus, contrary to liquid chromatography (LC) and supercritical fluid chromatography (SFC), the elution time cannot be controlled by changing the composition of the mobile phase. In GC the counterpart to gradient elution is to control the elution times by raising the column temperature either continuously or step-wise, so called temperature programming. For GC-MS applications, see Chapters 8, 9 and 16. Reference 6 provides a summary of recent fundamental developments in GC, and for an overview of various aspects of coupling of gas chromatography with mass spectrometry see Reference 1.

4.1.1.1. Gas-Solid Chromatography. In gas-solid chromatography (GSC) the stationary phase is a porous solid that interacts directly with the sample molecules. The adsorbent can be relatively nonspecific (e.g., graphitized carbon) or have a negatively or positively charged surface that interacts specifically with the sample molecules (e.g., aluminium oxide). In addition to specific and nonspecific adsorption, GSC (often with synthetic zeolites as adsorbent) can also be used for molecular sieving, that is, small molecules that enter into the adsorbent pores will be retained longer on the column than large molecules that cannot. An example is that straight-chain hydrocarbons can be separated from their branched-chain isomers with GSC molecular sieves. Today, the main applications of GSC are separation of permanent gases and low molecular weight hydrocarbons. For the rest of the GC applications GC is often the same as gas-liquid chromatography.

4.1.1.2. Gas-Liquid Chromatography. In gas-liquid chromatography (GLC) the stationary phase is a liquid. GLC capillary columns are coated internally with a liquid (WCOT columns) stationary phase. As discussed above, in GC the interaction of the sample molecules with the mobile phase is very weak. Therefore, the primary means of creating differential adsorption is through the choice of the particular liquid stationary phase to be used. The basic principle is that analytes selectively interact with stationary phases of similar chemical nature. For example, a mixture of nonpolar components of the same chemical type, such as hydrocarbons in most petroleum fractions, often separates well on a column with a nonpolar stationary phase, while samples with polar or polarizable compounds often resolve well on the more polar and/or polarizable stationary phases. Reference 7 is a "metabolomics" example of capillary GC-MS.

4.1.2. Liquid Chromatography

In liquid chromatography (LC) the mobile phase is a liquid and the stationary phase typically consists of small porous particles with a large surface area, for example, silica beads coated with different ligands. Sample molecules travel through the chromatographic system and interact with the surface of the stationary phase. On the whole, sample molecules that interact strongly with the stationary phase will take a longer time to travel through the chromatographic system, that is, will have longer retention times. Depending on the LC mode, different types of sample-ligand interactions are responsible for the retention process. The most commonly utilized retention mechanisms in LC are summarized below. The performance of a chromatographic system, for example, separation power, retention times, and loading capacity, is affected by many factors. In high performance liquid chromatography (HPLC), formerly high pressure liquid chromatography, a combination of packing materials with smaller and more

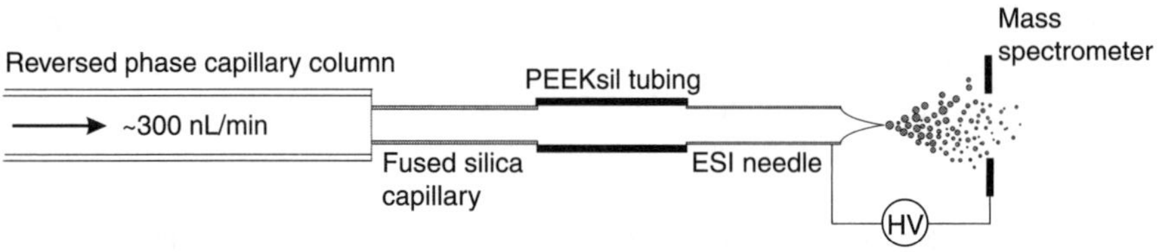

Figure 4.1. Schematic of a nanoflow LC-ESI-MS interface (not to scale).

uniform particle sizes, and increased flow pressure yields both higher resolving power and faster analyses. Recently, increased chromatographic performance for complex mixture separation has been demonstrated with so-called "ultra performance liquid chromatography" (UPLC) using a packing material with particle size less than 2 μm and a chromatographic system operating at very high pressures [8]. For LC-MS applications, see Chapter 8. Reference 9 is a review of LC techniques in proteomic analysis. Figure 4.1 shows a schematic of a nanoflow LC-MS interface.

4.1.2.1. Normal-Phase Chromatography. The classical LC method is the so-called normal-phase chromatography (NPC) or polar-phase chromatography. In NPC the stationary phase is highly polar (e.g., silica) and the mobile phase relatively nonpolar, (e.g., hexane). The retention time of the sample molecules increases with increasing polarity. By increasing the polarity of the mobile phase the retention time can be decreased. An application of NPC is separation of nonpolar compounds soluble only in nonpolar organic solvents. These compounds do not show strong interaction with a highly polar stationary phase and can be eluted from the column relatively quickly. For an example of coupling of NPC separation to MS, see Reference 10.

4.1.2.2. Reversed-Phase Chromatography. Reversed-phase chromatography (RPC) also relies on hydrophobic interactions but in this case the mobile phase is more polar than the stationary phase. In RPC columns, the packing material is often silica with covalently attached carbon chains. The retention time is longer for more hydrophobic molecules. Retention time is decreased by the addition of nonpolar solvent to the mobile phase. The hydrophobicity of the sample molecules is pH dependent. For this reason an organic acid, such as formic acid or trifluoroacetic acid, is often added to the mobile phase. The effect varies depending on application but generally improves the chromatography. RPC is the most widely used mode of LC. A closely related technique is hydrophobic interaction chromatography (HIC). The basic molecular interactions are very similar to RPC, but adsorbents for HIC are less highly substituted with hydrophobic ligands, which allows for mild (not denaturing) elution conditions, thus maintaining the biological activity of the sample molecules. For a recent comparison between nano(RP)LC-MS and CZE-MS, see Reference 11.

4.1.2.3. Ion Exchange Chromatography. Ion-exchange chromatography (IEC) is based on interactions between charged functional groups in the stationary phase and ions or polar molecules of opposite charge in the sample. When positive ions bind to negatively charged functional groups it is called cation exchange chromatography. Correspondingly, anion exchange chromatography uses positive immobilized groups that bind negative ions. The mobile phase is an aqueous buffer and the elution of sample molecules is controlled by adjusting the pH or ionic strength. IEC is often used for peptides and proteins, but can be used for almost any kind of charged molecules. Proteins and peptides are amphoteric molecules with both acidic and basic groups. The net charge of an amphoteric molecule can be altered by changing the pH of the mobile phase, thereby forcing them to elute from the stationary phase. Another method to control elution is changing the ionic strength of the mobile phase. Ions in the mobile phase interact with the sample molecules and the charged groups on the stationary phase shielding them from each other and thus weakening the interactions. Reference 12 provides an example of IEC coupled to ICP-MS.

4.1.2.4. Affinity Chromatography. Affinity chromatography utilizes specific biological interactions between a binding molecule, for example, an antibody, receptor, lectin, or metal, and a ligand (most often a protein, peptide, etc.) present in the sample mixture [13]. As the sample molecules flow through the chromatographic system, the ligand is selectively adsorbed to the binding molecule that is immobilized on the solid phase. By changing the properties of the mobile phase (e.g., acidification) the complex can be dissociated and the ligand eluted. Affinity chromatography is used both to selectively enrich and to selectively remove a certain biopolymer or group of biopolymers from a complex mixture. The technique can also be used to separate functionally different forms of the same biopolymer, for example active and inactive forms of an enzyme.

4.1.2.5. Size Exclusion Chromatography. Contrary to the methods described above, size exclusion chromatography (SEC), also referred to as gel filtration or gel permeation chromatography (GPC), does not rely on interaction between the sample molecules and the stationary or mobile phase. Instead, molecules are separated according to their size relative to pores in the stationary phase [14]. As the sample molecules travel down the column some enter into the pores. The larger the molecules, the less possibility for them to penetrate into the pores, and the faster the elution. Since both aqueous and organic solutions can be used as mobile phases, SEC is applied both to the fractionation of proteins and other large water-soluble molecules, as well as organic-soluble polymers. See Chapter 15 for applications.

4.1.3. Supercritical Fluid Chromatography

In supercritical fluid chromatography (SFC) the mobile phase is a supercritical fluid, such as carbon dioxide [15]. A supercritical fluid can be created either by heating a gas above its critical temperature or compressing a liquid above its critical pressure. Generally, an SFC system typically has chromatographic equipment similar to a HPLC, but uses GC columns. Both GC and LC detectors are used, thus allowing analysis of samples that cannot be vaporized for analysis by GC, yet cannot be detected with the usual LC detectors, to be both separated and detected using SFC. SFC is also in other ways a middle way between GC and LC. Due to the low viscosity of supercritical fluids, pressures and flows are similar to GC. However, the low diffusion gives efficiencies similar to HPLC methods. All three chromatographic methods have some way of controlling the elution by using a gradient. GC uses a temperature gradient and LC a solvent nature or polarity gradient. SFC uses temperature gradients and solvent nature/polarity gradients as well as pressure gradients. An important advantage with SFC is the possibility of combining the advantages of GC and LC, high (liquid-like) solubility, high (gas-like) flow rates, and nonselective gas-phase detection. A disadvantage is that due to the many tunable parameters development time is often increased. Another disadvantage is that since the technique is not as widespread as GC and LC not as many methods have been developed.

4.2. ELECTRIC-FIELD DRIVEN SEPARATIONS

In electric-field driven separations an electric field causes ions to travel through a matrix, such as a gas, liquid, or gel. The movement is retarded by frictional forces from interaction with the matrix and the ions almost instantly reach a steady-state velocity. This velocity depends on properties of both the sample molecules and the surrounding matrix. The two main types of electric-field driven separations are ion mobility spectrometry where the matrix is a gas and electrophoresis where the matrix is a liquid or gel.

4.2.1. Ion Mobility

In conventional ion mobility spectrometry (IMS, or drift tube ion mobility spectrometry, DTIMS) a discrete pulse of ions is driven through a neutral buffer gas by an electric field [16]. The time it takes an ion to travel through the drift tube to the detector depends on several factors, including the ion's charge, size, and shape. While DTIMS resembles the TOF mass analyzer (see Section 2.2.1) high-field asymmetric waveform ion mobility spectrometry (FAIMS) has more in common with the quadrupole filter [17] (see Section 2.2.4). As in the quadrupole filter, a continuous flow of ions is subjected to a combination of DC and high-frequency potentials. In FAIMS the high-frequency field is in the direction perpendicular to the flow of carrier gas and is characterized by a significant difference in voltage amplitude for the positive and negative polarities (set so that the ions are experiencing no net momentum perpendicular to the flow direction). A strong field causes more energetic collisions between the ions and the gas molecules than a weak field and this affects the mobility of the ion. How the mobility changes with field strength is a combined result of the properties of the ions and the buffer gas, and may increase in one bath gas yet decrease in another. Only ions where the DC voltage compensates for the drift caused by the high-frequency potentials avoid colliding with the electrodes and successfully reach the detector.

4.2.2. Electrophoresis

In electrophoresis ("carry across by electricity") ions migrate in a liquid or gel under the influence of an electric field. As in ion mobility, separations are caused by mobility differences due to frictional forces caused by differences in charge, size, and shape. As in chromatography, electrophoresis can be performed according to different physicochemical principles and the electrophoresis system can vary in size and shape. One example is capillary electrophoresis (CE), which is electrophoresis performed in a capillary tube. Advantages of CE include high sensitivity; low consumption of reagents, rapid separations, extremely high resolution, and possibility to automate. Below the basic principles of some common forms of electrophoresis are summarized. For an enjoyable review "encompassing ca. 65 years of history of developments in electrokinetic separations," see Reference 18.

4.2.2.1. Gel Electrophoresis. The matrix in gel electrophoresis (GE) is a colloid in a solid form, such as starch, agarose, or polyacrylamide. GE is used primarily for separation of large biopolymers, such as proteins, peptides, and nucleic acids. The pores in the gel act as a molecular sieve and the migration velocity of sample molecules depend on their charge, size, and shape. Nucleic acids carry one negative charge, provided by the phosphate group, per nucleotide. This charge is also independent of any pH used for electrophoresis. Hence, GE separates nucleic acids strictly according to size. A similar situation is created for proteins by performing GE under denaturing conditions. Denaturing conditions can be obtained by first reducing the protein's disulfate bridges and adding a denaturing agent such as sodium dodecyl sulfate (SDS). SDS binds to proteins in numbers proportional to the protein length and each SDS donates negative charge, that is, the number of charges will be proportional to the size of the protein. Hence, the migration velocity through the gel is determined only by size or molecular mass. By varying the pore size of the gel the conditions can be optimized for separation of biopolymers of different sizes. Reference 19 provides a comparison of the yeast proteome coverage between chromatographic fractionation, gas phase fractionation, protein gel electrophoresis, and isoelectric focusing.

4.2.2.2. Isoelectric Focusing. Isoelectric focusing (IEF) is used for the separation of amphoteric molecules, such as proteins and peptides. Amphoteric molecules contain both acidic and basic functional groups and due to protonation or deprotonation of these functional groups the net charge of the molecule is pH dependent. At a particular pH, the so-called isoelectric point (pI), the charge on the groups is balanced and the molecule is neutral. In IEF the sample mixture is deposited in a medium with a previously generated pH gradient and subjected to an electric field. Negatively charged molecules will migrate through the pH gradient toward the anode, and the positively charged ions will migrate to the cathode. As a molecule moves through the pH gradient it eventually reaches a pH that equals its pI. At this pH the molecule no longer has a net electric charge and stops moving. Hence, the amphoteric molecules are focused into narrow zones according to their pI. If a molecule diffuses away from its pI, it will regain its charge and migrate back. This focusing effect allows proteins to be separated based on very small charge differences. IEF is often used in combination with SDS-PAGE gel electrophoresis or so-called two-dimensional gel electrophoresis (2-DGE) [20, 21]. For an engaging account of the evolution of IEF see References 22 and 23.

4.2.2.3. Capillary Zone Electrophoresis. In capillary zone electrophoresis (CZE or often only CE) the sample is applied as a narrow zone (band), surrounded by a buffer (electrolyte). When the electric field is applied, both the buffer ions and the sample ions start to migrate towards the detector. The charge and the mobility of the sample ions as well as the velocity of the electroosmotic flow will determine the time required for a specific zone to reach the detector. A major advantage in ZE is that cations and anions can be separated simultaneously. Depending on the polarity of the electric field, the electroosmotic flow will help the migration of some ions, while hindering the migration of others (with the opposite charge). Neutral molecules are not separated in ZE, since all migrate at the velocity of electroosmotic flow. ZE is often performed as capillary ZE (CZE) and typically yields very high separation power. For a recent comparison between nano(RP)LC-MS and CZE-MS see Reference 11. Figure 4.2 shows schematics of two different CZE-MS interfaces.

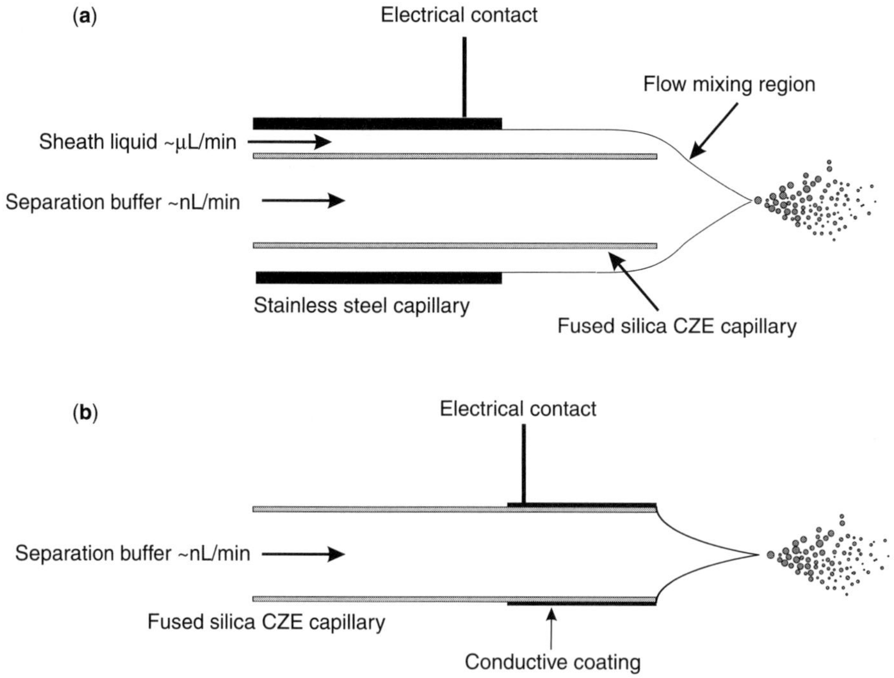

Figure 4.2. Schematics of two CZE-ESI-MS interfaces: (a) sheathflow interface and (b) sheatless interface (not to scale).

4.2.2.4. Micellar Electrokinetic Chromatography. An important advantage of micellar electrokinetic chromatography (MEKC) is that it is possible to separate both neutral and charged molecules at the same time. The matrix in MEKC is a micellar pseudo-stationary phase surrounded by aqueous buffer (electrolyte). Micelles are aggregates of surfactant molecules with hydrophobic heads that point at the core, protected from water by the surrounding hydrophilic tails. When the electric field is applied both the electrolyte and the anionic micelles start to migrate but with different rates of migration, thus creating two different phases. MEK separation is a result of ionic species and neutral molecules moving into and out of the micellar phase depending on their relative affinity for the two phases. Concentration and chemical composition of the surfactant and other additives to the buffer can be used to control the migration rates of sample molecules. Other factors that affect the MEK separation are temperature, pH, and ionic strength. Reference 24 is a review of the combination of MEKC and MS.

4.2.2.5. Electrochromatography. Electrochromatography (EC) is a hybrid separation method between CZE and LC where an electric field is used instead of hydraulic pressure to drive the sample ions through a chromatographic system. The stationary phase used in EC is typically standard reverse-phase LC packing material. Since there is no back pressure in EC it is possible to use smaller diameter particle packing material and thereby achieve very high efficiencies. Another advantage with EC is the plug-like flow profile created by the electric field. This is more efficient then the parabolic flow profile of the pressure driven system. Reference 25 provides an example of online hyphenation of capillary EC to ESI-MS and APCI-MS instruments.

4.2.2.6. Isotachophoresis. In isotachophoresis (ITP), or displacement electrophoresis or multizonal electrophoresis, the sample is inserted between two different buffers (electrolytes) without electroosmotic flow. The electrolytes are chosen so that one (the leading electrolyte) has a higher mobility and the other (the trailing electrolyte) has a lower mobility than the sample ions. An electric field is applied and the ions start to migrate towards the anode (anions) or cathode (cations). The ions separate into zones (bands) determined by their mobilities, after which each band migrates at a steady-state velocity and steady-state stacking of bands is achieved. Note that in ITP, unlike ZE, there is no electroosmotic flow and cations and anions cannot be separated simultaneously. Reference 26 provides a recent example of capillary isotachophoresis/zone electrophoresis coupled with nanoflow ESI-MS.

REFERENCES

1. J. Abian. The Coupling of Gas and Liquid Chromatography with Mass Spectrometry. *J. Mass Spectrom.*, **34**(1999): 157–168.
2. M. Tswett. Physikalische-chemische Studien über das Chlorophyll. Die Adsorptionen. *Ber. Deutschen Botan. Ges.*, **24**(1906): 316–323.
3. M. Tswett. Adsorption Analyse und Chromatographische Methode: Anwendung auf die Chemie des Chlorophylls. *Ber. Deutschen Botan. Ges.*, **24**(1906): 386.

4. L. S. Ettre. M. S. Tswett and the Invention of Chromatography. *LCGC N. Amer.*, **21**(2003): 458–467.
5. A. T. James and A. J. P. Martin. Gas-Liquid Partition Chromatography: The Separation and Micro-Estimation of Volatile Fatty Acids from Formic Acid to Dodecanoic Acid. *Biochem. J.*, **50**(1952): 679–690.
6. G. A. Eiceman, J. Gardea-Torresdey, F. Dorman, E. Overton, A. Bhushan, and H. P. Dharmasena. Gas Chromatography. *Anal. Chem.*, **78**(2006): 3985–3996.
7. J. A. J. Trygg, J. Gullberg, A. I. Johansson, P. Jonsson, H. Antti, S. L. Marklund, and T. Moritz. Extraction and GC/MS Analysis of the Human Blood Plasma Metabolome. *Anal. Chem.*, **77**(2005): 8086–8094.
8. I. D. Wilson, J. K. Nicholson, J. Castro-Perez, J. H. Granger, K. A. Johnson, B. W. Smith, and R. S. Plumb. High Resolution "Ultra Performance" Liquid Chromatography Coupled to oa-TOF Mass Spectrometry as a Tool for Differential Metabolic Pathway Profiling in Functional Genomic Studies. *J. Proteome Res.*, **4**(2005): 591–598.
9. I. Neverova and J. E. Van Eyk. Role of Chromatographic Techniques in Proteomic Analysis. *J. Chromatogr.*, **B815**(2005): 51–63.
10. A. Raffaelli and A. Saba. Atmospheric Pressure Photoionization Mass Spectrometry. *Mass Spectrom. Rev.*, **22**(2003): 318–331.
11. M. Pelzing and C. Neusüß. Separation Techniques Hyphenated to Electrospray–Tandem Mass Spectrometry in Proteomics: Capillary Electrophoresis versus Nanoliquid Chromatography. *Electrophoresis*, **26**(2005): 2717–2728.
12. L. S. Milstein, A. Essader, E. D. Pellizzari, R. A. Fernando, and O. Akinbo. Selection of a Suitable Mobile Phase for the Speciation of Four Arsenic Compounds in Drinking Water Samples Using Ion-exchange Chromatography Coupled to Inductively Coupled Plasma Mass Sectrometry. *Environ. Int.*, **28**(2002): 277–283.
13. Y. D. Clonis. Affinity Chromatography Matures as Bioinformatic and Combinatorial Tools Develop. *J. Chromatogr.*, **A1101**(2006): 1–24.
14. L. K. Kostanski, D. M. Keller, and A. E. Hamielec. Size-Exclusion Chromatography: A Review of Calibration Methodologies. *J. Biochem. Biophys. Meth.*, **58**(2004): 159–186.
15. R. M. Smith. Supercritical Fluids in Separation Science: The Dreams, the Reality and the Future. *J. Chromatogr.*, **A856**(1999): 83–115.
16. G. F. Verbeck, B. T. Ruotolo, H. A. Sawyer, K. J. Gillig, and D. H. Russell. A Fundamental Introduction to Ion Mobility Mass Spectrometry Applied to the Analysis of Biomolecules. *J. Biomol. Tech.*, **13**(2002): 56–61.
17. R. Guevremont. High-Field Asymmetric Waveform Ion Mobility Spectrometry: A New Tool for Mass Spectrometry. *J. Chromatogr.*, **A1058**(2004): 3–19.
18. P. G. Righetti. Electrophoresis: The March of Pennies, the March of Dimes. *J. Chromatogr.*, **A1079**(2005): 24–40.
19. L. Breci, E. Hattrup, M. Keeler, J. Letarte, R. Johnson, and P. A. Haynes. Comprehensive Proteomics in Yeast Using Chromatographic Fractionation, Gas Phase Fractionation, Protein Gel Electrophoresis, and Isoelectric Focusing. *Proteomics*, **5**(2005): 2018–2028.
20. A. Görg, W. Weiss, and M. J. Dunn. Current Two-Dimensional Electrophoresis Technology for Proteomics. *Proteomics*, **4**(2004): 3665–3685.
21. T. Rabilloud. Two-Dimensional Gel Electrophoresis in Proteomics: Old, Old Fashioned, But It Still Climbs Up the Mountains. *Proteomics*, **2**(2002): 3–10.

22. P. G. Righetti. The *Alpher*, *Bethe*, *Gamow* of Isoelectric Focusing, the Alpha-Centaury of Electrokinetic Methodologies. I. *Electrophoresis*, **27**(2006): 923–938.

23. P. G. Righetti. The *Alpher*, *Bethe* and *Gamow* of IEF, the Alpha-Centaury of Electrokinetic Methodologies. II. Immobilized pH Gradients. *Electrophoresis*, **28**(2007): 545–555.

24. L. Yang and C. S. Lee. Micellar Electrokinetic Chromatography-Mass Spectrometry. *J. Chromatogr.*, **A780**(1997): 207–218.

25. D. Norton, J. Zheng, N. D. Danielson, and S. A. Shamsi. Capillary Electrochromatography-Mass Spectrometry of Zwitterionic Surfactants. *Anal. Chem.*, **77**(2005): 6874–6886.

26. Y. An, J. W. Cooper, B. M. Balgley, and C. S. Lee. Selective Enrichment and Ultrasensitive Identification of Trace Peptides in Proteome Analysis Using Transient Capillary Isotachophoresis/Zone Electrophoresis Coupled with Nano-ESI-MS. *Electrophoresis*, **27**(2006): 3599–3608.

PART II

INTERPRETATION

INTRODUCTION

The majority of mass spectrometers are found today in laboratories that are not dedicated to studying and improving mass spectrometers. Instead, mass spectrometers are widely used as analytical tools in various fields of research (see Part III). As a consequence, most people who use mass spectrometric data in their daily work are not primarily experts on the instruments, but rather on the samples. Detailed knowledge of the sample is essential when it comes to MS data interpretation, but knowing the limits and strengths of the particular mass spectrometer is also vital. Interpreting mass spectra manually used to be a cherished art among mass spectrometrists. However, manual interpretation is getting more and more arduous with more rapid data acquisition and higher sample throughput. Not even the most dedicated graduate student can check the validity of each fragment in every tandem mass spectrum in a week-long series of LC-MS/MS runs from a commercial high throughput system. Even if someone was brave enough to attempt it, the risk of mistakes (or severe neck and arm pains) would be practically close to 100%. It has also been shown in the case of peak assignment in MALDI-TOFMS that manual interpretation even by experts (also by the same expert on different occasions) is not very reliable. Luckily, all commercial mass spectrometers come with the necessary software tools to control the instrument, acquire mass spectra, and aid in interpretation of the results. However, it is crucial to remember

Mass Spectrometry. Edited by Ekman, Silberring, Westman-Brinkmalm, and Kraj

that these software tools are man-made computer algorithms and the results must be in compliance with the chemistry and physics of the ion source, the fragmentation method, the mass analyzer, the detector, the acquisition electronics, and, of course, the sample itself. In other words, the result must make sense. At the end, it always comes back to the individual scientist. How sure are you that your interpretation of your data is correct? Are you willing to stake your scientific career on the identity of peak nr 37? [See *Medicine* Man (1992), starring Sean Connery and a GC with amazing software.] In this part of the book, we focus on mass spectra interpretation in two important fields of application. Chapter 5 begins by explaining some basic concepts in mass spectra interpretation and then continues with describing in detail how to interpret mass spectra in organic chemistry. The subject of Chapter 6 is how to use mass spectrometry as a tool for peptide sequencing. In Chapter 7 the authors discuss how to optimize sensitivity and specificity in mass spectrometric proteome analysis.

5

INTRODUCTION TO MASS SPECTRA INTERPRETATION: ORGANIC CHEMISTRY

Albert T. Lebedev

5.1. BASIC CONCEPTS

Mass spectrometry is an analytical technique to measure molecular masses and to elucidate the structure of molecules by recording the products of their ionization. The mass spectrum is a unique characteristic of a compound. In general it contains information on the molecular mass of an analyte and the masses of its structural fragments. An ion with the heaviest mass in the spectrum is called a molecular ion and represents the molecular mass of the analyte. Because atomic and molecular masses are simple and well-known parameters, a mass spectrum is much easier to understand and interpret than nuclear magnetic resonance (NMR), infrared (IR), ultraviolet (UV), or other types of spectra obtained with various physicochemical methods. Mass spectra are represented in graphic or table format (Fig. 5.1).

In the graphic form the abscissa represents the mass of ions (to be more precise, the mass-to-charge ratio, m/z), while the ordinate represents the relative intensity of these ions' peaks. Atomic mass units (unified atomic mass unit) or daltons are used as units to measure masses of ions, while intensity is represented in percent relative to the base peak in the spectrum or to the total abundance of all the ions in the spectra. The atomic mass unit (dalton) is equal to the mass of one-twelvth of the mass of a ^{12}C atom ($1{,}661 \times 10^{-27}$ g) (see Chapter 1).

Mass Spectrometry. Edited by Ekman, Silberring, Westman-Brinkmalm, and Kraj

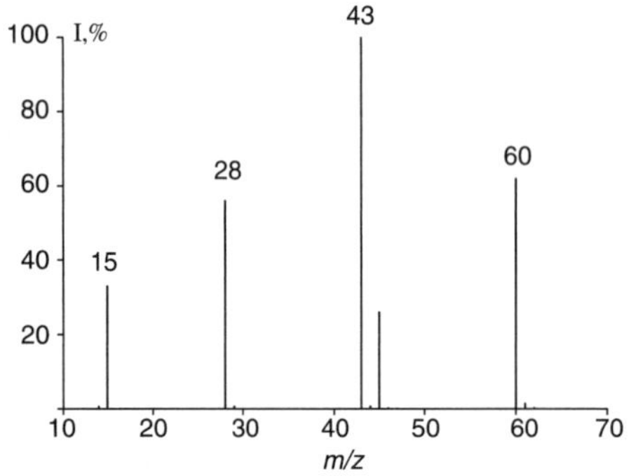

m/z	Intensity (%)
14	0.7
15	33.0
28	56.0
29	0.65
30	0.11
43	100
44	2.2
45	26.0
46	0.31
47	0.10
60	62.0
61	1.48
62	0.26

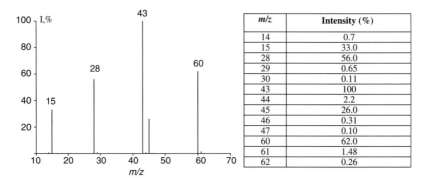

Figure 5.1. Mass spectrum of a compound $C_2H_4O_2$ in table and graphic format.

Unknown 1. Try to identify a compound with the spectrum represented in Fig. 5.1. The exact molecular mass of the compound is 60.0211 Da, which defines its elemental composition as $C_2H_4O_2$. At this stage pay attention only to the most abundant peaks in the spectrum: m/z 60 (molecular ion) and primary fragment ions of m/z 45, m/z 43, m/z 28, and m/z 15. Use the masses of elements from the periodic table of chemical elements.

To record a mass spectrum it is necessary to introduce a sample into the ion source of a mass spectrometer, to ionize sample molecules (to obtain positive or negative ions), to separate these ions according to their mass-to-charge ratio (m/z) and to record the quantity of ions of each m/z. A computer controls all the operations and helps to process the data. It makes it possible to get any format of a spectrum, to achieve subtraction or averaging of spectra, and to carry out a library search using spectral libraries. A principal scheme of a mass spectrometer is represented in Fig. 5.2. To resolve more complex tasks (e.g., direct analysis of a mixture) tandem mass spectrometry (see below and Chapter 3) may be applied.

There are certain rules determining fragmentation of organic compounds in a mass spectrometer. That is why on the basis of the fragmentation pattern it is possible to define the molecular mass, elemental composition, presence of certain functional groups, and often the structure of an analyte. There are a lot of similarities in the mass spectrometric behavior of related compounds. This fact facilitates manual interpretation of a mass spectrum, although it requires some experience. It is also worth mentioning that mass

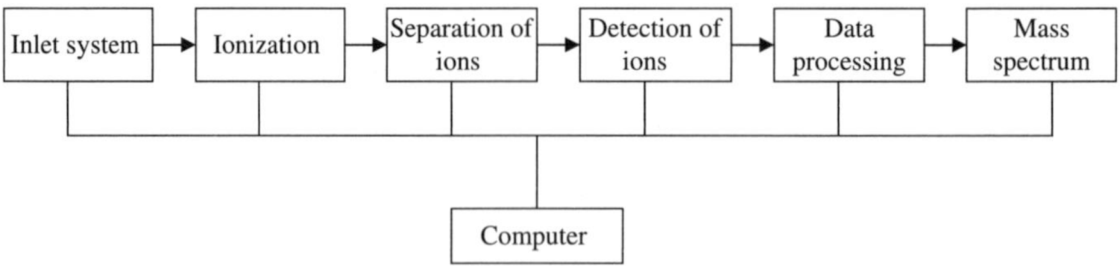

Figure 5.2. Principal scheme of a mass spectrometer.

spectrometry is the most sensitive method of analysis of organic compounds. Modern instruments require up to 10^{-21} M of a sample or even less to record a reliable mass spectrum.

5.2. INLET SYSTEMS

5.2.1. Direct Inlet

The most straightforward tool for the introduction of a sample into a mass spectrometer is called the direct inlet system. It consists of a metal probe (sample rod) with a heater on its tip. The sample is inserted into a crucible made of glass, metal, or silica, which is secured at the heated tip. The probe is introduced into the ion source through a vacuum lock. Since the pressure in the ion source is 10^{-5} to 10^{-6} torr, while the sample may be heated up to 400°C, quite a lot of organic compounds may be vaporized and analyzed. Very often there is no need to heat the sample, as the vapor pressure of an analyte in a vacuum is sufficient to record a reasonable mass spectrum. If an analyte is too volatile the crucible may be cooled rather than heated. There are two main disadvantages of this system. If a sample contains more than one compound with close volatilities, the recorded spectrum will be a superposition of spectra of individual compounds. This phenomenon may significantly complicate the identification (both manual and computerized). Another drawback deals with the possibility of introducing too much sample. This may lead to a drop in pressure, ion–molecule reactions, poor quality of spectra, and source contamination.

5.2.2. Chromatography-Mass Spectrometry

If you have a sample of a pure compound it is easy to analyze it using a direct inlet system. However, this system is used less and less frequently because a chromatographic column (LC or GC) is a much more efficient inlet system into a mass spectrometer. This combined method is called LC-MS or GC-MS (see Chapter 4), correspondingly. A combined method "chromatography-mass spectrometry" adds to the mass spectral information another important parameter, which is retention time. Due to this feature it is possible to analyze isomers, while their mass spectra are often indistinguishable. The combination of chromatography and mass spectrometry is quite logical as the sensitivity of both methods is rather similar and they complement each other with great efficiency. Chromatography is responsible for the separation of an analyzed sample, while mass spectrometry is responsible for the detection and identification of all the chemical compounds in the sample. The problems involving pressure restrictions were resolved in the 1970s for GC and in the 1980s for LC. In both variants (LC-MS or GC-MS) chromatographic separation of the sample takes place in the usual way, while the mass spectrometer serves as a detector to the chromatograph. The eluate leaving a column comes to the ion source of a mass spectrometer, where ionization is realized. The ions are separated, while the full mass spectrum is recorded repetitively in the required mass range. A resulting chromatogram (Fig. 5.3a) provides a sequence of peaks representing individual components of the analyzing sample. The abscissa shows the

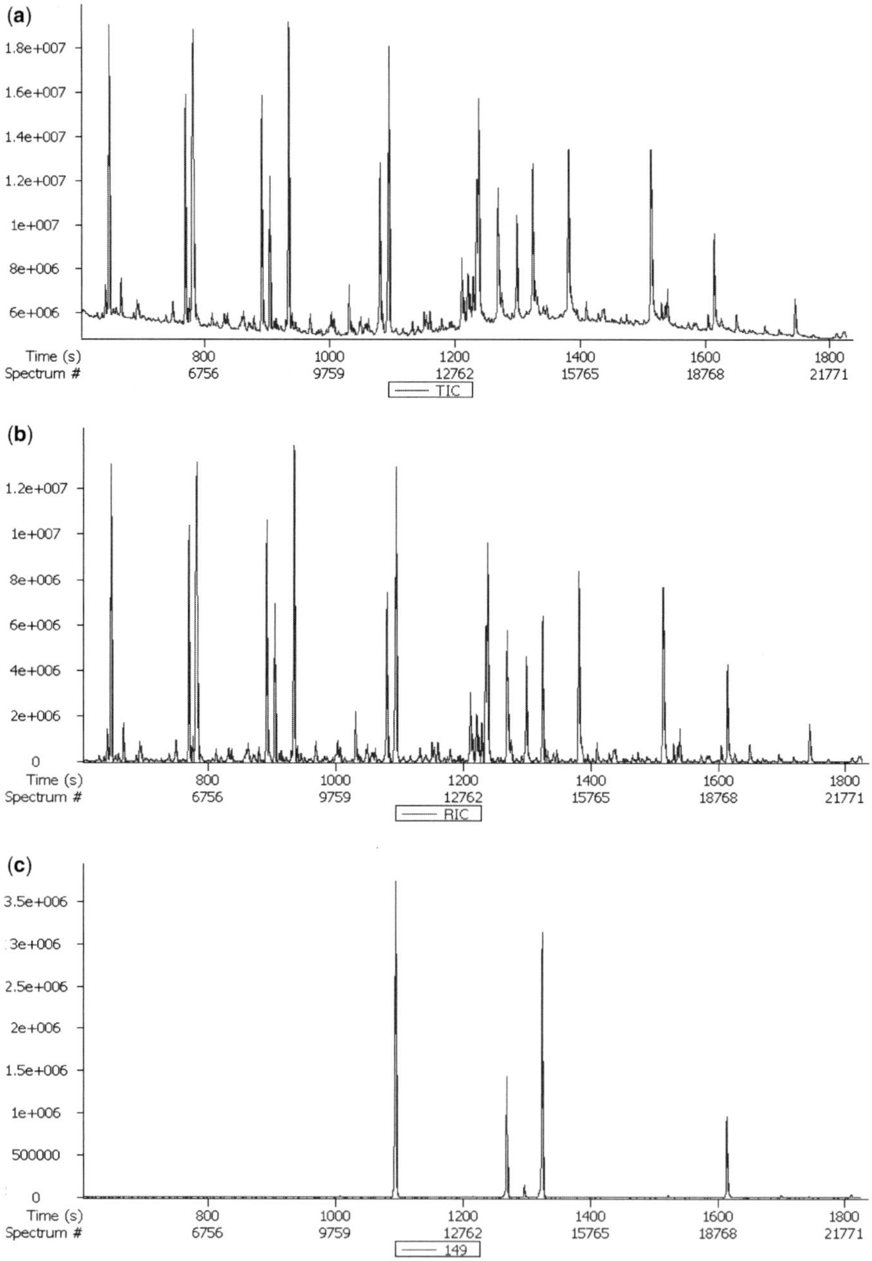

Figure 5.3. GC-MS analysis of organic pollutants in a sample of natural water. (a) Total ion current chromatogram; (b) reconstructed ion current chromatogram; (c) mass chromatogram based on the ion m/z 149 current.

retention times of the ingredients, while the ordinate gives a value of the total ion current at this particular time (the value of the current of all the ions, forming in the precise moment and registering in a single scan). The higher the ion current the higher the position of the scan signal on the ordinate scale. The area of a chromatographic peak corresponds with the quantity of this compound in the sample. The original chromatogram is called the total ion current chromatogram or TIC (Fig. 5.3a). Modern software automatically processes TIC to eliminate background peaks, solvent traces, artifacts, etc. As a result, a reconstructed ion current chromatogram or RIC is created (Fig. 5.3b). The ordinate again represents the total ion current. One can see that the majority of the peaks are resolved at the base line. Mass spectra are also of better quality when dealing with RIC, rather than TIC. A computer can produce chromatograms based on the current of a certain ion (e.g., m/z 149, Fig. 5.3c). The resulting picture is called a mass chromatogram and may be used successfully to detect a certain compound in the sample. For example,

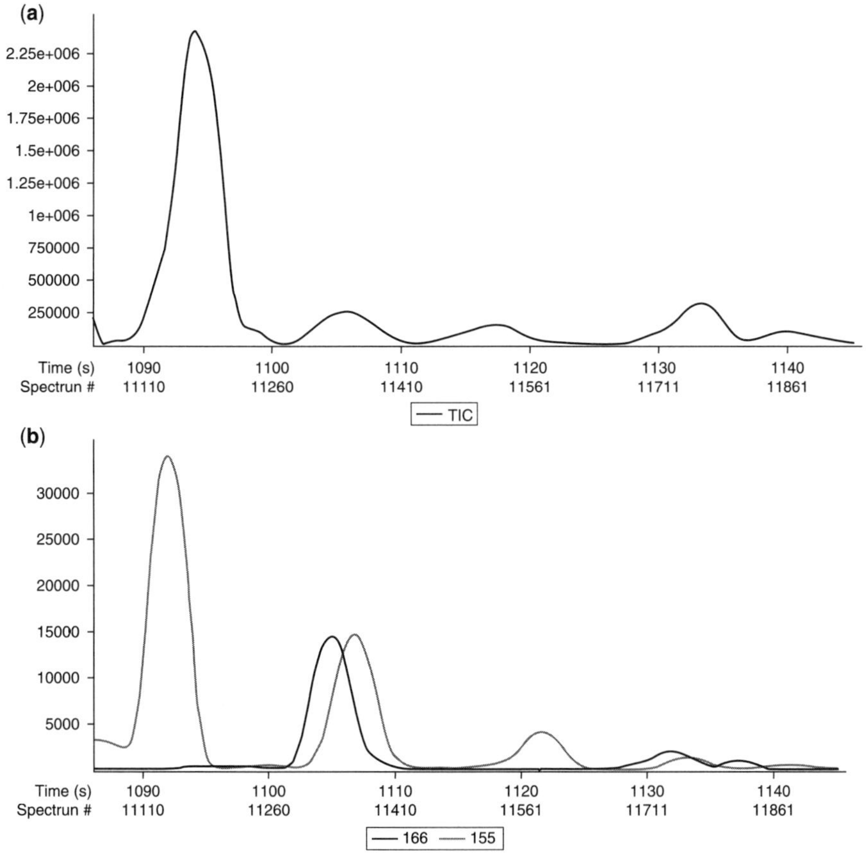

Figure 5.4. GC-MS analysis of organic pollutants in a sample of natural water. (a) One minute segment of TIC chromatogram; (b) mass chromatograms based on ions of m/z 166 (fluorene, black) and m/z 155 (trimethylnaphthalene, grey) current.

an ion of m/z 149 characterizes phthalates. The mass chromatogram reliably gives the position of the phthalate peaks.

Quantitative GC-MS and LC-MS analyses are also based on mass chromatograms (see below). The mass chromatograms are always less complex and much more selective. They permit detection of trace components hidden by peaks of major ingredients in the complex mixtures. In addition, mass chromatograms provide an opportunity to separate and identify individual coeluting compounds. A chromatographic peak with retention time of 1106 seconds (Fig. 5.4) is due to coelution of fluorene and trimethylnaphthalene (Fig. 5.4a). Mass chromatograms based on the current of ions of m/z 166 and 155 (Fig. 5.4b) show that the retention times of these compounds are 1005 and 1008 s, respectively. Mass spectra recorded at the top of each peak permit identification of the ingredients, while the area of the peaks allows measurement of the levels of these chemicals in the sample.

There is a version of GC-MS (LC-MS) analysis that does not record full mass spectra. In this mode the mass spectrometer is adjusted to monitor one or several ions with known m/z values. The method is called selected ion monitoring (SIM) or mass fragmentography and is used to detect and quantify analytes of interest in complex samples. In this case all the information about other components of the sample is lost. However, due to the fact that the detector working in this mode permanently measures only the current of the selected ions, integral sensitivity may be increased about two orders of magnitude. SIM is often used in ecological, forensic studies (see Chapter 16), doping control (see Chapter 8), etc., where it is necessary to detect ultra low traces of the known organic compounds. The appearance of a peak of the selected characteristic ion at the known retention time is used to confirm the presence of the compound in the sample, while the peak area is used for quantitation. Since the probability of

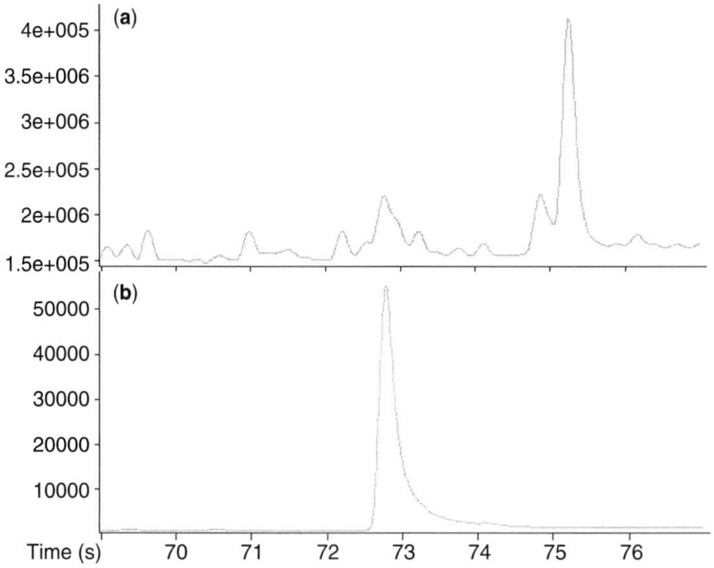

Figure 5.5. SIM in low resolution R = 500, m/z 74 (a) and high resolution R = 5000, m/z 74,0480 (b) modes of a sample containing *N,N*-dimethylnitrosoamine.

the presence of coeluting components giving ions of the same mass may be rather high in complex mixtures, SIM results are not as reliable as those obtained in the scan mode. To increase reliability simultaneous registration of several characteristic ions is used. Alternatively high resolution SIM is used (Fig. 5.5).

The scanning speed of a mass spectrometer should be adequate to the separation power of the chromatograph. If the width of the chromatographic peak is about 3 s and the mass spectrum scanning speed is 1 s we can get the situation represented in Fig. 5.6a to c. The intensities of the peaks in the resulting spectra of the same compound are different. There is a definite discrimination of low mass fragment ions in favor of high mass ions in Fig. 5.6a, and vice versa in Fig. 5.6b. This phenomenon is called spectral skewing. It may create problems in the library search and identification of analytes. Thus, spectra should be recorded with an adequate speed to get at least ten mass spectra during a chromatographic peak elution, while the best spectra are these recorded at the top of the chromatographic peak.

Computer software is used to improve spectral quality. The most widespread procedures deal with averaging and background subtraction. The averaging process is rather obvious. The intensities of ions' peaks at each m/z, recorded along the analyte chromatographic peak profile, are summed in several spectra and divided by the number of spectra used. Averaging minimizes, for example, spectral skewing problems.

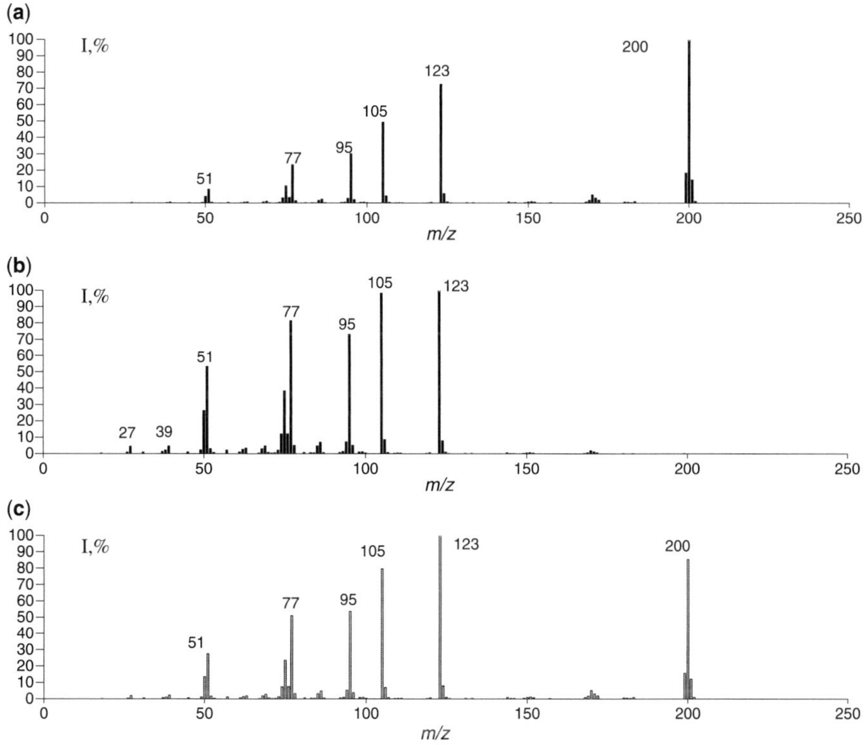

Figure 5.6. Mass spectra of *para*-fluorobenzophenone on the front side (a), on the back side (b), and at the top (c) of the chromatographic peak.

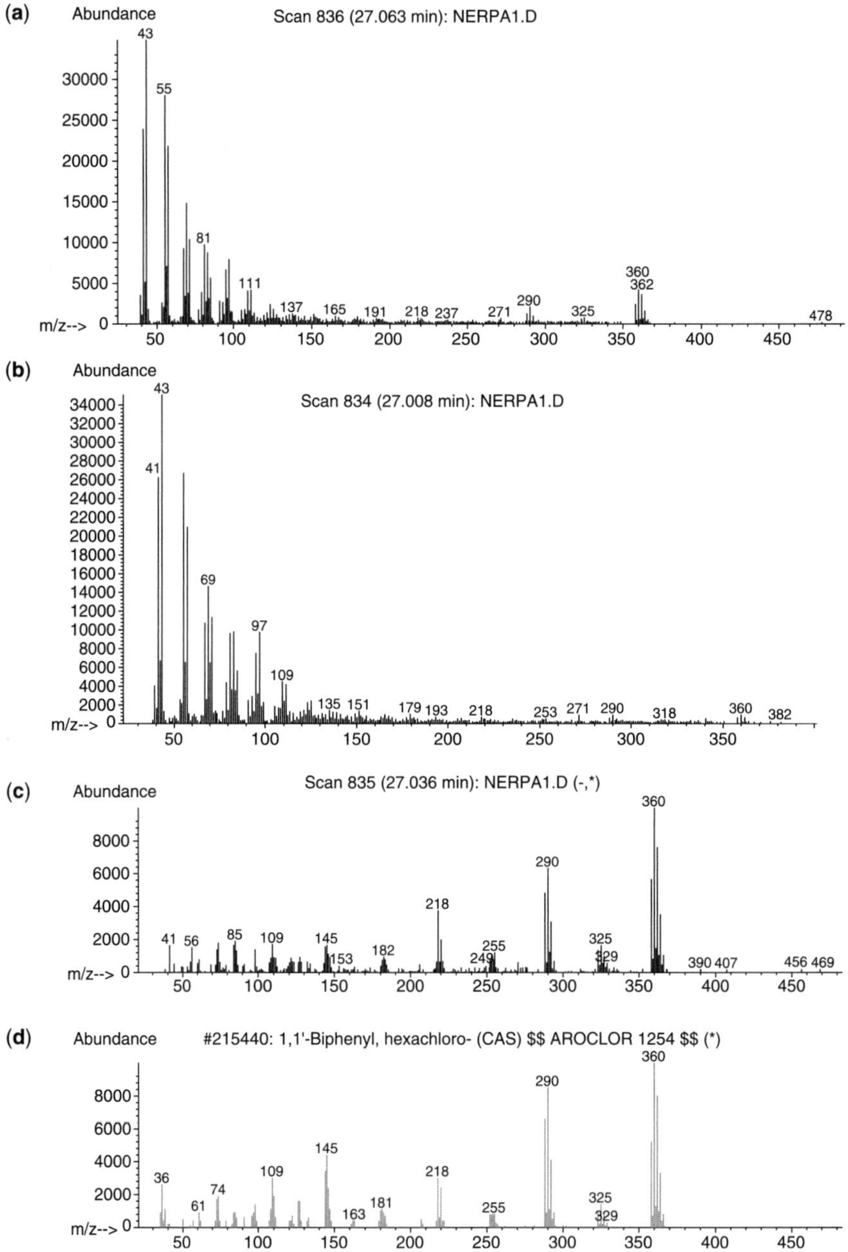

Figure 5.7. (a) Mass spectrum at the top of the chromatographic peak; (b) background spectrum; (c) analyte spectrum after background subtraction; (d) library spectrum of hexachlorobiphenyl (score 99%).

Figure 5.7 demonstrates the process of background subtraction. Background spectra should be selected near the peak of the analyte, a few scans to the left or right. The ion current at each mass of the background spectra should be subtracted from the corresponding ion current in the spectrum recorded at the top of the chromatographic peak. The resulting spectrum is free from background ions. A library search becomes much more efficient and reliable in this case. Sometimes very low analyte signals may be cleared using this procedure. To improve spectral quality it is also possible to average the background spectrum before subtraction. Then the abundance of ions' peaks at each m/z in the spectra recorded at the left and right sides of the chromatographic peak are summed and divided by two. The resulting averaged background spectrum is then subtracted from the analyte spectrum.

Figure 5.8 represents profiles of chromatographic peaks of the same amount of the same analyte recorded at various scan speeds. It is clear that besides problems with spectral skewing leading to poor library search, another drawback exists. Since quantitative GC-MS or LC-MS analysis is based on the measurement of a chromatographic peak area, low scan speed may lead to a significant error in the measured quantity of a component. In the case of very narrow chromatographic peaks an operator may even miss an important ingredient of the sample.

A very high speed of data acquisition with time-of-flight (TOF) instruments (see Chapter 2, Section 2.2.1) allows registration of several hundred full mass spectra per second, resulting in very high precision of quantitative analysis. Besides that, all the spectra are identical to one another and spectral skewing problems are eliminated. This is a significant advantage of this type of analyzer. Due to this feature the fast chromatography approach was created. The method involves fast temperature gradients and short chromatographic columns. Complete chromatographic separation is usually not achievable. However, the apexes of the peaks of individual components are separated by several milliseconds. A TOF instrument acquires 200 to 500 full mass spectra per second, allowing a computer to create mass chromatograms for each m/z value and then identify all m/z values reaching their maximum at particular scans. These m/z values and the intensities of the corresponding peaks constitute a reconstructed mass

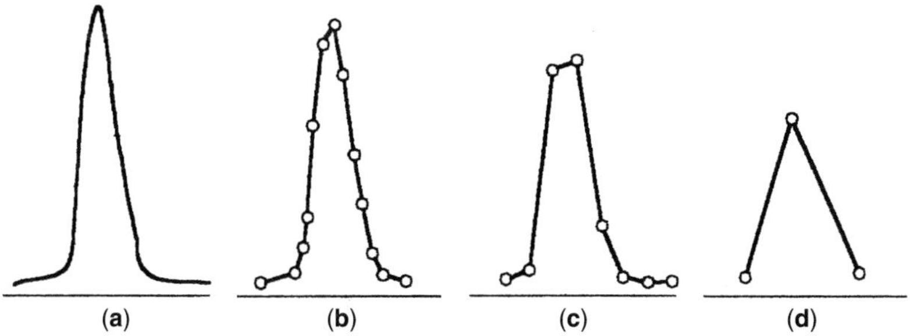

Figure 5.8. Shape and area of a chromatographic peak depending on the scan speed. (a) Analog signal; (b) 13 mass spectra are recorded during the elution of the analyte; (c) 7 mass spectra are recorded during the elution of the analyte; (d) 3 mass spectra are recorded during the elution of the analyte.

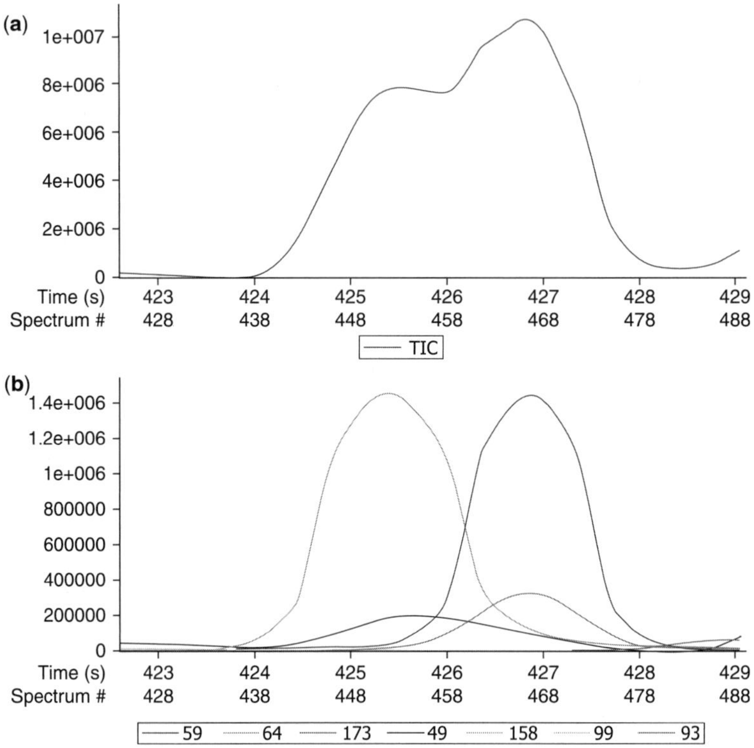

Figure 5.9. GC-MS analysis of organic pollutants in natural water (Pegasus III instrument, LECO). (a) Six-seconds segment of TIC chromatogram. (b) Mass chromatograms, reconstructed by software and based on the current of ions of m/z 59, 64, 173, 49, 158, 99, and 93.

spectrum, which is used for the library identification. If some of the fragment ions belong to several ingredients with close retention times (coeluting peaks) the software divides the abundances of their peaks according to the standard abundances of these fragments in the library spectra of the identified compounds. The final reconstructed spectra may be used for quantitative analysis. Figure 5.9 represents this approach. Four major individual compounds were detected at the 4 s (424 to 428 s) segment of a TIC chromatogram by the computer. The mass spectra were reconstructed by the software and allowed identification of the ingredients, while the corresponding peak areas allowed measurement of the levels of these chemicals in the sample.

5.3. PHYSICAL BASES OF MASS SPECTROMETRY

Because mass spectrometry deals with positive or negative ions it is necessary to ionize a sample after its introduction onto an ion source. There are several dozens of ionization methods. Some of them are used very commonly, others only in unique experiments. The popularity of a method may reach its peak at some stage and then with the appearance of new more efficient techniques it can decline (see Chapter 2, Section 2.1).

5.3.1. Electron Ionization

Electron ionization (earlier called electron impact) (see Chapter 2, Section 2.1.6) occupies a special position among ionization techniques. Historically it was the first method of ionization in mass spectrometry. Moreover it remains the most popular in mass spectrometry of organic compounds (not bioorganic). The main advantages of electron ionization are reliability and versatility. Besides that the existing computer libraries of mass spectra (Wiley/NIST, 2008) consist of electron ionization spectra. The fragmentation rules were also developed for the initial formation of a radical-cation as a result of electron ionization.

A certain amount of energy needs to be applied to a molecule to make it lose its electron. This energy is characteristic for each type of molecules and is called the ionization energy (IE). Since each molecular (atomic) orbital possesses its own energy, a molecule may have several values of ionization energies (a method of photoelectron spectroscopy deals with these features). In mass spectrometry the energy of the highest occupied molecular orbital (HOMO) is used to characterize a molecule. Another term still in use for this phenomenon is the first ionization potential. For the vast majority of organic compounds the IE value is in the range of 6 to 12 eV. If the energy of ionizing electrons is below this value ionization does not take place and a mass spectrum cannot be recorded. The standard value of the energy of ionizing electrons is 70 eV.

When a molecule loses its electron a molecular ion (radical-cation, $M^{+\bullet}$) arises. Not all the energy of the ionizing electrons transfers to the forming molecular ions. Usually these radical cations possess an excess internal energy in the range of 0 to 20 eV. This energy randomly spreads over all the chemical bonds in the ion. If this energy exceeds a certain bond energy this chemical bond ruptures, with elimination of a neutral species and formation of a fragment ion. The minimum energy required to form a fragment ion is called the appearance energy (AE) of this ion. Besides simple cleavages rearrangement processes may also occur. In this case the loss of a neutral molecule is accompanied by the formation of a rearrangement ion. It is worth mentioning that a fragment ion formed due to a simple bond cleavage is an odd electron particle (cation), while a rearrangement ion is an even electron particle (radical-cation). The higher the energy of the ionizing electrons the higher is the number of the fragmentation pathways of the initial $M^{+\bullet}$. If the internal energy of the fragment ions remains high enough, secondary processes of fragmentation take place. Considering that the AE of the primary fragment ions is quite similar, small changes in the energy of the ionizing electrons may lead to notable changes in the recording spectra. Thus, optimal conditions may be achieved at 50 to 70 eV, when the yield of ions reaches its maximum, the slope of the efficiency of the ion current curve is minimal, and all the possible $M^{+\bullet}$ fragmentation pathways are realized.

Together with singly charged ions doubly and multiply charged ions may also arise in the ionization process. However, the number of doubly charged and especially of multiply charged species is much lower. The yield of these ions depends on the structure of a molecule and on the experimental conditions. For example, polycyclic aromatic hydrocarbons give more ions of these types compared to aliphatic or monoaromatic compounds.

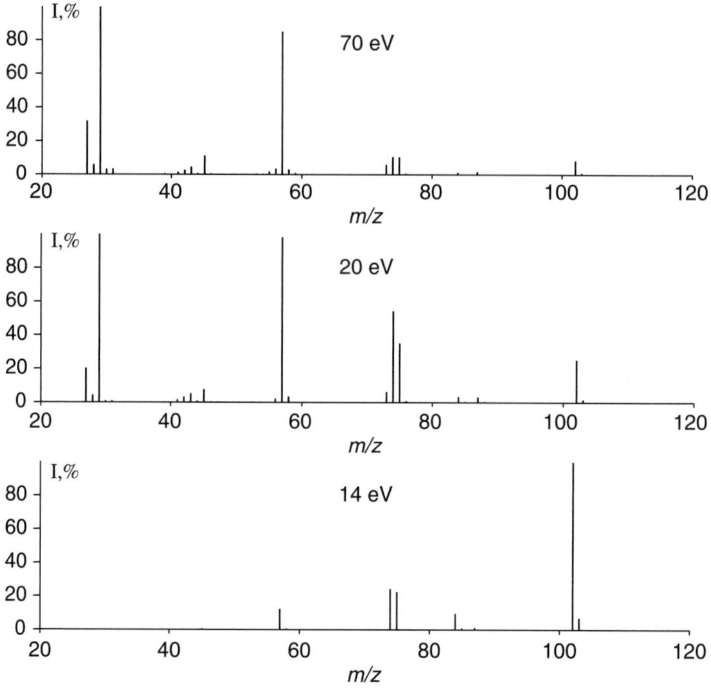

Figure 5.10. Electron ionization mass spectra of ethylpropionate (molecular mass 102 Da) registered at ionizing electron energies of 70, 20, and 14 eV.

When one wants to increase the relative intensity of the $M^{+\bullet}$ peak, electrons with 12 to 20 eV energies are used for the ionization. It is worth stressing that only the relative intensity of $M^{+\bullet}$ and of the rearrangement ions' peaks increases in comparison with the intensity of fragment ions' peaks, while the absolute intensity of all the peaks becomes lower. Besides that a loss of information may occur, as some fragmentation pathways are not realized in these conditions. Figure 5.10 illustrates this statement. An important conclusion may be formulated as follows: *if a molecular ion peak is absent in the spectrum recorded with* 70 eV *electrons, it will be absent at lower ionizing energies as well.* In this case it is possible to conclude that a molecular ion of this type is unstable. By the way the frequent absence of the $M^{+\bullet}$ peak in the spectrum is one of the major drawbacks of electron ionization.

Another important drawback involves the necessity to evaporate samples. Although at pressures of 10^{-5} to 10^{-6} torr and temperatures of 200 to 300°C many organic compounds evaporate, the method is unacceptable for thermolabile, polar, and high molecular weight compounds.

5.3.2. Basics of Fragmentation Processes in Mass Spectrometry

There are several more or less strict approaches to describe the processes of formation and fragmentation of molecular ions. Unfortunately none of them is able to predict the masses and peak abundances of all the ions in mass spectra of even relatively simple organic compounds. To avoid excessive delving into mathematics and physics only the basic qualitative approaches will be discussed here.

Any process of ionization generates a charged particle with an excessive internal energy. For example the most popular electron ionization leads to molecular ions with internal energies in the range 0 to 20 eV. Taking into account that 1 eV = 96.48 kJ, the "hottest" ions possess about 2000 kJ of excessive internal energy, that is, much higher than the particles in solution. The spread of the energies of $M^{+\bullet}$ may be represented by a probability function P(E) (see the upper part of Fig. 5.11). This curve resembles to some extent the photoelectron spectrum of the molecule. In solution the process of averaging of the molecular energies takes place due to their collisions. In mass spectrometry the method of ionization plays a key role. Thus, in the case of chemical ionization the ion source pressure is quite high (up to 1 torr). As a result multiple collisions of ions, radicals, and molecules are unavoidable. Therefore, averaging of energies of the initial molecular ions occurs. On the contrary, high vacuum conditions corresponding to electron ionization exclude particle collisions and the fate of a molecular ion depends exclusively on its initial energy acquired in the process of ionization. This energy does define that a certain portion of $M^{+\bullet}$ reaches the detector, the second portion fragments, the third rearranges, the fourth leads to the fragment ions of a second generation, etc. However, it is worth mentioning that since it deals with monomolecular reactions of isolated particles with very short leaving times the array and quantities of the reaction products are defined mainly by kinetic rather than thermodynamic factors.

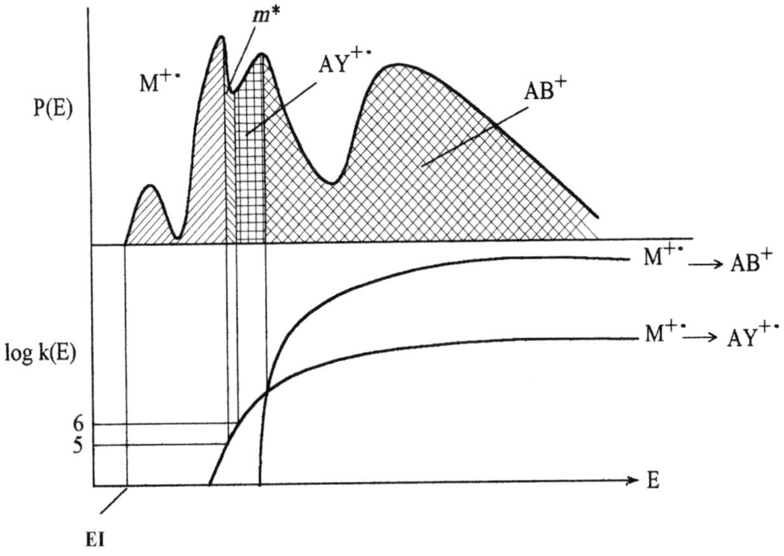

Figure 5.11. P(E) and k(E) functions of a molecular ion of organic compound ABXY; Wahrgaftig diagram.

Two theories deal with the interpretation of the gas-phase monomolecular reactions in high vacuum conditions. Rosenshtock proposed a physical description of the processes in the ion source of a mass spectrometer. The theory is called quasiequilibrium or QET, because it defines how the excessive internal energy spreads along all the bonds in an ion or, in other words, along all the energy states of this ion during 10^{-12} s. Thus, quasiequilibrium between these states is realized before the ion starts fragmenting. The probabilities of the fragmentation pathways depend only on the energy and structure of an ion. The ionization method, mechanism of formation, and the structure of the precursors (for fragment ions) do not influence this process. An analogous theory deals with similar processes for neutral molecules. It was named RRKM according to the names of its authors: Rays, Rampsberg, Kassel, and Marcus. Both theories are based on certain postulates that are beyond the scope of this manual.

Unfortunately both these mathematical theories (QET or RRKM) are not applicable to real organic molecules. With the exception of the simplest molecules these approaches require too much calculation time even using modern software. In addition significant errors may arise due to unknown structures of parent and daughter ions as well as transition states. The spread of energy along the chemical bonds is also not obvious. Actually the best achievement of the strict theories is the calculation of the spectra of light alkanes (propane, butane), that is, compounds with only one type of chemical bond. Nevertheless one can hope that the problems mentioned above may be overcome with the development of new generations of computers in the future.

Let us discuss electron ionization in more detail. The loss of an electron proceeds in 10^{-16} s. According to the Frank-Condon principle the molecular geometry remains unchanged during this process. The nuclei oscillations start in the range 10^{-12} to 10^{-13} s. The ionization process generates ions at all the possible levels of excited energy states (rotational, vibrational, electron). It is important to stress that an ion exists in its excited electron state for 10^{-8} s. During this time the ion reaches its ground electron state or fragments. Since ions stay in the electron ionization ion source for 10^{-6} s, about 99% of the time they are in the ground state. When you see a scheme of fragmentation with a sign of a charge and unpaired electron attached to a certain moiety of an ion, it does not mean that the electron was lost precisely from this position. It is just an attempt to represent a forming ion in its ground electron state.

The probability of some sort of transformation of $M^{+\bullet}$ (isomerization, fragmentation) depends on its internal energy and on the rate constant k(E) of a certain reaction (Fig. 5.11, bottom). A portion of $M^{+\bullet}$ with excessive internal energy below the critical energy value (E_0) for the lowest energy consuming fragmentation reaction will be registered as $M^{+\bullet}$.

Let us suppose that molecular ions of the ABXY molecule may fragment, with the rupture of the B–X bond, or rearrange, with formation of a new bond between A and Y, accompanied by the cleavage of A–B and X–Y bonds.

$$AY^{+\bullet} + B{=}X \longleftarrow ABXY^{+\bullet} \longrightarrow AB^{+} + XY^{\bullet} \qquad \text{(Scheme 5.1)}$$

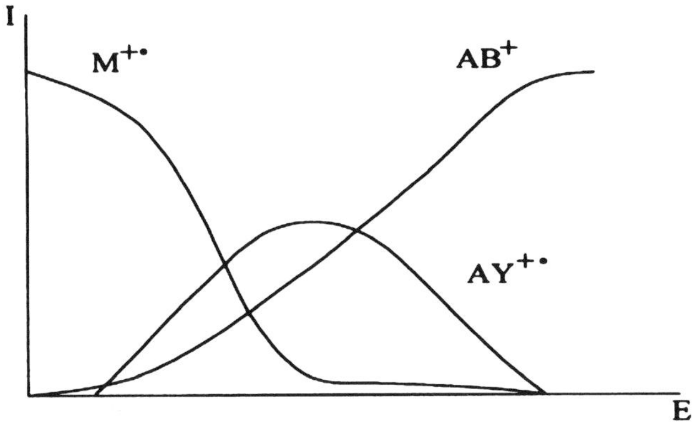

Figure 5.12. Intensities of ion peaks in mass spectra versus internal energy of $M^{+\bullet}$.

As one can see in Fig. 5.11, if $E(M^{+\bullet}) < E_0(AY^{+\bullet})$, the ion is stable and may be registered as $M^{+\bullet}$. Let us define as $E_{1/2}(AY^{+\bullet})$ a threshold energy level at which a half of $M^{+\bullet}$ fragments, with formation of $AY^{+\bullet}$, and another half remains as $M^{+\bullet}$. Similarly $E_{1/2}(AB^{+})$ corresponds to the energy level at which a half of $M^{+\bullet}$ fragments with formation of $AY^{+\bullet}$ and another half with formation of AB^{+}. Thus, for the case represented in Fig. 5.11 all $M^{+\bullet}$ with energies higher than $E_{1/2}(AB^{+})$ will fragment forming AB^{+}, as at this point the rate of formation of AB^{+} becomes higher than the rate of formation of $AY^{+\bullet}$. Roughly speaking the areas under the curves in Fig. 5.12 marked as $M^{+\bullet}$, $AY^{+\bullet}$, AB^{+} correspond to the quantity of these ions in the total ion current. However, it is worth mentioning that these are initial quantities. To calculate the real intensities of the ion peaks in the mass spectrum one should take into account all the further fragmentation processes possible for these primary ions.

Figure 5.12 represents another version of the graphical form of mass spectral information. Assuming that the same reactions (Scheme 5.1) are relevant for molecular ion $ABXY^{+\bullet}$ it is possible to observe the spectral changes depending on the internal energy of $M^{+\bullet}$. Any vertical cross section gives a mass spectrum of $M^{+\bullet}$ with corresponding excessive energy.

Figures 5.11 and 5.12 demonstrate that the most intense peak in the spectra is not necessarily due to a reaction with the lowest critical energy E_0. For the molecular ion $ABXY^{+\bullet}$ the reaction leading to ion AY^{+} has preferential enthalpy, while the entropy factor favors formation of ion AB^{+}. In the former case two bonds are cleaved (AB and XY) and two bonds are formed (AY and BX), that is, the energy losses are low. In the latter case one bond (BX) is cleaved and no new bonds are formed, that is, energy restrictions are tougher. On the other hand steric requirements are stricter in the first case. Ion AB^{+} arises as soon as BX bond acquires an appropriate amount of

$$AY^{+\bullet} + B{=}X \leftarrow ABXY^{+\bullet} \rightarrow AB^{+} + XY^{\bullet}$$

Scheme 5.1.

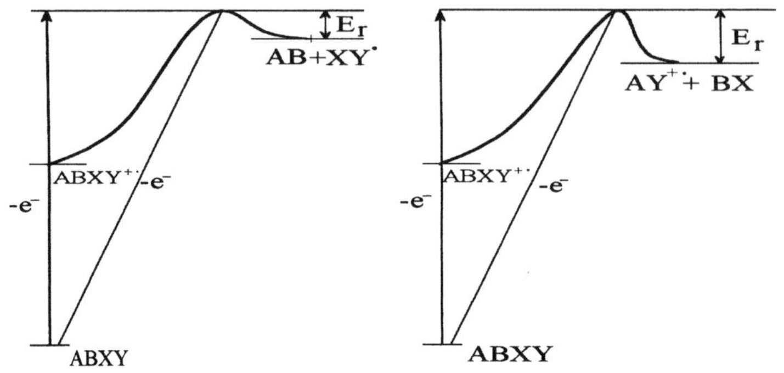

Figure 5.13. The variation of the potential energy along the reaction coordinate for the simple bond cleavage (left) and rearrangement (right).

vibrational energy, while formation of ion $AY^{+\bullet}$ requires approaching of A to Y at a certain distance with the distortion of the valent angles. This situation may be realized only for a small fraction of numerous conformations of $M^{+\bullet}$ possessing sufficient energy for this process. This simple example demonstrates the competitiveness of various fragmentation reactions and rationalizes the complexity of mass spectra of multiatomic molecules.

Variation of the potential energy along the reaction coordinate for the simple cleavage and rearrangement reactions is represented in Fig. 5.13. For simple cleavages (Fig. 5.13, left) the reverse reaction usually hardly requires any activation energy (E_r). For rearrangements (Fig. 5.13, right) the reverse process may require a certain activation energy.

Earlier, definitions of ionization energy (IE) and appearance energy (AE) were introduced. Measuring $AE(AY^{+\bullet})$ by increasing the energy of ionizing particles one would register at some stage the appearance of $AY^{+\bullet}$ and get the value of $E_{1/2}(AY^{+\bullet})$ rather than $E_0(AY^{+\bullet})$. Value $E_{1/2} - E_0$ is called the kinetic shift. Usually it is low (0.01 to 0.1 eV), reaching in some cases 2 eV. The measurements of $E_0(AB^+)$ are even more complicated since kinetic shift is accompanied by competitive shift due to the fact that at $E < E_{1/2}(AB^+)$ $M^{+\bullet}$ will fragment preferentially forming $AY^{+\bullet}$. Nevertheless, considering the known thermochemical characteristics of neutral particles mass spectral measurements of IE, AE, and electron affinity (EA) allow the calculation of the heats of formation of ions and chemical bond energies.

Measurement of the AE of fragment ions is an important method to study their structures. Assume we wish to identify the structure of ion AB^+, resulting due to fragmentation of $ABXY^{+\bullet}$. If the ions of the same composition being generated from $ABCD^{+\bullet}$, $AB_2Y^{+\bullet}$, etc., were studied earlier, while their heats of formation and structures were established, it is necessary to measure AE of ion AB^+, forming from $ABXY^{+\bullet}$, and calculate its heat of formation by Equation 5.1:

$$\Delta H(AB^+) = AE + \Delta H(ABXY) - \Delta H(XY) \qquad (5.1)$$

The coincidence of the resulting value of $\Delta H(AB^+)$ with that of ion AB^+ of known structure is a significant proof of their identical structures. It is clear that a notable difference between values $\Delta H(AB^+)$ and ΔH of all other ions AB^+, studied earlier, proves the ion of interest has a different structure. The development of quantum chemical calculations now allows checking the ion structures, comparing experimental and theoretical values of their heats of formation.

5.3.3. Metastable Ions

The kinetic shift in AE measurements is due to the fact that the fragment ions arise only from $M^{+\bullet}$ having internal energy adequate to realize the corresponding reaction with the rate constant not less than 10^6. However, there is a possibility to register processes with the rate constant 10^5. An ion abandons the ion source in 10^{-6} seconds, but it takes 10^{-5} seconds for this ion to reach the detector, that is, having an appropriate energy it can fragment between the source and the detector. The ions fragmenting this way, as well as the resulting fragment ions, are called "metastable" (see Chapter 3).

How can metastable ions be registered with a classic magnetic sector mass spectrometer? (See Chapter 2, Section 2.2.2) Let ion m_1^+ leave the ion source and after acceleration with accelerating voltage V fragment, with formation of ion m_2^+ and a neutral particle m_3^0 between the source and magnetic analyzer (first field-free region, 1 FFR).

$$m_1^+ \longrightarrow m_2^+ + m_3^0$$

Let v_n be the velocity of the ion with mass m_n. Then Equations 5.2 and 5.3 are valid for the field-free region:

$$m_1 v_1^2 = m_2 v_2^2 + m_3 v_3^2 \tag{5.2}$$

$$m_1 v_1 = m_2 v_2 + m_3 v_3 \tag{5.3}$$

As $m_1 = m_2 + m_3$, both equations may be true only in the case that $v_1 = v_2 = v_3$, that is, in the first approximation both products of fragmentation move in the same direction and with the same speed as their parent ion. Any magnetic analyzer is an analyzer of momentum. Thus, a metastable ion with real mass m_2 and velocity v_1 will be registered together with stable ion m^*, with velocity v^* (Equation 5.4) if their momentums are equal.

$$m^* v^* = m_2 v_1 \tag{5.4}$$

Equation 5.5 is valid for any stable ion.

$$\frac{m^* v^{*2}}{2} = \frac{m_1 v_1^2}{2} = eV \tag{5.5}$$

Combining Equations 5.4 and 5.5 one can get Equation 5.6 to calculate the mass with which the metastable ion will be detected:

$$m^* = \frac{m_2^2}{m_1} \tag{5.6}$$

Therefore, a metastable ion forming in the 1FFR of a single focus sector instrument may be detected with the mass calculated by Equation 5.6. Metastable ions are usually represented in the mass spectra by weak, wide peaks. The low abundance is due to the low amount of the parent ions with the energies favoring their fragmentation out of the ion source. Even the least energy consuming process (rearrangement with formation of ion $AY^{+\bullet}$, Fig. 5.11) results in a weak metastable ion peak. For more energy-consuming reactions formation of a stable ion due to a less energetic reaction is more favorable than the alternative formation of metastable ions. The peaks of the latter may be absent in the spectrum. The widening of a metastable peak is due to energy release during fragmentation. A portion of internal energy transfers into kinetic energy, while the vector of this additional kinetic energy may have any direction, from totally coinciding to totally opposite to the initial vector of movement of the fragmenting ion.

Registration of a metastable ion in the spectrum is rather useful, as it confirms realization of a certain fragmentation reaction. The fragmentation schemes are considered to be true if corresponding metastable peaks are detected. On the other hand, metastable peaks deteriorate spectral resolution. Depending on the amount of energy released, the forms of the metastable peaks may be quite different. These peaks are eliminated from the spectra as part of the computer deconvolution process.

Because an electrostatic analyzer separates ions according to their kinetic energy metastable ions having less energy compared to the stable ones do not pass through it. Nevertheless, metastable ions may be recorded with a direct geometry double focusing instrument if the fragmentation takes place in the 2FFR between the electrostatic and magnetic sectors.

Various instruments allow working in special regimes to detect only metastable ions (MI spectra). The conditions of experiments in this case are the same as for the MS/MS experiments, but without collision activation. Any sort of spectrum (daughter ions, parent ions, constant neutral losses) may be generated this way. These spectra are used to establish the pathways of fragmentation, to resolve structural problems. However, the abundance of the metastable signals and even their presence or absence in the spectrum depends on the energy of the parent ions. Therefore, in contrast to CID (see Chapter 3) spectra the difference in MI spectra of two parent ions does not confirm their different structures.

The following example demonstrates the usefulness of MI spectra. Ions $[M - N_2]^{+\bullet}$ of 1,2,3-triazols usually exist as azirines (Scheme 5.2). However, the daughter

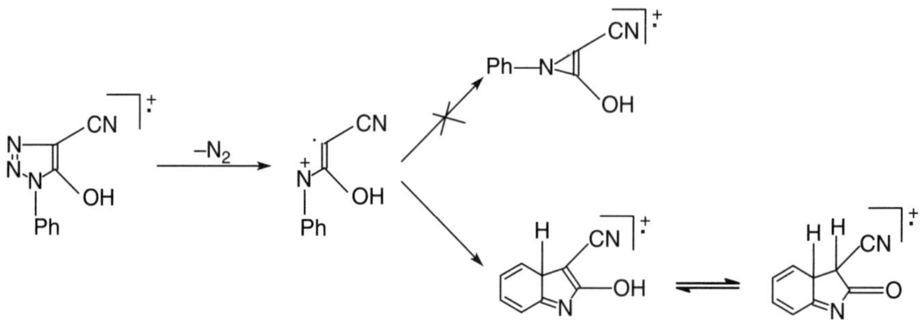

Scheme 5.2.

ions spectrum of these ions in the case of *N*-phenyl-4-cyano-5-hydroxy-1,2,3-triazol shows that one of the fragmentation pathways involves the loss of $NCO^{\bullet}$. This process cannot be realized from azirine and another transformation of the original $[M - N_2]^{+\bullet}$ ion must be proposed, for example, oxindol formation due to interaction of the radical center with a benzene ring (Scheme 5.2).

5.4. THEORETICAL RULES AND APPROACHES TO INTERPRET MASS SPECTRA

The accumulation of sufficient empirical information on the fragmentation pathways of organic compounds allowed understanding certain rules and creating qualitative theories of mass spectral fragmentation. The admissions of these theories do not camouflage the real sense and permit understanding the essence of the processes. These rules are quite useful to remember and to apply various mechanisms of fragmentation for the interpretation of mass spectra depending on the structure of compounds. Using this approach K. Biemann proposed the first classification of the principal fragmentation reactions as early as 1962 [1]. It was followed by the Benz classification [2].

The most accepted among the qualitative theories of mass spectral fragmentation are the conception of charge and unpaired electron localization and the estimation of ions and neutral particles stability. Despite their qualitative character these approaches are quite useful to work with mass spectra. Both theories use the principle of the minimal structural changes at each stage of fragmentation, while the structure of the molecular ion is considered to be the same as that of the initial molecule. Certain isomerization processes of $M^{+\bullet}$ before the fragmentation are usually a matter of special study.

These two qualitative concepts are not opposed to each other. For example, criticizing the first approach one could mention that the electrons in any molecule are socialized, while it is impossible to establish the exact sites of localization of the charge and unpaired electron. It is not necessary to suppose that a certain trigger mechanism except accumulation of a certain amount of energy in a certain chemical bond is required to start fragmentation. Anyway both theories are qualitative and their joint usage allows successful interpretation of the spectra of various organic compounds.

5.4.1. Stability of Charged and Neutral Particles

Let us discuss again the primary fragmentation pathways of ion $ABXY^{+\bullet}$:

$$AY^{+\bullet} + B{=}X \longleftarrow ABXY^{+\bullet} \longrightarrow AB^{+} + XY^{\bullet}$$

It is known that in the vast majority of cases the activation energy E_r of the reverse reaction is very small or even negligible. Using Hammond's postulate [3], it is possible to assume that in the case of endothermic fragmentation the transition state will be much closer to the products than to the initial particle (Fig. 5.14). Thus, the stability of the products influences significantly the efficiency of fragmentation. It is important to consider stability of both products: a neutral and a daughter ion.

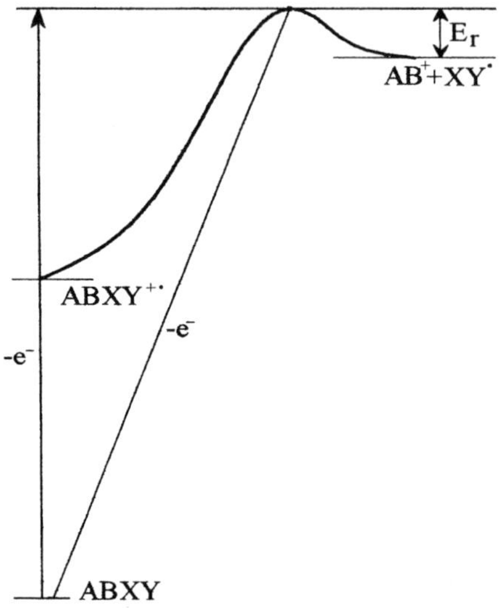

Figure 5.14. Energetic diagram of formation of ion AB^+ from ABXY molecule.

Estimating stability it is possible to apply criteria commonly used in organic chemistry. Tertiary alkyl carbocation is more stable than the secondary one which is in its turn more stable than the primary one. For the carbon ions of this type the row of the stability is reversed. Allyl and benzyl cations are stable due to the resonance stabilization. The latter having four resonance structures may rearrange to be energetically favorable in the gas phase tropilium cation possessing seven resonance forms (Scheme 5.3).

On the contrary sp^2-hybride phenyl and vinyl carbocations are less stable in comparison with their sp^3-hybride analogues. The resonance notably increases ion stability in the case of participation of heteroatomic n electrons (Scheme 5.4).

The fragmentation may occur due to favorable delocalization of the unpaired electron in the neutral particle forming. Again the rules of classic organic chemistry

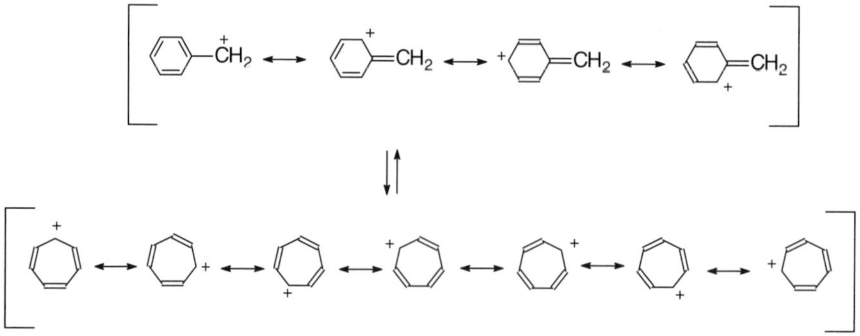

Scheme 5.3.

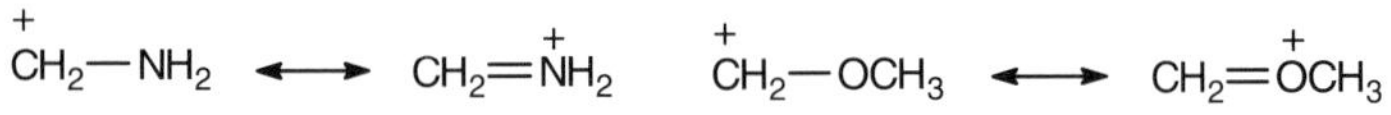

Scheme 5.4.

are applicable. Electron delocalization due to the resonance effect increases the stability of allyl and benzyl radicals. The involvement of α-CH bonds leads to the increased stability of tertiary radicals, etc. Nevertheless, it is worth emphasizing that charge stabilization is more valuable than that of the unpaired electron. For example, if *tert*-$C_4H_9^+$ is 1.5 eV more stable than *n*-$C_4H_9^+$, *tert*-$C_4H_9^\bullet$ is only 0.4 eV more stable than *n*-$C_4H_9^\bullet$.

Usually ions fragment losing small neutral molecules: hydrogen, methane, water, carbon monoxide and dioxide, hydrogen chloride, hydrogen sulfide, etc. The high negative enthalpy of formation of these molecules favors realization of the corresponding pathways of fragmentation. The alternative processes involving the losses of less stable particles become less pronounced while the corresponding peaks may be absent in the mass spectra. As a result the presence of certain functional groups in the analyte may be camouflaged.

5.4.1.1. The Loss of the Largest Alkyl Group. The preferential loss of the largest alkyl radical is the most important exception, as the abundance of the alternative analogous daughter ions decreases with the increase of their thermodynamic stability.

This effect is called the loss of the largest alkyl. The rule does not have exceptions and often helps in spectra interpretation. Scheme 5.5 demonstrates the fragmentation of tertiary amines as an example. At the first stage of fragmentation $M^{+\bullet}$ loses one of three hydrocarbon radicals ($R^{1\bullet}$, $R^{2\bullet}$, or $R^{3\bullet}$). Considering three alternatives the most intense peak belongs to the fragment ion formed with the loss of the largest radical, while the lowest peak will belong to the ion formed due to the loss of the smallest radical.

Similarly among the primary fragment ions of alcohol's ion $[M - R_i]^+$, forming due to the loss of the maximal radical, is represented by the maximal peak (Scheme 5.6). An important consequence of this rule involves the fact that if an intense peak of

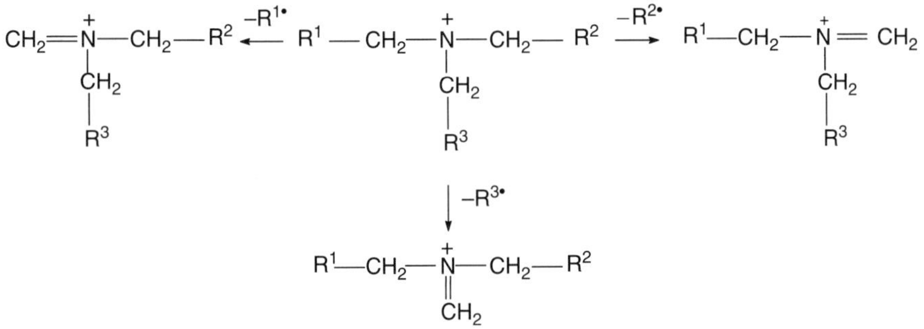

Scheme 5.5.

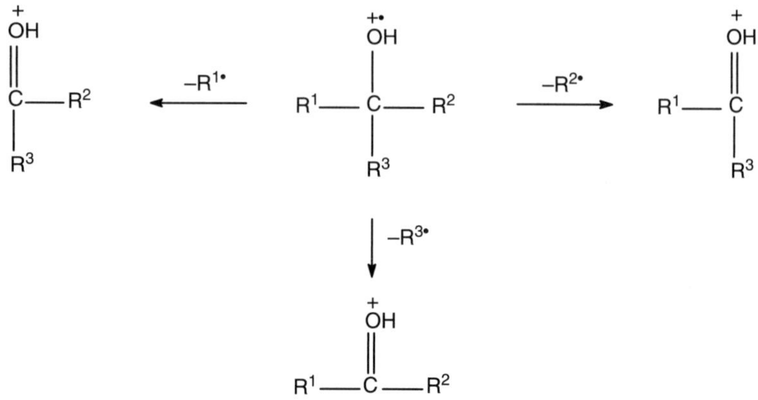

Scheme 5.6.

$[M - H]^+$ ion is observed in the mass spectrum, the carbon atom linked to this hydrogen is not linked to any alkyl group.

The rule of the largest alkyl loss may be quantifiable. Zahorszky [4], studying fragmentation of alcohols and amines, showed that the peak intensities of $[M - Alk_i]^+$ ions may be calculated, as the ratio of intensities of these peaks is inversely proportional to the ratio of the masses of the corresponding ions. The intensities of the peaks of secondary ions should be added to the intensity of the corresponding primary ion peak. This rule is not applicable to the loss of methyl radical.

For example fragmentation of $M^{+\bullet}$ of 4-ethyloctanol-4 (Scheme 5.7) starts with the alternative losses of ethyl, propyl, or butyl radicals.

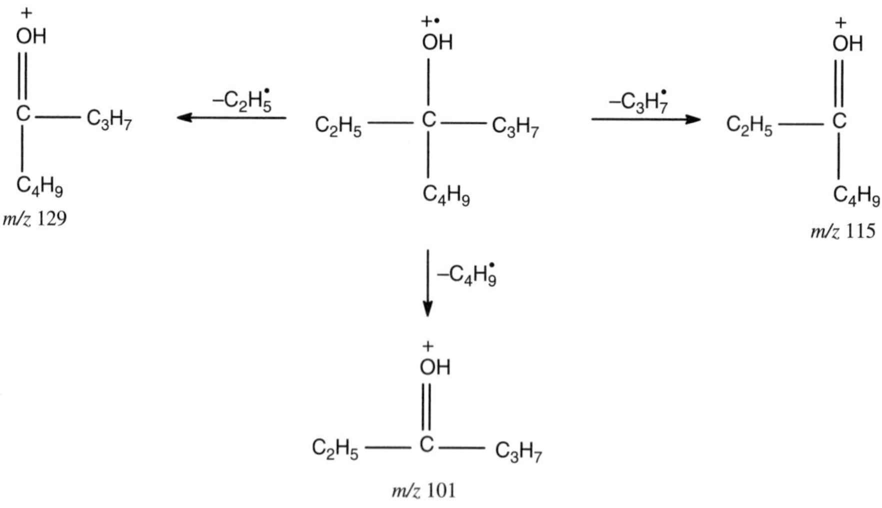

Scheme 5.7.

If the energy of the ionizing electrons is 13 eV (to minimize the secondary fragmentation processes), the intensities of the primary ions peaks with m/z values 129, 115, and 101 will be 78%, 85%, and 100%, correspondingly. Theoretically calculated intensities should be $(101/129) = 0.78$ (78%) and $(101/115) = 0.88$ (88%). As one can see, the resulting values are fairly close to the theoretical ones. The estimation of the intensity of an ion peak resulting due to the loss of methyl radical has a value two to three times higher than the experimental one.

Task 1. Try to estimate the intensities of the primary ion peaks forming due to the loss of alkyl radicals from the molecular ions of the compounds listed below.

A) Butylethylmethylamine
B) Butylethyl-*sec*-pentylamine
C) 3-Methylheptanol-3
D) 3,4,5-Trimethyloctanol-4

5.4.1.2. Stevenson's Rule (Stevenson–Audier). A simple cleavage of a chemical bond in an odd electron ion may result in two pairs of ions and neutrals:

$$AB^+ + XY^\bullet \longleftarrow ABXY^{+\bullet} \longrightarrow AB^\bullet + XY^+$$

Stevenson in 1951 was the first to answer the question which of two pieces of the original molecule would retain the charge [5]. The initial rule was formulated mathematically as follows: if $IE(AB) < IE(XY)$, then $AE(AB^+) = IE(AB) + D(AB - XY)$, but if $IE(AB) > IE(XY)$, then $AE(AB^+) > IE(AB) + D(AB - XY)$, that is, in the second case the formation of ion AB^+ requires an excess of energy. The modern form of the rule is stricter. *The fragment with higher IE will preferentially retain an unpaired electron. Hence the probability of formation of an ion with lower ionization energy will be higher* [6]. Since the corresponding ion is usually more stable its peak will be more intense in the spectrum than the peak of the alternative ion. The experimental results confirming Stevenson's rule are summarized in Table 5.1 [7].

TABLE 5.1. IE of the Fragments and Peak Intensities of the Complementary Ions

Compound AB–XY	IE (AB)	Abundance of AB^+ (% to max.)	IE (XY)	Abundance of XY^+ (% to max.)
$HOCH_2–CH_2NH_2$	~7.6	2.3	~6.2	100
$(CH_3)_2CH–CH_2OH$	7.55	100	~7.6	67
$(CH_3)_2CH–CH(OH)CH_3$	7.55	14.5	~6.9	100
$(CH_3)_3C–CH_2OH$	6.93	100	~7.6	7.4
$(CH_3)_3C–CH(OH)CH_3$	6.93	100	~6.9	79.2
$(CH_3)_3C–CH_2NH_2$	6.93	7.7	~6.2	100
$ClCH_2–CH_2OH$	9.3	4.0	~7.6	100
$BrCH_2–CH_2OH$	8.6	15.2	~7.6	100

If the difference in the IE values of the alternative radicals exceeds 0.3 eV, the ion peak with the lower IE dominates in the spectrum. If the difference in IE values is less than 0.3 eV, both peaks are observed in the spectrum with close intensities.

Stevenson's rule is applicable for the rearrangement processes as well. In this case a radical cation and a molecule are formed, that is, two molecules compete for the charge. The molecule with lower IE becomes the radical cation.

$$AY^{+\bullet} + BX \longleftarrow ABXY^{+\bullet} \longrightarrow AY + BX^{+\bullet}$$

This postulate may be confirmed by the results on McLafferty rearrangement [8] in the $M^{+\bullet}$ of aldehydes (Scheme 5.8 [9]), summarized in Table 5.2.

A 1,5-sigmatropic shift of the hydrogen atom at the first stage of the process (Scheme 5.8a) leads to a distonic intermediate. The cleavage of the C–C bond at the second stage results in formation of a pair of molecules (alkene and enol) one of which is neutral and the other is a radical cation. Sometimes this rearrangement is represented as a concerted process of 1,5-sigmatropic shift and C–C bond cleavage (Scheme 5.8b); however, numerous studies have demonstrated that in the majority of cases the process proceeds stepwise [10].

In the cases of butanal and pentanal the IE of the enol is lower than that of the alternative alkene. As a result the intensities of the enol peaks are five to ten times higher. For the additional homologues the IE of the terminal olephines forming are in the range 9.4 to 9.6 eV, that is, approximately equal to the IE of enol. Thus, the intensities of the peaks belonging to the alternative enol and alkene ions are similar. The methyl group in position 2 (2-methylpentanal) decreases the IE of the enol, resulting

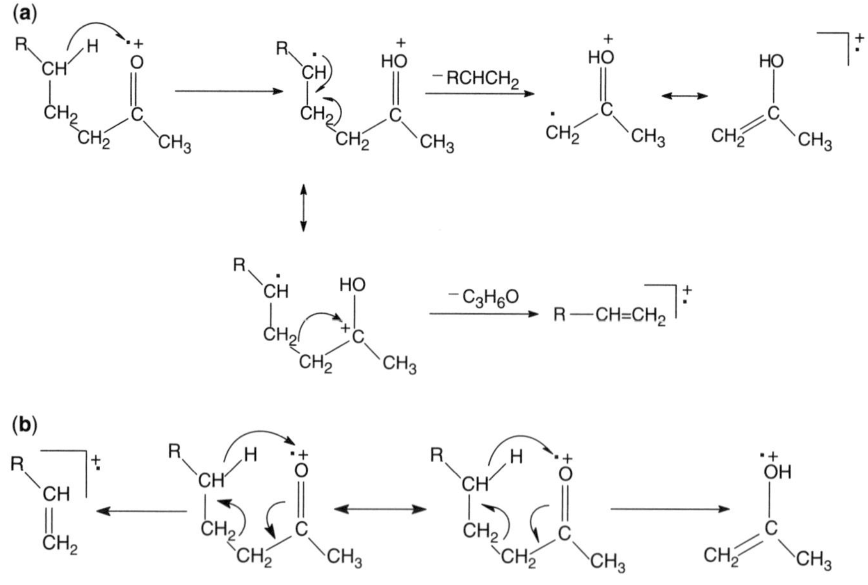

Scheme 5.8.

TABLE 5.2. IE of the Complementary Fragments and the Intensities of the Corresponding Peaks in the Mass Spectra of Aldehydes

Compound	Fragment—IE (eV)	Abundance (% in Σ_{27})	Fragment—IE (eV)	Abundance (% in Σ_{27})
Butanal	C_2H_4–10.5	3.4	C_2H_4O–9.5	16.5
Pentanal	C_3H_6–9.8	2.4	C_2H_4O–9.5	26.4
Hexanal	C_4H_8–9.6	16.1	C_2H_4O–9.5	15.0
2-Methylpentanal	C_3H_6–9.8	1.7	C_3H_6O–9.0	25.1
3-Methylpentanal	C_4H_8–9.3	27.0	C_2H_4O–9.5	5.9

Source: Date from Harrison, A. G. *Org. Mass Spectrom.*, 1970, 3, 549.
Σ_{27}—total ion current, that is, the sum of abundances of all the ions in the range from m/z 27 to $M^{+\bullet}$.

in the increased intensity of its peak. It is an order of magnitude higher in comparison with the alternative peak of the olephine ion. On the contrary, a substitute in position 3 (3-methylpentanal) leads to the formation of alkene with an internal double bond and lower IE (9.3 eV for butane-2). As a result the peak of the alkene ion becomes four times more intense than the alternative enol peak.

It is worth mentioning that Stevenson's rule deals with reactions with thermodynamic control. If the reaction is kinetically controlled, the rule is inapplicable.

5.4.1.3. The Rule of Even Electron Ions.

A rule formulated in 1980 by Karni and Mandelbaum [11] states that even electron ions usually fragment eliminating molecules rather than radicals, that is, cations rather than radical cations are formed. This observation involves the fact that a chemical bond cleavage with formation of two radicals consumes significantly more energy. Taking into account that the stability of radicals is lower than that of molecules or cations it is obvious that these processes cannot compete successfully with alternative elimination of molecules. The charge in this reaction of fragmentation of an even electron ion may be retained by the same atom or migrate to another one. The rule does not mean that these unfavorable reactions are totally banned. They proceed, although the peaks of resulting fragment ions are usually very low. However, there are exceptions. For example, it is possible to mention consecutive losses of methyl radicals by trimethylsilyl derivatives of amines, alcohols, and acids or consecutive losses of chlorine and bromine atoms from the corresponding polyhalogenated derivatives.

An analog of Stevenson's rule for even electron ions is proposed by Field [12] and Bowen et al. [13]. Let us discuss a reaction of fragmentation of an even electron ion:

$$AB^+ \longrightarrow A^+ + B$$

The expedience of the reverse reaction, proceeding almost without any activation energy, depends on the value of affinity of molecule B to cation A^+. The higher this value is, the more favorable is the reaction. Field [12] supposed that cation affinity should vary in parallel with proton affinity. Thus, the probability of fragmentation of various ions AB^+, with formation of identical ion A^+, will be inversely proportional to the proton affinity values of the corresponding molecules B. To confirm this statement let us compare the losses of ammonia, water, and hydrogen chloride molecules by MH^+ of the corresponding amines, alcohols, and alkylchlorides. MH^+ of amines practically never lose ammonia (PA = 9.1 eV). The peaks of the ions formed by the loss of water molecule (PA = 7.7 eV) from MH^+ of alcohols are quite pronounced, while the loss of HCl (PA = 6.4 eV) from MH^+ of alkylchlorides is the dominant process.

Bowen et al. [13] also used proton affinity values to establish the fragmentation pathways of even electron ions:

$$A + B^+ \longleftarrow AB^+ \longrightarrow A^+ + B$$

The charge will be retained on the fragment with higher PA value. For example, in MI spectra of cation $C_2H_5O{=}CH_2{}^+$ the ion CH_2OH^+ (*m/z* 31) peak is the base one, while the alternative $C_2H_5{}^+$ (*m/z* 29) ion peak is absent. The reason involves the fact that ethylene (PA = 7.3 eV) loses to formaldehyde (PA = 7.9eV) their competition for the proton. The Bowen rule becomes less pronounced with increase in the precursor energy. Thus, in the CID spectra of $C_2H_5O{=}CH_2{}^+$ ion the intensity of the $C_2H_5{}^+$ ion peak is only six times lower than that of CH_2OH^+. In a regular mass spectrum, where all mechanisms of fragmentation are triggered, the intensities of the corresponding alternative peaks may be completely reversed.

Similarly to Stevenson's rule, Field's and Bowen's rules are applicable only to the reactions with thermodynamic control. Any reaction with kinetic control may lead to their violation.

5.4.1.4. The Energy of Chemical Bonds.

The energy of formation of a fragment ion from a neutral molecule may be represented as the sum of the energy of homolytic cleavage of a certain bond in the molecule plus ionization energy of the radical formed. The atomic bonding energies in organic molecules are in the range 2 to 4 eV, while ionization energies are higher, 6 to 12 eV. Hence IE is more important. The influence of IE on the peak intensity is postulated by Stevenson's rule (Section 5.4.1.2). However, if competing processes lead to ions with similar stability, the bond energies become a decisive factor.

For example, fragmentation of $M^{+\bullet}$ of 1-chloro-3-bromopropane (Fig. 5.15) leads to two alternative fragment ions $Hal(CH_2)_3{}^+$. The IE of the corresponding radicals

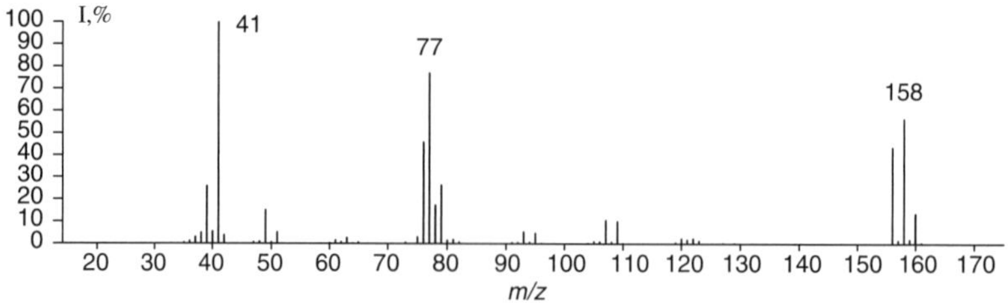

Figure 5.15. EI mass spectrum of 1-bromo-3-chloropropane.

should be very close; however, since the C–Br bond is less stable in comparison with the C–Cl bond, ion $[M - Br]^+$ (m/z 77 and 79) peaks are eight times more intense than ions $[M - Cl]^+$ and $[M - HCl]^+$ (m/z 120–123) peaks. Similarly in the mass spectra of α-bromo-ω-iodoalkanes the intensity of $[M - I]^+$ peaks is always higher than that of $[M - Br]^+$ peaks.

5.4.1.5. Structural and Stereochemical Aspects. High ion energies in mass spectrometry often devaluate the differences in fragmentation of isomeric molecules. Happily this is not always true. The steric factors influence most significantly the structure of the activated complex and may be clearly observed in rearrangement processes. As a result the competitiveness of rearrangement reactions is often notably lower than one can assume on the basis of their critical energy values. That is why an increase of the internal energy of $M^{+\bullet}$ leads to a decrease in formation of rearrangement products. As a consequence, spectra peaks that are intense in MI often become hardly visible in regular mass spectra. A decrease of molecular ion energy (e.g., using 12 to 20 eV ionizing electrons) on the contrary leads to an increase of intensities of the rearrangement ions. Comparing mass spectra of an organic compound recorded at normal and low energies of ionizing electrons one can easily distinguish between fragment and rearrangement ions (Fig. 5.10).

A spectacular example of the influence of steric factors on the fragmentation pattern involves the behavior of chloro- and bromoalkanes. If the hydrocarbon chain is unbranched, peaks of $C_4H_8Hal^+$ ions dominate in the homologous series of fragment ions $C_nH_{2n}Hal^+$ (Scheme 5.9). Any substitute in the carbon chain sharply decreases the competitiveness of this process.

Figure 5.16 represents mass spectra of 1-chlorooctane and 2-chlorooctane. One can see that the base peak of m/z 91 in the first spectra is absent in the second one. The intensity of the m/z 105 ion peak (an analog of the cyclic ion for 2-chlorooctane) constitutes not 100%, but only 6% (Fig. 5.16b). Similar processes are characteristic for alkylmercaptanes (5-membered cycles) and alkylamines (6-membered cycles).

The cycle size plays an important role in the transition state during elimination of simple molecules (HHal, H_2O, H_2S) from the $M^{+\bullet}$ of monofunctional alkyl derivatives. Five-member transition state (1,4-sigmatropic shift of a hydrogen atom) is preferential for the elimination of HCl and HBr. For thiols this process gives 40% of the eliminating H_2S, while the preferential one (60%) involves 1,4-elimination (1,5-sigmatropic shift). The latter process (1,4-elimination) completely dominates in the case of alcohols [14]. Another example of steric effects involves a McLafferty-type

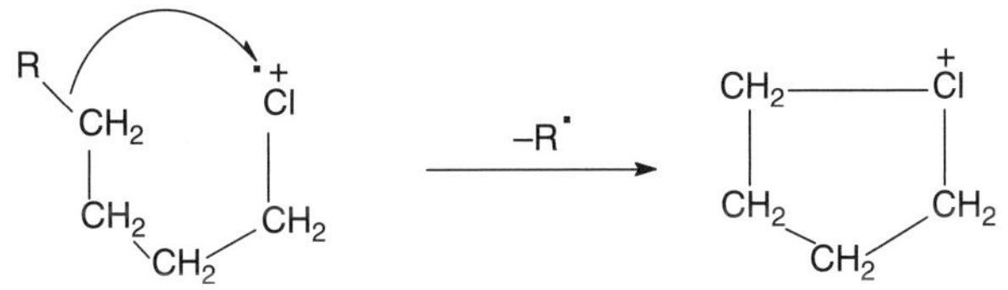

Scheme 5.9.

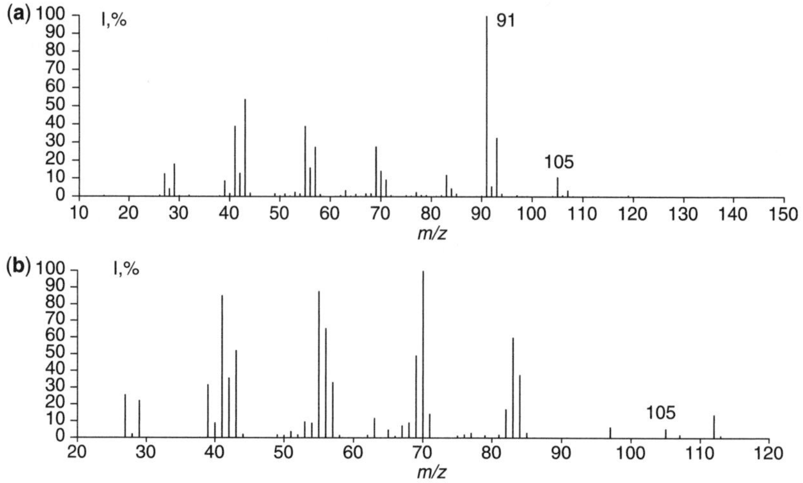

Figure 5.16. EI mass spectra of 1-chlorooctane (a) and 2-chlorooctane (b).

rearrangement process in alkylbenzenes (Scheme 5.10). Being very intense in the case of monosubstituted benzenes the corresponding peaks are hardly visible in the spectra of alkylbenzenes, with substitutes in both *ortho*-positions.

Stereochemical aspects in mass spectrometry have aroused more and more interest. EI mass spectra of stereoisomers are practically indistinguishable. However, the use of "soft" ionization methods (chemical ionization, field ionization, etc.) accompanied by tandem mass spectrometry allows important and reliable conclusions on the molecular structures to be drawn.

Sometimes even EI mass spectra may be informative. Thus, D-labeling shows that 1,4-elimination of hydrogen chloride and water from $M^{+\bullet}$ of chlorocyclohexane and cyclohexanol, respectively, when the heteroatom and eliminating hydrogen atom are 1.7 Å from one another (conformation "bath"), is stereospecific [15]. 1,3-Elimination of HCl (diaxial conformation of chlorine and hydrogen atoms) is also stereospecific. However, the losses of H_2O и HOD are completely equiprobable in case of water elimination from $M^{+\bullet}$ of cyclohexanol labeled with deuterium in position 3. As the minimal distance between the axial substitutes in positions 1 and 3 is 2.3 Å, it is insurmountable in the case of cyclohexanol. Its molecular ions isomerize with the cycle cleavage before the elimination. The 0.4 Å longer C–Cl bond allows 1,3-elimination in the original form of $M^{+\bullet}$ of chlorocyclohexane.

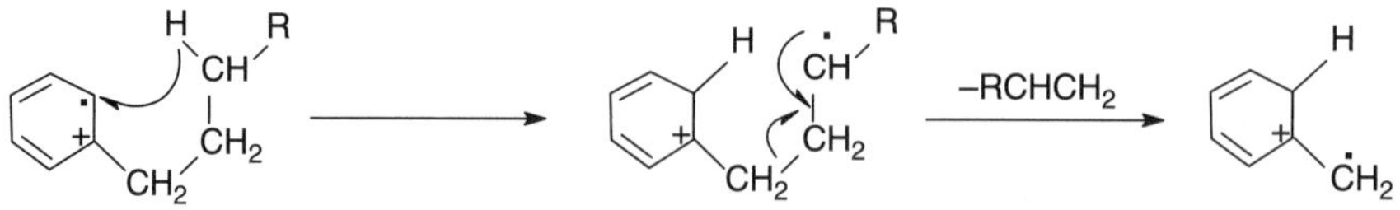

Scheme 5.10.

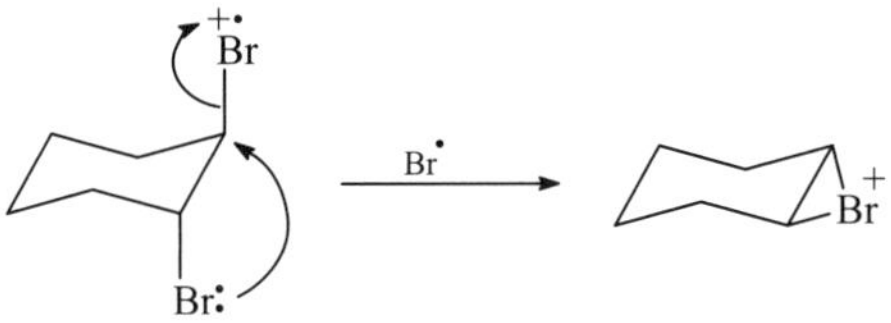

Scheme 5.11.

The ratio of intensities of ion peaks $[M - Br]^+/M^{+\bullet}$ for *cis*-1,2-dibromocyclohexane is 1 to 55, while for its *trans*-isomer the ratio is 64 to 5. The difference is due to anchimeric assistance of the second bromine atom (Scheme 5.11) in diaxial conformation) [16].

Important stereochemical results may be obtained using chemical ionization with chiral reagent gases. For example enantiomeric α-amino acids may be distinguished by CI with optically pure 2-methylbutanol [17].

It is important to emphasize that stereochemical aspects in mass spectrometry are a matter of special studies. Those who are interested in this field should read Green [18], Mandelbaum [19, 20], and Splitter and Turecek [21].

5.4.1.6. Ortho-Effect. The *ortho*-effect is one of the most widely known structural phenomena in organic chemistry. It is widely used in organic chemistry for synthetic purposes. The mass spectra of the majority of *ortho*-substituted aromatic compounds possess significant differences in comparison with the spectra of their *meta*- and *para*-isomers. A classic example of the *ortho*-effect in mass spectrometry involves fragmentation of alkylsalicylates. The intense peaks of $[M - ROH]^{+\bullet}$ ions dominate in the EI spectra of these compounds. These peaks are absent in the spectra of their *meta*- and *para*-isomers. The reaction leading to these ions may be represented by Scheme 5.12.

A detailed review of the *ortho*-effect in mass spectrometry was published by H. Schwarz [22]. He classified the processes related to the *ortho*-effect, gave examples of unusual elimination reactions, processes of intramolecular cyclization, exchange and reduction processes.

Certain analogies in the processes due to the *ortho*-effect in solution and in the ion source of a mass spectrometer allow realization of prognoses on the possibility of

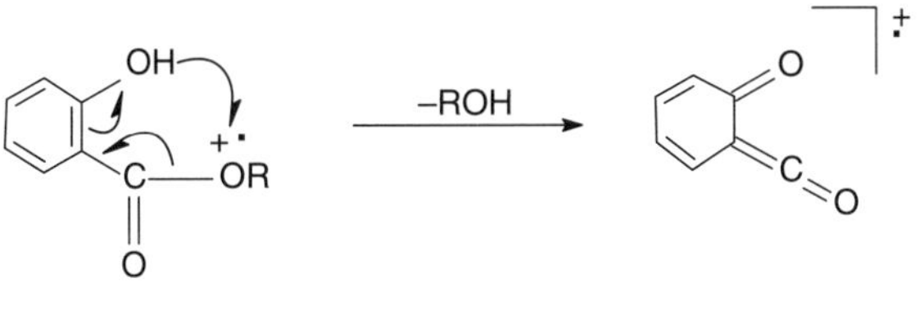

Scheme 5.12.

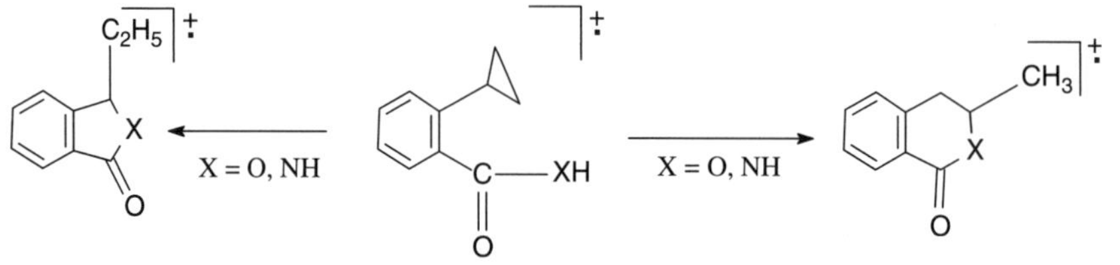

Scheme 5.13.

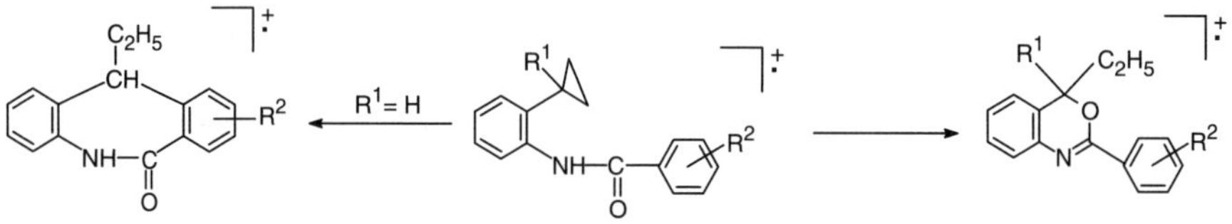

Scheme 5.14.

synthesis of organic compounds based on the mass spectral behavior of their precursors. Thus, molecular ions of *ortho*-carboxy and *ortho*-carboxamidophenylcyclopropanes [23] fragment with initial isomerization into 5- and 6-membered heterocycles (Scheme 5.13).

A similar reaction takes place with sulfuric acid. It takes 2 to 3 months to achieve thermodynamic equilibrium between 5- and 6-membered heterocycles. However, the ratio of the products may be easily predicted in several minutes measuring the ratio of intensities of peaks of $[M - CH_3]^+$ and $[M - C_2H_4]^{+\bullet}$ ions in an EI mass spectrum of the initial substituted arylcyclopropane [23].

Tandem mass spectrometry has been used to demonstrate that $M^{+\bullet}$ as well as MH^+ of substituted *N*-(*ortho*-cyclopropylphenyl)benzamides isomerizes before the fragmentation, with formation of 3-aryl-1-ethyl-1H-benzoxazines and 5-ethyl-2-oxodibenzoazepines (Scheme 5.14). The methyl group in *N*-[ortho-(1-methylcyclopropyl)-phenyl]benzamides quenches the latter process, leaving the formation of benzoxazines as the only cyclization reaction. A subsequent chemical experiment in solution confirmed the mass spectral predictions [24]. A similar study confirmed the analogy of cyclization of substituted *N*-(*ortho*-cyclopropylphenyl)-*N′*-aryl ureas and *N*-(*ortho*-cyclopropylphenyl)-*N′*-aryl thioureas in the ion source of mass a spectrometer and in solution [25].

5.4.2. The Concept of Charge and Unpaired Electron Localization

An extremely useful approach was proposed by Budzikievicz and coworkers [14] and later developed by McLafferty and Turecek [26]. The concept is based on the postulate that reactions of fragmentation of the molecular ions of complex organic molecules are initiated by the charge or unpaired electron, localized at a certain moiety. Despite its limitations this approach is very convenient to remember the different reactions of various particles.

Dealing with a molecular ion it is necessary to identify its ground state, that is to remove an electron from the highest occupied molecular orbital (HOMO). The most favorable sites for the charge and unpaired electron localization may be established by taking away an electron with minimal ionization energy. The energy requirements in this case are similar to these known in UV-spectroscopy for the electron transitions: $\sigma < \pi < n$.

Ionizing electrons with 70 eV energies may knock out an electron from any molecular orbital with smaller (than 70 eV) energy. However, due to radiating and nonradiating processes the initial electronically excited ion should pass to its ground state in 10^{-8} s. As an ion stays in the EI ion source 10^{-6} s, it exists there mainly in its ground state. The narrower the dispersion of the excessive energy of $M^{+\bullet}$ (function P(E)) is, the fairer becomes the concept of the charge and unpaired electron localization, since the quota of the low energy processes with $\log k(E) < 8$ increases in this case.

The charged and radical centers may coincide (e.g., elimination of n-electron) or may be separated (e.g., ionization of π-bond). These centers may be quite far away from one another in the fragment ions. For example, SO_2 loss from the molecular ion of tetrahydrothiophendioxide initially leads to a linear chain of four methylene groups, while the charge and the unpaired electron are located at the opposite ends of the chain. Certainly such an ion may further isomerize into a more stable structure.

Sometimes such distonic ions may be more stable than their isomers, with localization of the charge and the unpaired electron at the same atom. For example, the ion $CH_3OH^{+\bullet}$ is notably less stable than isomeric $^{\bullet}CH_2OH_2^+$, generated during ethylene glycol fragmentation. Calculations and experiments have shown that these two ions are on the bottom of the potential energy wells separated by a rather high barrier. This does not permit them to isomerize to one another [27].

Radical site reaction initiation (α-cleavage) involves the tendency for electron pairing. The unpaired electron participates in the formation of a new bond to an adjacent atom. Another bond of this α-atom cleaves (α-cleavage). Three general variants of α-cleavage (Scheme 5.15) are illustrated with real examples (in parentheses).

The tendency for the fragmentation initiation with the radical site is parallel to the donor properties of this site. The most spectacular examples involve the processes triggered by the removal of a nitrogen n-electron. Halogens are the least active in these reactions.

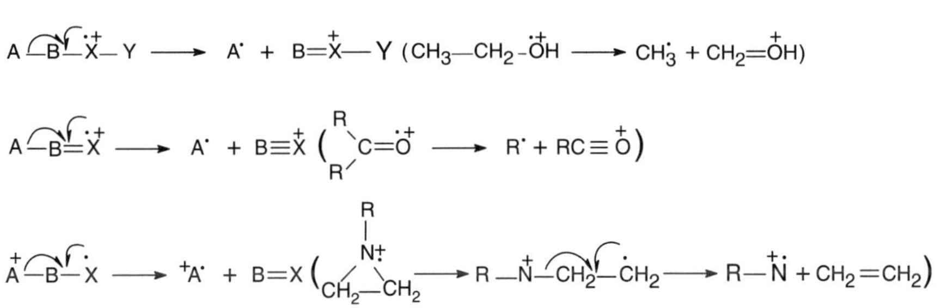

Scheme 5.15.

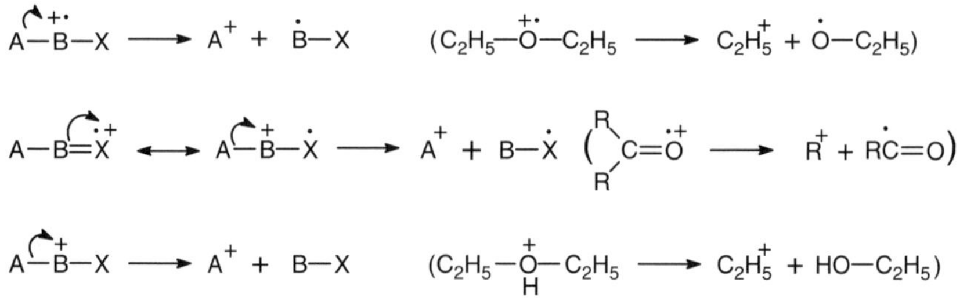

$$A\overset{\frown}{\overset{+\cdot}{-}B}-X \longrightarrow A^+ + \dot{B}-X \qquad (C_2H_5-\overset{+\cdot}{O}-C_2H_5 \longrightarrow C_2H_5^+ + \dot{O}-C_2H_5)$$

$$A-B\overset{\frown}{\overset{\cdot+}{=}X} \longleftrightarrow A\overset{\frown}{\overset{+}{-}B}-\dot{X} \longrightarrow A^+ + B-\dot{X} \quad \left(\overset{R}{\underset{R}{>}}C\overset{\cdot+}{=}\dot{O} \longrightarrow R^+ + R\dot{C}=O\right)$$

$$A\overset{\frown}{\overset{+}{-}B}-X \longrightarrow A^+ + B-X \qquad (C_2H_5-\overset{+}{\underset{H}{O}}-C_2H_5 \longrightarrow C_2H_5^+ + HO-C_2H_5)$$

Scheme 5.16.

The driving force of the charge center initiation of fragmentation is the attraction of an electron pair. The tendency in this reaction for heteroatoms is opposite to that mentioned above for the radical center. Halogens are the most active, sulfur and oxygen are less active, while nitrogen is the last in this row. Actually the favorability of this process increases with the increase of the atom's electronegativity. Three general variants of the reaction (Scheme 5.16) are illustrated with real examples (in parentheses).

The second reactions represented in Schemes 5.15 and 5.16 deal with the same initial ion and the same products. The essence consists in the question, which of two complementary particles retains the charge? To estimate the intensity of the alternative peaks one should use Stevenson's rule (Section 5.4.1.2). To propose a mechanism of the process (in terms of the described concept) it is necessary to take into account that stabilization of the charge is more important than stabilization of the radical (see above). As the reactions initiated by the charge involve charge migration, they are usually less favorable in comparison with the alternative reactions initiated by the radical center.

Unknown 2. Propose the reactions initiated by the charge and radical centers, which will take place during fragmentation of the $M^{+\bullet}$ of the following compounds:

A) Diethylsulfide (n-electron of S atom is lost)
B) Butanone-2 (n-electron of O atom is lost)
C) Butanol-2 (n-electron of O atom is lost)
D) Hexene-3 (π-bond is ionized)
E) Iso-butylchloride (n-electron of Cl atom is lost)
F) Triethylamine (n-electron of N atom is lost)

Rearrangement processes may also be divided into the same two types. The most common example of a rearrangement initiated by the radical center is the McLafferty rearrangement (Scheme 5.17). A hydrogen atom migrates to the oxygen atom (tendency for electron pairing) through a six-member transition state (1,5-sigmatropic shift). The radical center at the γ-carbon atom initiates the fragmentation reaction in the forming intermediate.

The secondary process of amine fragmentation (1,3-sigmatropic shift) may be used as an example of a process initiated by the charge center (Scheme 5.18).

Woodword-Hoffman theory states that there is high critical energy for the 1,3-sigmatropic shifts. However, theoretical calculations [28] and experimental data

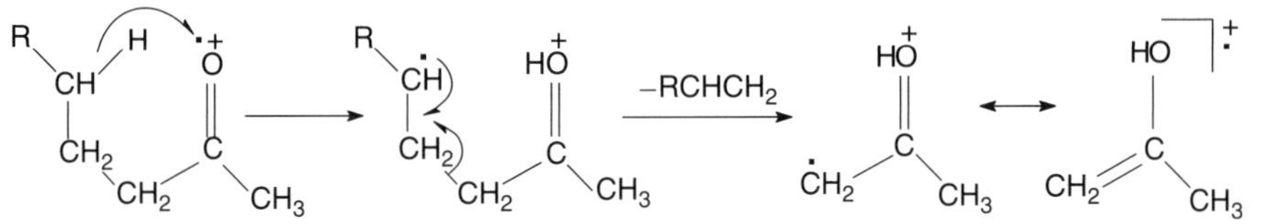

Scheme 5.17.

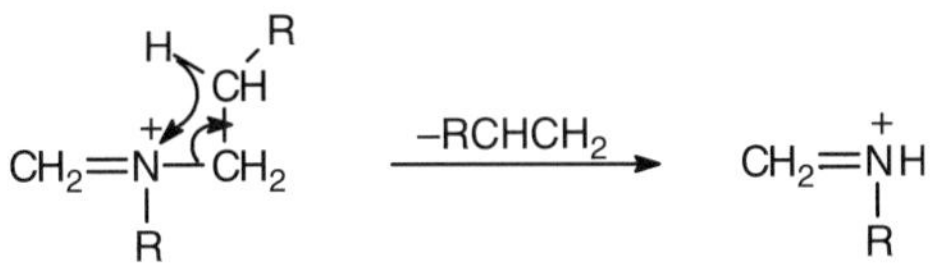

Scheme 5.18.

demonstrate that this rule is not applicable for the charged particles. In this case the hydride anion migrates through a four-member transition state to the nitrogen atom. This process is accompanied with heterolytic C–N bond cleavage and elimination of an alkene molecule. The nitrogen atom retains the charge. The forming ion is analogous to its precursor and, having enough energy and an appropriate length of R, may lose another olephine molecule by the same mechanism. These reactions have low activation energy and require an alkyl chain containing at least two carbon atoms. Therefore, the competitiveness of these secondary and subsequent rearrangements may be very high, while the intensity of the primary ions peaks, calculated on the basis of k(E) and P(E) functions for $M^{+\bullet}$, notably lower.

Unknown 3. What ions will form as a result of McLafferty rearrangement for the compounds listed below? Estimate the intensities of olephine and enol peaks using Stevenson's rule.

A) Pentanal
B) Decanone-4
C) 3-Methylpentanone-2
D) 3-Phenylbutanal
E) 5-Phenylpentanone-2

Unknown 4. Why is McLafferty rearrangement suppressed in the compounds listed below?

A) Butanone-2
B) 1-Phenylpentanone-3
C) Hepten-5-one-2
D) Hexene-4-one-2
E) Octen-4-one-3

5.4.3. Charge Remote Fragmentation

There are certain processes in which fragmentation proceeds without reliable participation of the site bearing the charge [29–33]. The most appropriate example, involving anions of fatty acids, is represented in Scheme 5.19.

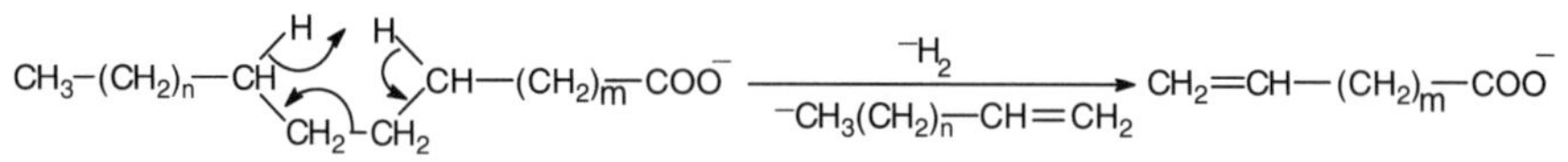

$$CH_3-(CH_2)_n-CH\overset{H\;H}{\underset{CH_2-CH_2}{\diagdown\diagup}}CH-(CH_2)_{\overline{m}}-COO^- \xrightarrow[-CH_3(CH_2)_{\overline{n}}-CH=CH_2]{-H_2} CH_2=CH-(CH_2)_{\overline{m}}-COO^-$$

Scheme 5.19.

The spectra of labeled compounds and the results of the neutralization-reionization experiments confirmed that the reaction proceeds according to the proposed mechanism [29], with the loss of hydrogen and alkene molecules. An experiment with a cholesterol derivative demonstrated that the twisting of a molecule with the shortening of the distance between the charge site and the cleaving bonds in space did not play any role, that is, charge remote fragmentation is realized. This process requires a carbon chain containing at least four atoms, while an increase in the chain length favors the reaction. Similar processes are reported for the anions of alkylsulfates, alkylsulfonates, N-acylated amino acids, quaternary alkyl ammonium salts, etc. [29].

5.5. PRACTICAL APPROACHES TO INTERPRET MASS SPECTRA

So you have a mass spectrum of an organic compound in your hands. How should you start the process of identification? By this time it is very useful to have all available information on the sample. Any information may be relevant (method of synthesis and isolation, the nature of precursors and solvents, the presence of impurities, etc.). Sometimes even knowing what was synthesized in the same flask earlier may be of crucial importance for the final decision.

First of all it is helpful to check the presence of similar spectra in the available spectral databases. Ideally the task may be completely resolved at this stage. The library search may help to refer the sample to a certain class of organic compounds, to get some clues on the presence of heteroatoms and functional groups.

Further, one should pay attention to the general appearance of the spectrum (parameters of the mass spectrometer, the most intense peaks, and characteristic groups of peaks). Thus, if there are many peaks of fragment ions, while their intensity increases toward the low m/z values, the sample most probably belongs to the aliphatic compounds. On the contrary, rare intense peaks indicate the aromatic nature of the sample. You should start the detailed interpretation with identification of the molecular ion peak.

5.5.1. Molecular Ion

One can get an enormous amount of information from studying the region of the molecular ion in a mass spectrum. The mass of $M^{+\bullet}$ is the molecular mass of the analyte. The ratio of the isotopic peaks (see below) allows one to roughly establish the elemental composition, while accurate mass measurements using high resolution mass spectrometry give exact elemental composition. The relative intensity of the $M^{+\bullet}$ peak provides certain clues as to the structure of the analyte and its relationship to one or another class of organic compounds. Thus, the relative abundance of $M^{+\bullet}$ in the total ion current in the case of hydrocarbons increases with the increase in the rate of unsaturation. The following order of the abundances of $M^{+\bullet}$ for hydrocarbons with 10 carbon atoms proves this statement: decane, 1.0%; decene-1, 1.1%; butylcyclohexane, 4.2%; decadiene–4,6, 4.7%; 2,6-dimethyloctatriene–2,4,6, 7.5%; butylbenzene, 10.2%; 1-phenyl-butene-2, 11.0%; 1-methyl-1H-indene, 19.9%; naphthaline, 58.7%.

Unfortunately many compounds do not have a peak for their $M^{+\bullet}$ in electron ionization mass spectra, as it is unstable. That is why it is of primary importance to be able to identify $M^{+\bullet}$ correctly. There are four necessary but not sufficient conditions to name an ion peak molecular. If at least one of these conditions fails the ion is not molecular. If all four statements are fulfilled, the ion can be molecular.

1. The ion should have the maximum m/z value in the spectrum;
2. The ion should be odd electron;
3. The ion should be able to generate the most valuable high m/z primary fragment ions due to the reactions with the losses of real neutral particles;
4. The ion should contain all the elements present in the fragment ions.

Let us consider these statements in detail. The first one is obvious, as the mass of the entire molecule is definitely higher than that of any of its fragment.

The number of electrons in an ion may be checked using a formula for the calculation of unsaturation degree (rings-plus-double-bonds, Equation 5.7):

$$R = x - \frac{1}{2}y + \frac{1}{2}z + 1 \tag{5.7}$$

where R is the degree of unsaturation (rings-plus-double-bonds) of an ion; x, y, and z are indexes in the general formula of ion $C_xH_yN_zO_n$. It is clear that the degree of unsaturation may be calculated only if the elemental composition of the ion is known. If the ion contains other elements, indexes x, y, n, z are the sums of atoms with the same valence (C and Si, 4 – x, N and P, 3 – z, O and S, 2 – n, H and Hal, 1–y). It is important to remember that if the elements mentioned are in other oxidation states, the formula may give an erroneous result. For example, using standard values for S and P in SO_2 or PO_4^{3-} groups (2 and 3 correspondingly) one will get an incorrect value.

Calculating R one gets, besides degree of unsaturation, another important parameter. If R has an integer value, the corresponding ion is odd-electron, that is, it can be molecular. If R has a noninteger value, the corresponding ion is even-electron, that is, it cannot be molecular.

As an example, let us calculate degree of unsaturation for ions $C_5H_9N_3O_2ClBr$ and $C_{12}SiH_{11}PSBr_3$. In the first case $x = 5$, $y = 9 + 1 + 1 = 11$, $z = 3$, $n = 2$. Then $R = 5 - 11/2 + 3/2 + 1 = 2$, that is, the ion has two double bonds, or one triple bond, or two cycles, or one cycle and one double bond. Besides, since R has an integer value, this ion, being odd-electron, can be molecular.

In the second case, $x = 12 + 1 = 13$, $y = 11 + 3 = 14$, $z = 1$, $n = 1$, that is, $R = 13 - 7 + 1/2 + 1 = 7.5$. Thus, the degree of unsaturation of the ion is 7, while the decimal point indicates that the ion is even-electron and cannot be molecular.

An alternative method of calculation of unsaturation involves substitution of heteroatoms with hydrocarbon moieties. All the elements with valence 1 are substituted for CH_3 groups, all the elements with valence 2 for CH_2 groups, all the elements with valence 3 for CH groups, and all the elements with valence 4 for C. The formula obtained is compared with that of the alkane with the same number of carbon atoms. The difference between the number of hydrogen atoms in the alkane and in the sample divided by 2 gives the *R* value. For the above examples one will get:

$$C_5H_9N_3O_2ClBr = C_5H_9 + 3CH + 2CH_2 + 2CH_3 = C_{12}H_{22}.$$

The general formula of the corresponding alkane (dodecane) is $C_{12}H_{26}$. Thus, $R = (26 - 22)/2 = 2$.

$$C_{12}SiH_{11}PSBr_3 = C_{12}H_{11} + C + CH + CH_2 + 3CH_3 = C_{18}H_{23}.$$

The general formula of the corresponding alkane (octadecane) is $C_{18}H_{38}$. Thus, $R = (38 - 23)/2 = 7.5$.

The third essential requirement allows checking the correctness of $M^{+\bullet}$ on the basis of primary fragment ions. Usually $M^{+\bullet}$ easily loses CO, CO_2, H_2O, C_2H_4, HHal molecules; $Alk^{\bullet}$, $H^{\bullet}$, $Hal^{\bullet}$, $OH^{\bullet}$ and other radicals. One should remember that *the losses of neutrals of 5 to 14 and 21 to 25 mass units from* $M^{+\bullet}$ *resulting in an intense peak of a fragment ion are highly unlikely.* Rather suspicious are also the mass losses of 37 and 38 units. If corresponding peaks are present in the spectrum, $M^{+\bullet}$ was selected incorrectly or there are impurities in the sample. For example, if the heaviest ion in the mass spectrum of a pure compound was recorded at m/z 120, while the next one was recorded at m/z 112, the former is not a molecular, but fragment, that is, in this case $M^{+\bullet}$ is unstable and was not recorded in the spectrum.

If the elemental composition is known additional clues appear. For example, the loss of 33 mass units is possible only for the $SH^{\bullet}$ elimination. Similarly the presence of chlorine allows elimination of 35 mass units, which is otherwise hardly possible.

The fourth requirement is also quite obvious. It may be illustrated by the following example. If there are certain peaks, possessing isotopic ratio characteristic for bromine, chlorine, etc. (see below), in the middle part of the spectrum of a pure compound, while a possible $M^{+\bullet}$ does not contain these elements, the latter is not the molecular ion.

Unknown 5. Can the ion of the highest mass be the molecular ion if the following series of fragment ions is detected in the high m/z region of the spectrum?

A) 128, 127, 120, 113, 100 . . .
B) 178, 177, 176, 152, 151 . . .
C) 154, 153, 152, 151, 150, 127, 126. . .
D) 143, 142, 128, 125, 119, 115 . . .
E) 124, 123, 111, 109, 107, 106, 96, 95 . . .
F) 179, 178, 161, 150, 136 . . .
G) 100, 99, 85, 81, 57 . . .

Unknown 6. Can the ion of the highest mass be the molecular ion if the following series of fragment ions is detected in the high m/z region of the spectrum?

A) $C_{10}H_{14}$, $C_{10}H_{13}$, $C_{10}H_{11}$, C_9H_{11}, C_9H_9, C_8H_9, C_7H_7 ...
B) C_8H_{12}, C_8H_{11}, C_7H_9, C_6H_{12}, C_6H_7 ...
C) $C_{13}H_{15}NO_2$, $C_{13}H_{14}NO_2$, $C_{12}H_{12}NO_2$, $C_9H_6NO_2$, $C_{13}H_{15}NO$, $C_{13}H_{15}O$, $C_{12}H_{15}$ ...
D) $C_{10}H_8$, $C_{10}H_7$, $C_{10}H_6$, $C_{10}H_5$, C_8H_6, C_8H_5 ...
E) $C_{10}H_{12}N$, $C_{10}H_{11}N$, C_9H_9N, C_8H_7N, C_9H_{11} ...
F) C_7H_7ClO, C_7H_6ClO, C_6H_6Cl, C_7H_7O, C_7H_6O ...
G) $C_{10}H_7Cl$, $C_{10}H_6Cl$, $C_{10}H_5Cl$, $C_{10}H_7$, $C_{10}H_6$, C_9H_7 ...
H) $C_8H_{13}N_3O_2$, $C_8H_{12}N_3O_2$, $C_7H_{10}N_3O_2$, $C_8H_{13}N_2O$, $C_7H_{10}NO_3$, $C_7H_{13}N_2$ ...

5.5.2. High Resolution Mass Spectrometry

One of the most important spectral parameters is called resolution. The resolving power of an instrument R may be defined as its ability to record ions with masses m and $(m + \Delta m)$ as separate peaks (Fig. 5.17). The value R formally corresponds to the ratio $m/\Delta m$. The ideal peak shape is rectangular, while the real one is Gaussian. In magnetic sector mass spectrometers, peaks are usually defined to be separated down to a 10% valley (Fig. 5.17). Fourier transform ion cyclotron resonance and time-of-flight mass spectrometers use the same resolving power definition as magnetic sector mass spectrometers. However, it is common to use a 50% valley definition for the resolution of these analyzers. If $\Delta m = 1$, then R is a theoretical mass limit of an instrument when ion separation is still possible. Unit resolution means that you can separate each integer mass from the next integer mass. That is, you can distinguish mass 1000 Da from mass 1001 Da. This definition is commonly used when it deals with resolution on quadrupole and ion trap mass spectrometers. For a single peak in a mass spectrum made up of singly charged ions at mass m, the resolution may be expressed also as $m/\Delta m$, where Δm is the width of the peak at a height which is a specified fraction of the maximum peak height. It is recommended that one of three values 50%, 5%, or 0.5% should always be used.

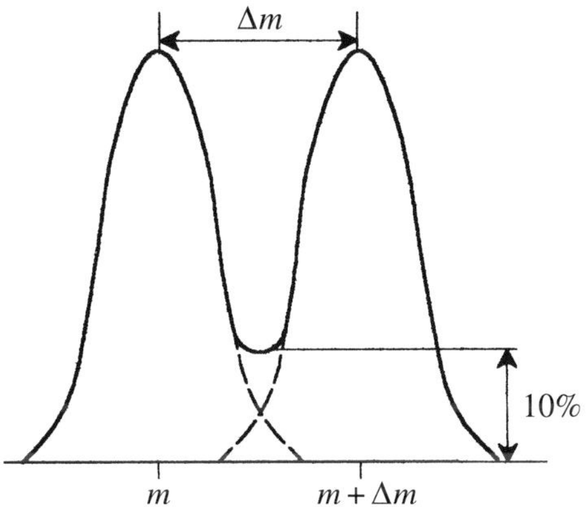

Figure 5.17. Definition of the spectral resolution.

Nevertheless, a resolution of 80,000 for a sector instrument does not mean that this mass spectrometer allows one to obtain separate peaks of the singly charged ions of masses 79,999 and 80,000. The registered mass is proportional to B^2/V. An unlimited increase of magnetic field B is technically unreasonable, while decrease of accelerating voltage V decreases the resolving power.

Taking into account the information above the question arises, what does the resolving power of many thousands or even millions mean? The answer deals with the fact that every isotope of any chemical element has a unique mass. Since carbon isotope ^{12}C (12.000000 Da) was accepted as a standard (the unified atomic mass unit is defined as $1/12$ of the mass of ^{12}C), the masses of all other isotopes have nonintegral values. The difference between the exact mass of an isotope and the nearest integral mass is called the mass defect. Thus, the mass of the main hydrogen isotope ^{1}H is 1.00782506 Da, that of ^{14}N is 14.00307407 Da, ^{16}O is 15.99491475 Da, etc. The mass defect may be positive or negative, depending on whether the exact mass value is below or above the closest integer number. Actually it is negative for the majority of isotopes of chemical elements.

The exact mass of an ion (4 to 6 decimal points) reliably defines its elemental and isotopic composition, while the method is called high resolution mass spectrometry. The measurements are conducted manually or automatically (computerized). Manual measurements are based on the parallel acquisition of the peak of interest with the closest peak of an ion with the known composition. Any compound with an intense ion peak with m/z value in the region $\pm 10\%$ may serve as a marker. The most widespread markers are perfluorokerosene, perfluorotributylamine, and other polyfluorinated compounds. The use of these compounds is based on their volatility, as well as on the fact that fluorine is a monoisotopic element. In the spectra of these compounds intense ion peaks randomly cover all the range between m/z 19 and $M^{+\bullet}$.

The simplest example of the importance of accurate mass measurements deals with the separation of a multiplet of m/z 28. Three compounds with this molecular mass are always present in the background spectrum. They are nitrogen, carbon monoxide, and ethylene. If the resolving power of an instrument is below 500, the molecular ions of these compounds are registered as a single peak (Fig. 5.18a). An increase in the resolving power significantly changes the picture (Fig. 5.18b, c, d).

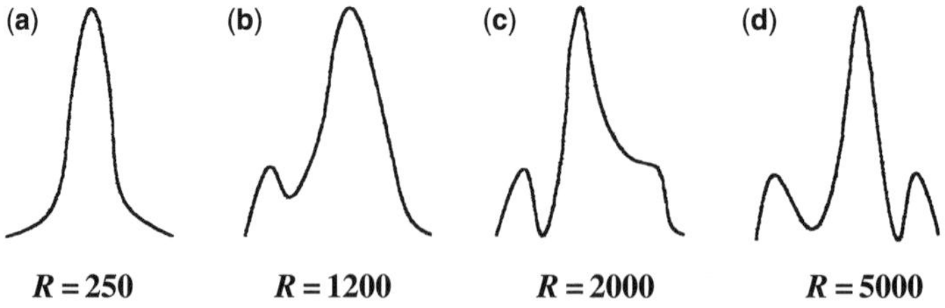

Figure 5.18. The peak shape of the background ions of m/z 28 versus resolving power R of a mass spectrometer.

Actually, to get separate images of the ions with composition CO and C_2H_4 with masses 27.994915 and 28.03300 Da, respectively, one needs resolving power $R = 28/(28.03300 - 27.994915) = 770$, while to separate CO and N_2 peaks (the mass of nitrogen molecule is 28.006148 Da) $R = 2500$.

Taking into account that the number of ions with compositions with the same nominal mass sharply increases with the increase of mass, the importance of high resolution mass spectrometry becomes obvious. Establishing the elemental composition of a compound containing any chemical elements in a few minutes is a notable advantage of mass spectrometry over laborious determination of the constituent elements by means of classic elemental analysis. Nevertheless, it is worth mentioning that elemental analysis provides information about the sample in general (including impurities), while mass spectrometry establishes the elemental composition of individual compounds, amenable to the analysis. Thus, to establish elemental composition of a compound it is better to use mass spectrometry, while to check sample purity one should select elemental analysis.

High resolution mass spectrometry becomes indispensable for the analysis of biomolecules with ESI (see Chapter 2, Section 2.1.15) and MALDI (see Section 2.1.22) techniques. In these cases very high resolving power and accuracy of measurements are required to measure reliably the real masses of the sample molecules.

The modern sector instruments have resolving power about 60,000 to 80,000, and in some cases up to 150,000. There are reports of resolution about 70,000 in the case of ion traps. The newly developed (2005) Orbitrap instruments demonstrated resolving power up to 200,000. Super high resolution (dozens of millions) may be achieved using FTMS (FTICR).

It is not true that an increase in the resolving power always simultaneously leads to an increase in the accuracy of mass measurement. As soon as the resolving power allows peaks separation it is useless to increase it more as the signal intensity will decrease with subsequent decrease in accuracy of measurements.

The results of the accurate mass measurements are often represented as a table, where measured mass, calculated mass, and elemental composition of ions are presented. Usually the table contains several ion compositions with the masses close to the experimental one. An example of the presentation is illustrated in Table 5.3. Five ions have their masses very close to the experimental one. Since *a priori* it was

TABLE 5.3. The Results of Accurate Mass Measurements

Measured Ion Mass (M_{exp})	Calculated Ion Mass (M_{cal})	Discrepancy (ppm) $M_{exp} - M_{cal}$	Elemental Composition
163.9497	163.9503	0.6	$C_6N_2S_2$
	163.9488	−0.9	C_8HSCl
	163.9521	2.4	$C_5H_5S_2Cl$
	163.9537	4.0	$C_3H_4N_2S_3$
	163.9447	−5.0	$C_3HN_2O_2SCl$

known that the sample did not contain nitrogen or hydrogen atoms the only possible composition is $C_6N_2S_2$.

5.5.3. Determination of the Elemental Composition of Ions on the Basis of Isotopic Peaks

As was mentioned above, accurate mass measurements may establish exact elemental composition of ions. However, as the majority of chemical elements exist as a mixture of several isotopes, the elemental composition may be established more or less reliably using low resolution mass spectra. The presence of isotopes may be demonstrated by spectra of carbon disulfide (Fig. 5.19a), chloroethane (Fig. 5.19b) and bromoethane (Fig. 5.19c).

Ions of m/z 76 (Fig. 5.19a), 64 (Fig. 5.19b), and 108 (Fig. 5.19c) are molecular ones, while those 2 Da heavier are due to heavy natural isotopes of sulfur (^{34}S), chlorine (^{37}Cl), and bromine (^{81}Br), respectively.

Table 5.4 represents natural isotopic abundances of the most important (for organic compounds) elements. This table is a fragment of the complete table of isotopes of

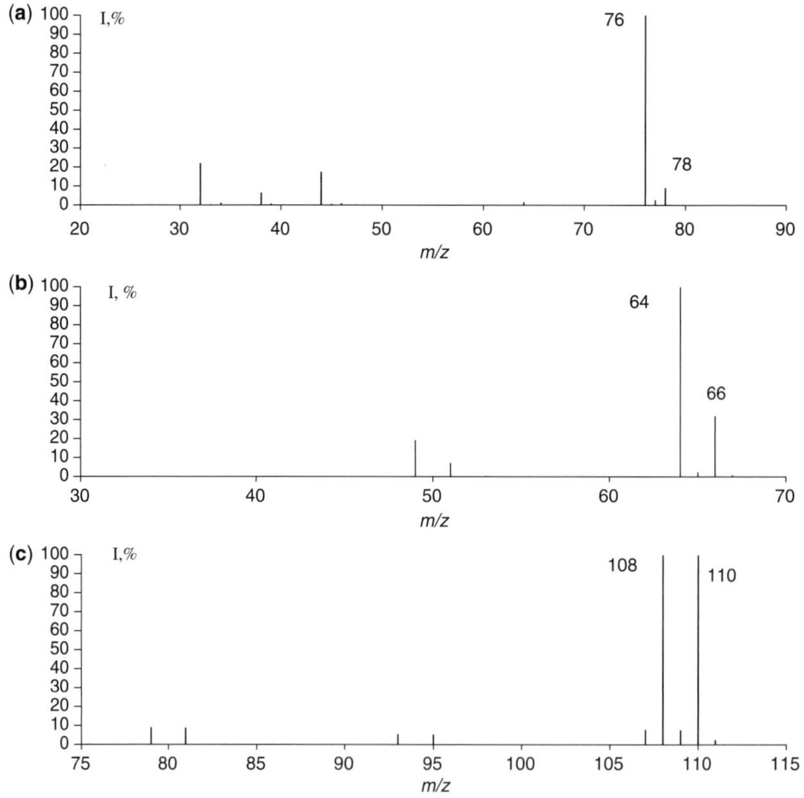

Figure 5.19. Mass spectra of carbon disulfide (a), chloroethane (b), bromoethane (c).

TABLE 5.4. Natural Isotopic Abundances of Widespread Chemical Elements

Element	Isotope	Isotope Type	Abundance to All the Isotopes (%)	Abundance to the Main Isotope (%)	Element Type
Hydrogen	^{1}H	A	99.985	100.00	A[a]
	^{2}D	A + 1	0.015	0.02	
Carbon	^{12}C	A	98.89	100.00	A + 1
	^{13}C	A + 1	1.11	1.12	
Nitrogen	^{14}N	A	99.64	100.00	A + 1
	^{15}N	A + 1	0.36	0.37	
Oxygen	^{16}O	A	99.76	100.00	A + 2
	^{17}O	A + 1	0.04	0.04	
	^{18}O	A + 2	0.20	0.20	
Fluorine	^{19}F	A	100.00	100.00	A
Silicon	^{28}Si	A	92.18	100.00	A + 2
	^{29}Si	A + 1	4.71	5.11	
	^{30}Si	A + 2	3.12	3.38	
Phosphorus	^{31}P	A	100.00	100.00	A
Sulphur	^{32}S	A	95.02	100.00	A + 2[a]
	^{33}S	A + 1	0.75	0.79	
	^{34}S	A + 2	4.21	4.44	
	^{36}S	A + 4	0.11	0.11	
Chlorine	^{35}Cl	A	75.40	100.00	A + 2
	^{37}Cl	A + 2	24.60	32.63	
Bromine	^{79}Br	A	50.57	100.00	A + 2
	^{81}Br	A + 2	49.43	97.75	
Iodine	^{127}I	A	100.00	100.00	A

[a]Despite the presence of the A + 1 isotope for hydrogen and the A + 4 isotope for sulfur these elements are considered A and A + 2 correspondingly, as the natural abundances of these isotopes are very low and may be detected only in the case of a large number of H and S atoms in a sample molecule.

chemical elements. All the elements mentioned may be divided into three groups (A, A + 1, and A + 2) depending on the mass of the heavier isotope.

Determination of the elemental composition should be started with the M + 2 peak. Chlorine, bromine, sulfur, and silicone are easily detected due to characteristic signal multiplicity for each of these elements. There is a simple rule to check the presence of the main A + 2 elements. *If the intensity of the* M + 2 *peak constitutes less than 3% of the intensity of the M peak, the compound does not contain chlorine, bromine, sulfur, or silicon atoms*. This rule is valid for the fragment ions as well, while its applicability is confirmed by the data summarized in Table 5.4.

Figures 5.19b and 5.19c represent the mass spectra of compounds with single chlorine and bromine atoms, respectively. Roughly, in the case of chlorine the ratio of intensities of A and A + 2 peaks is 3 to 1, and in case of bromine, 1 to 1. One should take into account that the presence of several atoms of A + 2 elements in the molecule results in the appearance of intense peaks M + 4, M + 6, etc., that is, the presence of two or more atoms of A + 2 elements in an ion again gives a unique ratio of isotopic peaks in the multiplet. Due to this fact the nature and number of A + 2 elements may be easily established.

$N + 1$ peaks separated by 2 mass units from one another will characterize an ion containing n atoms of A + 2 element. The intensities of these peaks may be calculated by binominal formula (Equation 5.8), where a and b represent the ratio of natural isotopic abundances for the corresponding element (for Cl, 1 and 0.325; for Br, 1 and 0.98; for S, 1 and 0.044), and n is the number of such atoms in the ion. For example, if an ion contains four chlorine atoms, the intensity of the A + 4 peak (third in the multiplet) in comparison to A will be: $4 \times 3 \times 1^2 \times 0{,}325^2/2 = 0{,}635$, that is, 63.5%.

$$(a+b)^n = a^n + na^{n-1}b + \frac{n(n-1)a^{n-2}b^2}{2!} + \frac{n(n-1)(n-2)a^{n-3}b^3}{3!} + \frac{n(n-1)(n-2)(n-3)a^{n-4}b^4}{4!} + \dots \quad (5.8)$$

To simplify calculations it is convenient to use orbed values (Cl, 3 : 1; Br, 1 : 1; S, 25 : 1). Thus, for a compound with three chlorine atoms there will be four peaks: M, M + 2, M + 4 and M + 6, with relative intensities $(3 + 1)^3 = 27 + 27 + 9 + 1$.

Figure 5.20 represents partial spectra (high m/z region) of tetrachlorobiphenyl and tetrabromobiphenyl. Isotopic clusters with 4, 3, and 2 atoms of chlorine (bromine) are clearly visible.

If an ion contains two different A + 2 elements, calculation of the peak intensities requires matrix multiplication. Thus, a multiplet of dibromodichloro-benzene $C_6H_2Br_2Cl_2$ (excluding carbon isotopes) will contain five peaks separated by 2 mass units from one another. To calculate the ratio of intensities of the peaks in this multiplet the following procedure should be applied:

Two chlorine atoms will give a triplet $(3 + 1)^2 = 9 : 6 : 1$.
Two bromine atoms will give a triplet $(1 + 1)^2 = 1 : 2 : 1$.

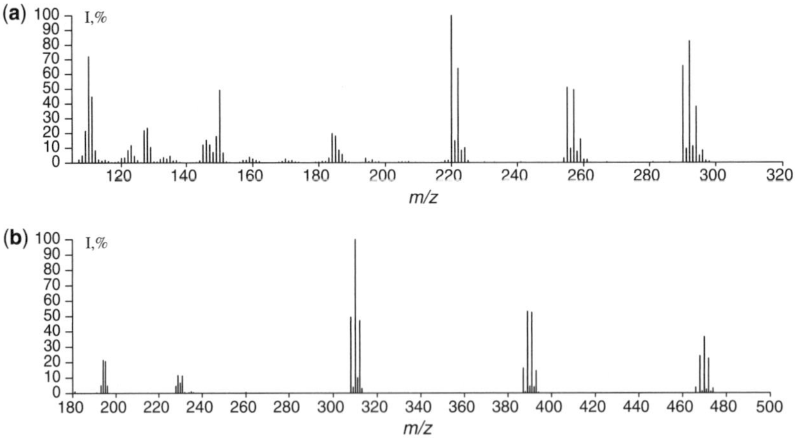

Figure 5.20. Partial spectra (high m/z region) of tetrachlorobiphenyl (a), and tetrabromobiphenyl (b).

Matrix multiplication of these triplets will give:

$$
\begin{array}{l}
(9:6:1) \times (1:2:1) = \\
\qquad\qquad 9:6:1 \\
\qquad 18:12:2 \\
9:6:1 \\
9:24:22:8:1
\end{array}
$$

Therefore, the relative intensities of the peaks in the multiplet will be: 9 : 24 : 22 : 8 : 1, or after normalization, 37.5 : 100 : 92 : 33 : 4. This example demonstrates that the intensity of the M peak may be notably lower than that of the isotopic peaks.

Unknown 7. Calculate the intensity of the M + 8 peak relative to M in the mass spectrum of a compound containing:

A) 5 chlorine atoms
B) 6 bromine atoms
C) 5 sulfur atoms (at this stage exclude the contribution of ^{33}S and ^{36}S isotopes)

Unknown 8. Calculate the intensity of the M + 6 peak relative to M in the mass spectrum of a compound containing:

A) 4 chlorine atoms
B) 7 bromine atoms
C) 6 sulfur atoms (at this stage exclude the contribution of ^{33}S and ^{36}S isotopes)
D) 2 chlorine atoms and 3 bromine atoms
E) 3 bromine atoms and 2 chlorine atoms

Formally oxygen is also an A + 2 element. However, the natural abundance of the ^{18}O isotope is only 0.2% of that of the main isotope ^{16}O. That is why reliable calculation of the number of oxygen atoms based on the intensity of the isotopic peaks is hardly possible. Nevertheless, it is quite possible to estimate this number. Thus, if the intensity of the M + 2 peak in the mass spectrum of a sample with a low number of carbon atoms is higher than 0.5% relative to $M^{+\bullet}$, this compound may contain one or more oxygen atoms.

When A + 2 elements are defined one should deal with A + 1 elements, that is, study the intensity of the M + 1 peak (F + 1 for the fragment ions). Carbon, nitrogen, and hydrogen are A + 1 elements. However, hydrogen may be excluded from this group as the natural abundance of deuterium is very low and may be reliably recorded only in the spectra of compounds with a very large number of hydrogen atoms. In the case of fragment ions one should consider that the F + 1 peak might be represented besides the isotopic ion with an ion of different elemental composition.

Carbon is the most important element comprising organic compounds. Depending on the origin of the sample the abundance of the ^{13}C isotope may be in the range 1.08% to 1.12% relative to ^{12}C. There is a special branch of mass spectrometry dealing with precise measurements of the abundances of isotopes in samples. This field is called isotopic mass spectrometry. It allows, for example, one to establish the origin of a sample according to the exact value of the ^{13}C isotope (Section 5.5.5). However, for the sake of interpretation of mass spectra usually a ratio of ^{13}C/^{12}C of 1.1% is used.

The presence of one carbon atom in a molecule of carbon dioxide results in registration of the molecular ion peak of m/z 44 and of the A + 1 isotopic peak of m/z 45. The intensity of the latter is 1.1% of that of $M^{+\bullet}$. It appears due to the presence of $^{13}CO_2$ molecules. An increase in the number of carbon atoms in a molecule leads to an increase of the intensity of the M + 1 ion peak to 1.1n% of $M^{+\bullet}$, where n is the number of carbon atoms in the molecule. To calculate the number of carbon atoms in a molecule using a mass spectrum one should divide the intensity of the M + 1 peak as a percentage of M by 1.1. The result defines the maximum possible number of carbon atoms. One should remember that calculations may be more complicated if an $[M - H]^+$ ion peak is present.

A large number of carbon atoms in a molecule leads to an increase in the possibility of the simultaneous presence of two or more ^{13}C atoms, resulting in a higher intensity of the isotopic peaks [M + 2], [M + 3], etc. (Table 5.5). For each atom of another element

TABLE 5.5. Isotopic Contributions of Carbon

Number of Carbon Atoms in an Ion	A + 1	A + 2	A + 3	A + 4	A + 5
1	1.1	0	0	0	0
2	2.2	0.01	0	0	0
3	3.3	0.03	<0.01	0	0
4	4.4	0.07	<0.01	<0.01	0
5	5.5	0.12	<0.01	<0.01	<0.01
6	6.6	0.18	<0.01	<0.01	<0.01
7	7.7	0.25	<0.01	<0.01	<0.01
8	8.8	0.34	<0.01	<0.01	<0.01
9	9.9	0.44	0.01	<0.01	<0.01
10	11.0	0.54	0.02	<0.01	<0.01
11	12.1	0.67	0.02	<0.01	<0.01
12	13.2	0.80	0.03	<0.01	<0.01
13	14.3	0.94	0.04	<0.01	<0.01
14	15.4	1.10	0.05	<0.01	<0.01
15	16.5	1.27	0.06	<0.01	<0.01
16	17.6	1.45	0.07	<0.01	<0.01
17	18.7	1.65	0.09	<0.01	<0.01
18	19.8	1.86	0.11	<0.01	<0.01
19	20.9	2.07	0.13	<0.01	<0.01
20	22.0	2.30	0.15	<0.01	<0.01
30	33.0	5.26	0.54	0.04	<0.01
40	44.0	9.44	1.32	0.13	0.01
50	55.0	14.8	2.54	0.32	0.03
60	66.0	21.4	4.55	0.71	0.09
70	77.0	29.2	7.29	1.34	0.19
80	88.0	38.2	10.9	2.32	0.39
90	99.0	48.5	15.6	3.74	0.68
100	110	59.9	21.5	5.74	1.21

Note: The intensity of the A peak is 100%.

present in the sample molecule the intensity of the A + 1 peak will increase 0.37% (for nitrogen), 0.04% (for oxygen), 5.1% (for silicon), 0.8% (for sulfur); and the intensity of the A + 2 peak 0.2% (for oxygen), 3.4% (for silicon), 4.4% (for sulfur), 32.5% (for chlorine), 98.0% (for bromine).

The theoretic intensity of isotopic peaks due to the presence of several ^{13}C isotopes may be easily calculated for any molecule. The intensity of the A + 2 peak (as a percentage of the A peak) for a compound with n carbon atoms may be established by the following formula:

$$I(\%) = 100C_n^2(0.011)^2 \tag{5.9}$$

The intensity of the A + 3 peak (as a percentage of the A peak) for a compound with n carbon atoms may be established by the following formula:

$$I(\%) = 100C_n^3(0.011)^3 \tag{5.10}$$

The intensity of the A + 4 peak (as a percentage of the A peak) for a compound with n carbon atoms may be established by the following formula:

$$I(\%) = 100C_n^4(0.011)^4 \tag{5.11}$$

The general formula to calculate the intensity of the A + m peak (as a percentage of the A peak) for a compound with n carbon atoms is as follows:

$$I(\%) = 100C_n^m(0.011)^m \tag{5.12}$$

When the number of carbon atoms is calculated on the basis of the M + 1 peak one should return to the M + 2 peak, to exclude contributions of ^{13}C isotopes and estimate the number of oxygen atoms with greater precision. Let us consider as an example the mass spectrum represented in Fig. 5.21.

The molecular ion peak (m/z 94) is the most intense in the spectrum. The relative intensity of the M + 2 ion peak is 0.4% of that of $M^{+\bullet}$. Thus, the sample does not contain chlorine, bromine, sulfur or silicon atoms, while there may be one or two

m/z	**Abund.(%)**	*m/z*	**Abund.(%)**
37	8.3	64	4.4
38	14.6	65	40.5
39	51.6	66	53.3
40	20.3	67	2.35
47	6.5	74	2.8
49	2.1	75	1.5
50	9.7	76	1.2
51	8.3	77	1.6
53	5.5	79	1.2
55	13.2	94	100
62	5.4	95	6.6
63	9.4	96	0.4

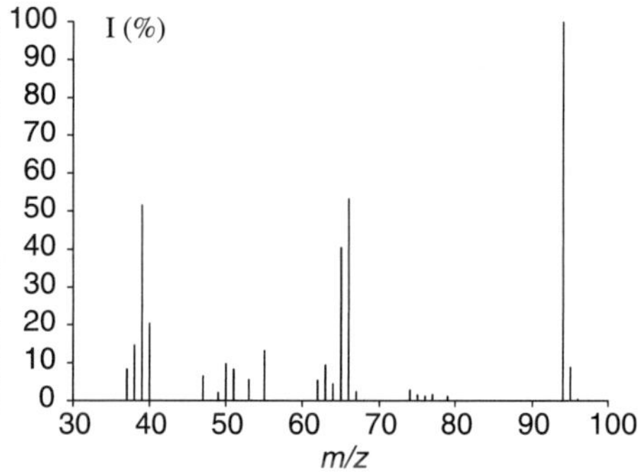

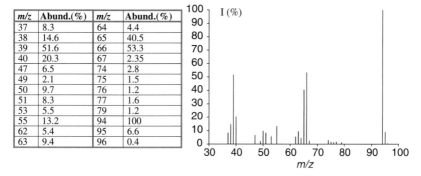

Figure 5.21. EI mass spectrum of a compound with molecular mass 94 Da.

oxygen atoms. As the relative intensity of the M + 1 ion peak is 6.6%, there are six carbon atoms in the molecule ((6.6/1.1) = 6). Table 5.5 lists the contribution of ^{13}C isotopes to the intensity of the M + 2 peak. For six carbon atoms this contribution is 0.18%. Therefore the molecule cannot contain more than one oxygen atom (0.4 − 0.18 = 0.22). The mass of six carbon atoms is 72 Da and the mass of one oxygen atom is 16 Da. The total mass of these elements is 88 Da. The final 6 mass units (94 − 88 = 6) may be only due to hydrogen. Thus, the composition of this compound is C_6H_6O. Other variants of chemically sensible compositions (with fluorine or nitrogen atoms) rationalize neither the mass nor the isotopic abundances. Fig. 5.21 represents the mass spectrum of phenol.

5.5.4. The Nitrogen Rule

There is a remarkable coincidence for the main elements comprising organic molecules. Two values (the valence and the mass of the most abundant isotope) are either odd-numbered or even-numbered. The principal exception involves nitrogen. This fact constitutes the basis for the *nitrogen rule*, which seems rather strange at first glance: *if a compound does not contain any nitrogen atoms or contains an even number of nitrogen atoms, its molecular ion will be at an even mass number; if a compound contains an odd number of nitrogen atoms, its molecular ion will be at an odd mass number.* This rule is applicable not only to molecular ions, but to the fragment ions as well. In this case the rule is phrased differently: *an odd-electron ion will be at an even mass number if it contains an even number of nitrogen atoms; an even-electron ion will be at an even mass number if it contains an odd number of nitrogen atoms.*

Nitrogen is an A + 1 element, while the natural abundance of ^{15}N isotope constitutes about 0.4% of ^{14}N. Since besides nitrogen only carbon contributes notably to the intensity of the A + 1 peak, it is possible to estimate the number of nitrogen atoms in the molecule on the basis of the nitrogen rule and the differences in the abundances of ^{13}C и ^{15}N isotopes.

The formulas analogous to that used for carbon may be applied to calculate the intensities of the isotopic peaks due to the presence of nitrogen atoms. The general formula for the intensity of the A + m peak (as a percentage of the A peak) for a compound containing *n* nitrogen atoms is represented by Equation 5.13:

$$I(\%) = 100C_n^m(0.0037)^m \qquad (5.13)$$

When the nature and number of A + 2 and A + 1 elements in the molecule is known it is easy to derive conclusions about monoisotopic elements. As the masses of these elements are rather unique, it is not a difficult task.

Establishing the elemental composition based on the isotopic peaks may be problematic if, for example, the sample contains impurities with the masses in the region of the molecular ion cluster. In the EI mass spectra of amines, alcohols, acids, and some other classes of organic compounds there is often a peak of $[M + H]^+$ ion. It distorts the isotopic picture. It is worth mentioning as well that in real experimental conditions the peak intensity may vary slightly in each spectrum. Thus, to establish the exact ratio of isotopic peaks it is better to record several spectra and to apply the averaging procedure.

Unknown 9. Estimate the number of carbon atoms in the following compounds using the intensities of the isotopic peaks. In the majority of cases the spectra are represented as m/z value (intensity to the base peak in the spectrum, %). The molecular peak is the first in the row.

A) $M^{+\bullet}$ – 100%, M + 1 – 7.8%, M + 2 – 0.2%
B) $M^{+\bullet}$ – 100%, M + 1 – 0%, M + 2 – 0%
C) $M^{+\bullet}$ (m/z 128) – 100%, M + 1 (m/z 129) – 11.2%, M + 2 (m/z 130) – 0.5%
D) 133 (100), 134 (9.3), 135 (0.3)
E) 156 (100), 157 (2.3), 158 (0)
F) 120 (100), 121 (8.6), 122 (4.7)

Unknown 10. Establish the elemental composition of the molecules on the basis of the intensities of the isotopic peaks. In the majority of cases the spectra are represented as m/z value (intensity to the base peak in the spectrum, %). The molecular peak is the first in the row.

A) $M^{+\bullet}$ (m/z 108) – 100%, M + 1 (m/z 109) – 7.7%, M + 2 (m/z 110) – 0.4%
B) 79 (100), 80 (5.9), 81 (0.1)
C) 98 (100), 99 (6.3), 100 (4.5)
D) 98 (100), 99 (2.2), 100 (66.0), 101 (1.4), 102 (11.0), 103 (0.2)
E) 85 (75), 86 (3.4), 87 (3.3)
F) 158 (100), 159 (7.4), 160 (5.2), 161 (0.2)

Unknown 11. Identify the compounds on the basis of their isotopic peaks. The spectra are represented as m/z value (intensity to the base peak in the spectrum, %). The molecular peak is the first in the row.

A) 94 (100), 95 (1.1), 96 (98), 97 (1.1)
B) 64 (100), 65 (2.2), 66 (33), 67 (0.7)
C) 64 (100), 65 (0.9), 66 (4.8)
D) 67 (100), 68 (4.8), 69 (0.1)
E) 84 (75), 85 (3.9), 86 (3.3)
F) 88 (100), 89 (1.1), 90 (0)

Unknown 12. Calculate the intensities of the molecular ion isotopic peaks for the compounds listed below:

A) Dibromomethane
B) Chloroform
C) Carbon disulfide
D) Tetrabromobenzene
E) Bromoform
F) Dibromodichloromethane

Unknown 13. Identify the molecular ion and establish its elemental composition. The spectra are represented as m/z value (intensity to the base peak in the spectrum, %).

A) 94 (1.9), 95 (7.1), 96 (100), 97 (6.5), 98 (0.2)
B) 160 (0.8), 161 (1.0), 162 (100), 163 (10.8), 164 (32.9), 165 (3.6), 166 (0.2)
C) 93 (1.1), 94 (100), 95 (3.8), 96 (8.8), 97 (0.26)

D) 134 (2.1), 135 (0.4), 136 (26.0), 137 (2.6), 138 (0.16)
E) 126 (6.0), 127 (9.8), 128 (100), 129 (11.0), 130 (0.5)
F) 127 (1.8), 128 (16.0), 129 (100), 130 (10.0), 131 (0.5)

5.5.5. Establishing the ^{13}C Isotope Content in Natural Samples

The natural cycle of carbon involves compounds of the atmosphere, hydrosphere, lithosphere, and biosphere. A certain difference in the ^{13}C isotope content exists between the samples, depending on their origin. To estimate the deviation from the average value of ^{13}C isotope contents δ(‰) scale is used. The deviation may be calculated by Equation 5.14:

$$\delta^{13}C = \left[\frac{(^{13}C/^{12}C)_o}{(^{13}C/^{12}C)_s} - 1\right] \times 1000 \qquad (5.14)$$

Indexes ***o*** and ***s*** define the ratio of the carbon isotopes in the sample and in the standard. A lithospheric carbonate material was accepted as standard. The closest to this zero point value belongs to standard sample NBS-19 (1.95‰). There are some other standard samples: NBS-22 oil (−29.74‰), NBS-18 calcium carbonate (−5.01‰). Usually $\delta^{13}C$ values for plants are in the range (−15‰) to (−30‰), and for oil (−20‰) to (−36‰). Atmospheric methane has the lowest content of ^{13}C. Its $\delta^{13}C$ value is approximately −47‰.

Isotopic mass spectrometry is used to establish $\delta^{13}C$ values. The sample is burned to CO_2 and the intensities of the ion peaks of m/z 44, 45, and 46 are measured. Then correction to eliminate the influence of ^{17}O isotope is achieved. If using sector magnetic instrument and three detectors (for each mass) the standard deviation of the results will be better than 0.001%. The high accuracy of the measurements allows valuable results to be obtained. The ancient Europeans and Americans may be distinguished by the analysis of their remnants. The reason involves the fact that wheat constituted the basic food ration in Europe, while corn played the same role in America. The difference in the isotopic composition of these plants forms several units of $\delta^{13}C$ scale [34].

5.5.6. Calculation of the Isotopic Purity of Samples

Mass spectrometry provides a unique opportunity to estimate the presence of heavy isotopes in a sample molecule. A chemist faces this problem rather often, for example, studying reaction mechanisms. Usually ^{2}H, ^{13}C, ^{14}C, ^{15}N, and ^{18}O isotopes are used to synthesize the labeled compounds. Other isotopes are also used.

Electron ionization is a perfect method for the analysis of labeled molecules as in this case ion–molecular reactions are suppressed. It is better to use for the calculations the most intense spectral peaks with the highest m/z values. Molecular ion is the best choice. However, if notable $[M + H]^+$ or $[M - H]^+$ peaks are present in the spectrum of the unlabeled compound the correct calculation will be problematic. To eliminate $[M + H]^+$ peaks it is helpful to record a spectrum with the minimum quantity of sample. To consider interference with $[M - H]^+$ ions one should know from what position the hydrogen atom is lost and whether deuterium could be in this position.

TABLE 5.6. Calculation of the Abundance of Isotopic Peaks in the Mass Spectrum of Deuterated Acetophenone

Isotopomer	*m/z* 120	*m/z* 121	*m/z* 122	*m/z* 123	*m/z* 124	*m/z* 125
d_0	100	8.8	0.54	–	–	–
d_1	–	100	8.8	0.54	–	–
d_2	–	–	100	8.8	0.54	–
d_3	–	–	–	100	8.8	0.54
10% d_0	10.0	0.88	0.05	–	–	–
10% d_1	–	10.0	0.88	0.05	–	–
20% d_2	–	–	20.0	1.76	0.11	–
60% d_3	–	–	–	60.0	5.28	0.32
Σ d_0–d_3	10.0	10.88	20.93	61.81	5.39	0.32
Normalized Σd_0–d_3 (%)	**16.2**	**17.6**	**33.9**	**100**	**8.7**	**0.5**

If the answer to these questions is not obvious, one can use for the calculations another ion of known composition, for example, $[M - CH_3]^+$, $[M - CO]^{+\bullet}$, or $[M - Cl]^+$. It is relevant to avoid the presence of fragment ions 1 to 2 Da lighter or heavier than the selected ion.

Let us calculate the intensities of the molecular ion cluster of an acetophenone sample, which contains 10% of unlabeled molecules, 60% of molecules with three deuterium atoms, 20% with two deuterium atoms, and 10% with one deuterium atom. Let us create a table (Table 5.6), taking into account that the molecular ion cluster of pure unlabeled acetophenone is as follows: 120 (100%), 121 (8.8%), 122 (0.54%). The calculations with the help of Table 5.6 are quite straightforward, while the final line represents the labeled sample spectrum.

Usually the reverse problem is relevant. Recording a spectrum of a sample one should calculate the percentage of molecules with various numbers of heavy isotopes. Let us suppose that the following mass spectrum was recorded for a sample of labeled acetophenone: 120 (20.0%), 121 (42.0%), 122 (100%), 123 (14.0%), 124 (1.0%), 125 (<0.03%). The best way to resolve the problem is to compose the table shown here as Table 5.7.

TABLE 5.7. Calculation of the Percentage of Acetophenone Molecules with Various Numbers of Deuterium Atoms

Isotopomer	*m/z* 120	*m/z* 121	*m/z* 122	*m/z* 123	*m/z* 124	*m/z* 125
Spectrum	20.0	42.0	100	14.0	1.0	0.02
d_0	20.0	1.76	0.11	–	–	–
Result 1	–	40.24	99.89	14.0	1.0	0.02
d_1	–	40.24	3.54	0.2	–	–
Result2	–	–	96.35	13.8	1.0	0.02
d_2	–	–	96.35	8.5	0.5	–
Result 3	–	–	–	5.3	0.5	0.02
d_3	–	–	–	5.3	0.47	0.02
Result 4	–	–	–	–	<0.03	–

An ion of m/z 120 may appear only due to a molecule of unlabeled acetophenone. Since the intensity of the corresponding peak is 20%, one may consider that there are 20 parts of the unlabeled molecules in the sample. On the basis of this conclusion let us calculate the relative intensities of m/z 121 and 122 ion peaks due to the unlabeled molecules. The following cluster of molecular ions may characterize the pure unlabeled compound: 120 (100%), 121 (8.8%), 122 (0.54%). Then 20 parts of this compound lead to the following spectrum: 120 (20.0%), 121 (1.76%), 122 (0.11%). These figures are presented in the second row of Table 5.7. The third row (Result 1) represents the spectrum of the sample without the contribution of the unlabeled compound. The next step involves estimation of the quota of monodeuterated acetophenone. The intensity 40.24 of the m/z 121 ion peak is due to the $M^{+\bullet}$ of this compound. Let us calculate the intensities of the isotopic peaks for 40.24 parts of this product. The following figures appear: 121 (40.24%), 122 (3.54%), 123 (0.2%). Subtraction of these values from Result 1 leads to Result 2, representing the mass spectrum due to the molecules with two and three deuterium atoms. Repeating this procedure two times more we get Result 4 with only one peak of m/z 124 with intensity 0.03%, which is at the level of experimental error. Therefore, we get the following row of results: d_0, 20 parts; d_1, 40.24 parts; d_2, 96.36 parts; d_3, 5.3 parts. Assuming that all isotopomers constitute 100%, the percentage of each of them will be: d_0, 12.3%; d_1, 24.9%; d_2, 59.5%; d_3, 3.3%.

Unknown 14. Calculate the percentage of isotopomers in the sample of acetophenone with the following mass spectra:

A) 120 (15.0%), 121 (32.0%), 122 (50.0%), 123 (100%), 124 (8.7%), 125 (0.5%)
B) 120 (40.0%), 121 (40.0%), 122 (100%), 123 (50%), 124 (3.5%), 125 (0.25%)

Unknown 15. Calculate the percentage of isotopomers in the sample of *para*-methoxyphenol with the following mass spectra:

A) 123 (0.21%), 124 (30.0%), 125 (60.0%), 126 (80.0%), 127 (100%), 128 (7.7%), 129 (0.6%)
B) 124 (0.2%), 125 (1.2%), 126 (100%), 127 (28.0%), 128 (1.8%), 129 (0.05%)

Unknown 16. Calculate the percentage of isotopomers in the sample of 4-fluoro-2-methoxyanisole with the following mass spectra:

A) 155 (0.12%), 156 (20.0%), 157 (100%), 158 (12.0%), 159 (0.8%), 160 (0.02%)
B) 155 (0.06%), 156 (10.0%), 157 (23.0%), 158 (42.0%), 159 (56.0%), 160 (100%), 167 (59.2%), 168 (3.7%), 169 (0.37%)

Unknown 17. Calculate mass spectrum in the region of the molecular ion for the sample of 4,7-difluoronaphthylmethylketone containing 10% of unlabeled, 20% d_1, 30% d_2 and 40% d_3 labeled molecules.

5.5.7. Fragment Ions

When all the available information is extracted from the molecular ion cluster it is possible to pass to the fragment ions. As mentioned above the fragment ions arise due to simple cleavages of the chemical bonds and to rearrangements. For the interpretation of mass spectra it is useful to use another classification. All fragment ions may be divided into three types:

1. The heaviest ions due to the losses of small neutral particles from the $M^{+\bullet}$, that is without notable transformations of the initial molecule.
2. The ions represented with the most intense peaks.
3. Homologous (14 Da from one another) ion series.

One should take into account that the intensities of the peaks of the first and third types may be insignificant. It is very important not to lose them.

5.5.7.1. Homologous Ion Series. The peaks of high m/z ions are important to establish the primary fragmentation reactions, the presence of labile groups. Homologous ion series provide important information on the degree of unsaturation, on the presence of heteroatoms, and finally give a chance to refer the analyzed sample to a certain class of compounds. It is much easier to elucidate the elemental composition of these ions (usually even-electron) in comparison to the heavy unique fragment ions with many possible isobaric compositions. Homologous ion series of the main classes of organic compounds are listed in Table 5.8.

The alkane series is present in mass spectra of any compound containing an alkyl group. In case of isobaric series (e.g., alkanes and ketones) one should pay attention to the intensities of the isotopic peaks. Thus, for the isobaric ions of m/z 43 (CH_3CO and C_3H_7) the abundance of the isotopic peak (m/z 44 ion) will be 2.2% and 3.3%, respectively. The situation is very simple with A + 2 elements. In this case there are two homologous ion series due to A and A + 2 ions. It is worth emphasizing that

TABLE 5.8. Homologous Ion Series of the Main Classes of Organic Compounds

Class of Compounds	Formula	m/z values
Alkanes	$C_nH_{2n+1}{}^+$	15, 29, 43, 57, 71, 85, ...
Alkenes, naphthenes	$C_nH_{2n-1}{}^+$	27, 41, 55, 69, 83, ...
Alkynes, dienes	$C_nH_{2n-3}{}^+$	25, 39, 53, 67, 81, ...
Alcohols, ethers	$C_nH_{2n+1}O^+$	31, 45, 59, 73, 87, ...
Aldehydes, ketones	$C_nH_{2n-1}O^+$	29, 43, 57, 71, 85, ...
Acids, esters	$C_nH_{2n-1}O_2^+$	45, 59, 73, 87, 101, ...
Thioles, sulfides	$C_nH_{2n+1}S^+$	47, 61, 75, 89, 103, ... (по ^{32}S)
Amines	$C_nH_{2n+2}N^+$	30, 44, 58, 72, 86, 100, ...
Alkylchlorides	$C_nH_{2n}Cl^+$	35, 49, 63, 77, 91, 105, ... (по ^{35}Cl)
Alkylfluorides	$C_nH_{2n}F^+$	19, 33, 47, 61, 75, ...
Alkylbromides	$C_nH_{2n}Br^+$	79, 93, 107, 121, ... (по ^{79}Br)
Alkyliodides	$C_nH_{2n}I^+$	127, 141, 155, 169, ...
Nitriles	$C_nH_{2n-2}N^+$	40, 54, 68, 82, 96, ...
Alkylbenzenes		38–39, 50–52, 63–65, 75–78, 91, 105, 119, ...

usually there are several homologous series in the mass spectra. For example, in the spectrum of an aliphatic alcohol there will be alkane (29, 43, 57, ...), alkene (27, 41, 55), and alcohol (31, 45, 59, ...) series.

A notable homologous low mass ion series will be absent in the spectra of compounds with high degree of unsaturation or with many functional groups. Alkylbenzenes may be considered a convenient example. Although there are certain characteristic ions, a homologous series of low mass ions is absent in their spectra. Nevertheless aromatic series may be observed being shifted to the high mass region (Table 5.8).

If an alkyl group is bonded to a substitute of high ionization energy one can observe a homologous series (sometimes quite abundant) of even-electron ions with general formula $C_nH_{2n}{}^+$.

The correct analysis of the homologous ion series has certain limitations. Low abundances of peaks in some series require the attention and experience of a researcher. Usually alkane series are dominated in the mass spectra of the most various compounds. Fragmentation initiated by one functional group may completely suppress or notably camouflage other reactions of polyfunctional substances. In the latter case it is useful to consider IR-spectroscopy data in mass spectral interpretation.

Unknown 18. Figure 5.22 represents the mass spectra of three monofunctional compounds with a similar array of fragment ions. Molecular ion peaks are absent in two cases. Try to refer the substances to widespread classes of organic compounds. Take into account that the intensities of the peaks of some homologous series may be very low.

Unknown 19. Figure 5.23 represents the spectra of three monofunctional compounds. Could you refer them to certain classes of organic compounds?

5.5.7.2. Neutral Losses. The simplest but important considerations may be done on the basis of the losses of neutral particles from $M^{+\bullet}$. These processes lead to the formation of high mass ions: $[M-1]^+$, $[M-15]^+$, $[M-18]^{+\bullet}$, $[M-20]^{+\bullet}$, etc. The eliminating neutrals represent common particles (radicals or molecules) with small masses (H_2O, CO, $CH_3^\bullet$, $Cl^\bullet$, H_2S, CH_3OH, etc.). As mentioned above, *the losses of neutrals of 5 to 14 and of 21 to 25 mass units from $M^{+\bullet}$ resulting in an intense peak of a fragment ion are highly unlikely*. In this region of a mass spectrum all the peaks are important; even if their intensity is below 1%. They are very useful for the confirmation of the correct identification of the molecular ion.

5.5.7.3. The Most Intense Peaks. It is not so easy to extract valuable information dealing with the most intense peaks in mass spectra. In contrast to other physico-chemical methods (IR, NMR, UV), registration of an ion peak of an integer m/z value does not provide an unequivocal identification of its structure or even composition. Even accurate mass measurements (high resolution mass spectrometry) defining the elemental composition of an ion does not establish its structure, as it could be formed directly from the $M^{+\bullet}$, with minimal distortion of the authentic structure, or as a result of numerous rearrangement processes.

However, if the majority of small peaks may be ignored without notable effect on the result of interpretation, it is highly advisable to establish the exact composition of the

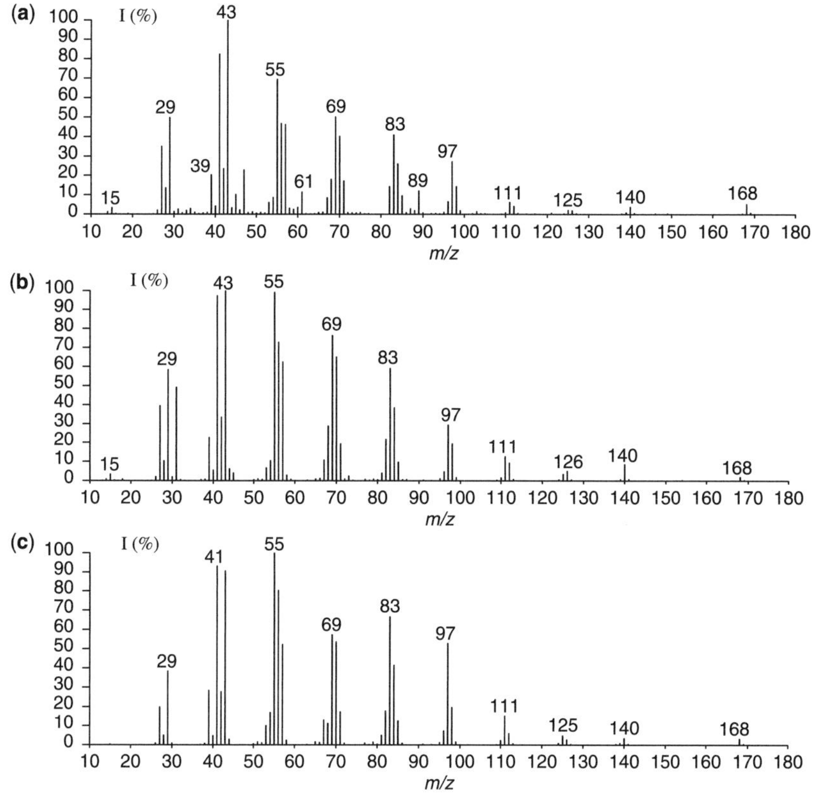

Figure 5.22. Mass spectra of monofunctional compounds referring to three different classes.

ions represented by the dominant peaks in the mass spectrum. These ions are usually present at the fragmentation schemes demonstrating the principal pathways of fragmentation of the analyte. These ions may be of high mass and may have many possible isobaric compositions. That is why application of high resolution mass spectrometry (accurate mass measurements) and tandem mass spectrometry (identification of parent and daughter ions) may be necessary. If a researcher works with a certain group of compounds for a prolonged period, the composition and structure of the main fragment ions are usually well known, significantly facilitating interpretation of mass spectra of the related substances. The task is much more difficult when dealing with new samples belonging to the most different classes.

Several books on mass spectrometry contain tables linking the m/z value of a fragment ion with certain moieties in the structure of the initial molecule. These tables may be helpful but should be used cautiously. For example, registration of the ion $C_2H_3O^+$ often may be interpreted as acetyl fragment (CH_3CO), referring the sample to methylketones. However, the CH_3CO^+ ion arises as a result of fragmentation (McLafferty rearrangement) of various ketones with appropriate length of the alkyl chain. An

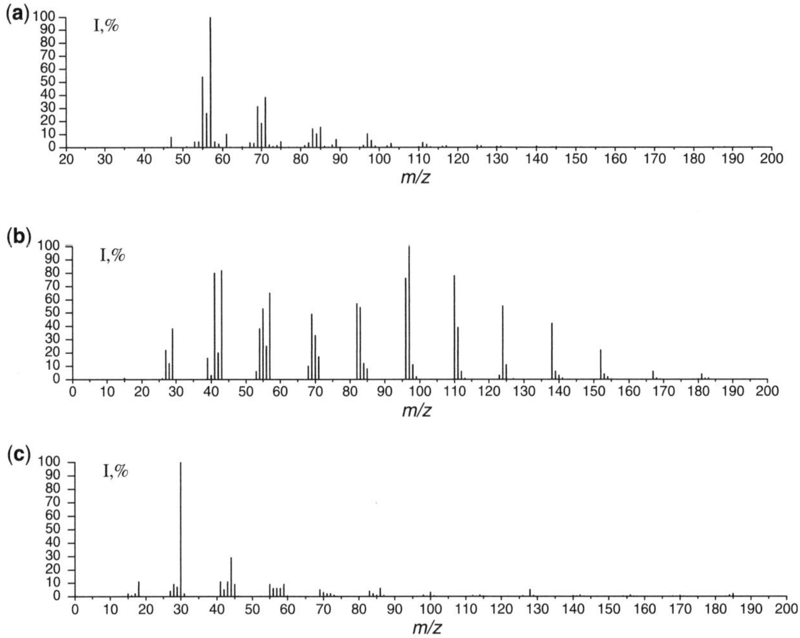

Figure 5.23. Mass spectra of three monofunctional aliphatic compounds.

intense peak of m/z 30 $(CH_2{=}NH_2)^+$ does not allow referring the analyte to primary amines as it readily arises from secondary or tertiary amines with an adequate length of their alkyl radicals.

Nevertheless, some specific ions being recorded in the mass spectrum may provide a definite clue concerning structural features: m/z 77, phenyl; m/z 91, C_7H_7 tropilium (benzyl); m/z 30, CH_2NH_2 amino group; m/z 105, benzoyl-PhCO. An ion of m/z 149 dominating in a mass spectrum almost unequivocally means that the sample belongs to the phthalates. By the way compounds of this class are widely used as plasticizers all over the world. They are rather stable and are treated as environmental priority pollutants in many countries. Now it is difficult to find an environmental sample (water, soil, air, food, beverages) without the presence of one or several phthalates. Thus, phthalates are quite often impurities (sometimes rather unexpected) in the most different samples.

Working in GC/MS mode a researcher often encounters chromatographic peaks due to periodic elution of compounds with abundant ions of m/z 73, 147, 207, 281, 355, etc. These ions represent the fragments of the widespread silicon phases used in the chromatographic columns.

When all available spectral information is processed the task involves creation of the most probable structure. If there are some facts indicating one or another structure *a priori*, one should try to draw theoretical possible pathways of its fragmentation and compare the result with the experimental spectrum.

5.5.8. Mass Spectral Libraries

Nowadays there are quite a number of atlases of mass spectra (mainly electron ionization mass spectra) although computer libraries are much more convenient. NIST and WILEY are the main commercial mass spectral libraries available for users. Both libraries contain electron ionization mass spectra as the vast majority of organic compounds have been analyzed with this method. The intensive fragmentation allows reliable comparison of the spectra. This is an important advantage over alternative chemical ionization or field ionization spectra. Another benefit of the electron ionization mass spectra involves standard conditions of registration of the spectra. As a result mass spectra recorded with different instruments are amenable to reliable comparison. The modern software allows, besides routine comparison of the spectra, various types of selection of the database spectra according to elemental composition, molecular mass, intensities of certain peaks, etc. Working with these selected spectra a researcher may use his or her experience, apply additional information, and ignore the peaks known to be due to some impurities. In this case the reliability of the correct identification may be notably increased.

Before the comparison of the recorded spectra with those in a library it is worth estimating its quality and making possible improvements. For example, working in GC/MS mode it is always helpful to subtract the background.

The spectral libraries definitely facilitate the process of identification. However, a computer cannot completely substitute a qualified operator. Even if a spectrum of the analyte is present in the library, it is not obvious that the computer will detect it correctly or will place it as the first choice of the proposed list of candidates. At this stage "manual" comparison of the selected variants with the newly recorded spectrum is very useful. If there are no spectra of the analyte in the library, the computer nevertheless generates a list of possible candidates "from its point of view." This information definitely helps an operator, providing some clues to refer the sample to one or another class of organic compounds. Several laboratories are working now on algorithms of the new approaches for the computer interpretation of mass spectra. These programs require creative thinking of a highly qualified operator allowing impressive results, that is, elucidation of the most probable structure for an analyte, which spectrum is absent in the library.

5.5.9. Additional Mass Spectral Information

Quite often a normal electron ionization mass spectrum appears insufficient for reliable analyte identification. In this case additional mass spectral possibilities may be engaged. For example, the absence of the molecular ion peak in the electron ionization spectrum may require recording another type of mass spectrum of this analyte by means of "soft" ionization (chemical ionization, field ionization). The problem of impurities interfering with the spectra recorded via a direct inlet system may be resolved using GC/MS techniques. The value of high resolution mass spectrometry is obvious as the information on the elemental composition of the molecular and fragment ions is of primary importance.

At first glance the necessity to use collision activation or metastable ion spectra is less evident. However, these two methods allow, first of all, recording pure mass spectra,

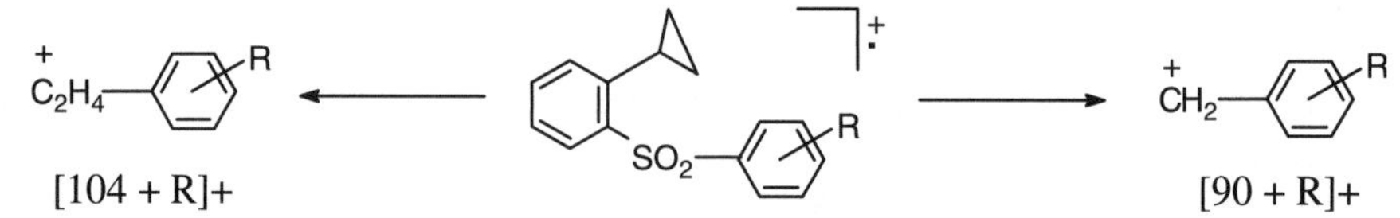

Scheme 5.20.

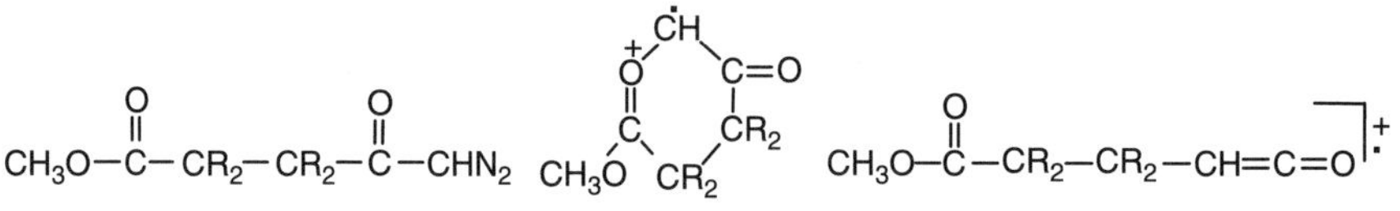

4-1 4-2 4-3

Scheme 5.21.

without interferences due to sample impurities. Besides they provide information on the pathways of fragmentation linking certain parent and daughter ions. These methods also help in obtaining valuable structural information.

For example, the fact that ions of m/z $[90 + R]^+$ and $[104 + R]^+$ arise directly from the molecular ions of sulfones (Scheme 5.20) confirms a transformation with a new C–C bond formation between carbon atoms of the small ring and of the second benzene ring prior to the fragmentation of the $M^{+\bullet}$. In this case the cyclopropyl moiety (maybe isomerized) retains the charge and unpaired electron and attacks the second aromatic ring by a nucleophilic or radical mechanism.

Sometimes it is impossible to elucidate the structure of an ion by means of any possible types of mass spectra. In this case it is useful to obtain spectra of labeled analogs of these compounds. This approach is widely used in other physicochemical methods, as well as in classic chemistry.

The following example demonstrates the usefulness of this approach for the solution of certain mass spectrometric tasks. Ions $[M - N_2]^{+\bullet}$ of diazo ketones 4-1 fragment by several pathways, which require their preliminary isomerization into some other structures. The most probable among them are 4-2 and 4-3 (Scheme 5.21).

To select between these two alternative structures it was necessary to synthesize a labeled analog. Three hydrogen atoms of the methyl moiety of the ester group were substituted for deuterium. One of the principal pathways of fragmentation of $[M - N_2]^{+\bullet}$ ions involves the loss of CH_3 radical. Since all R substitutes in diazo ketones 4-1 were also methyls it was important to detect what group exactly is eliminated from the $[M - N_2]^{+\bullet}$ ion. The spectrum of deuterated sample has confirmed that the methyl radical of the ester moiety leaves the parent ion. As a result the cyclic structure 4-2 was selected as the most probable. The ketene structure 4-3 is hardly able to trigger this process, while for heterocyclic ion 4-2 it is highly favorable (Scheme 5.22).

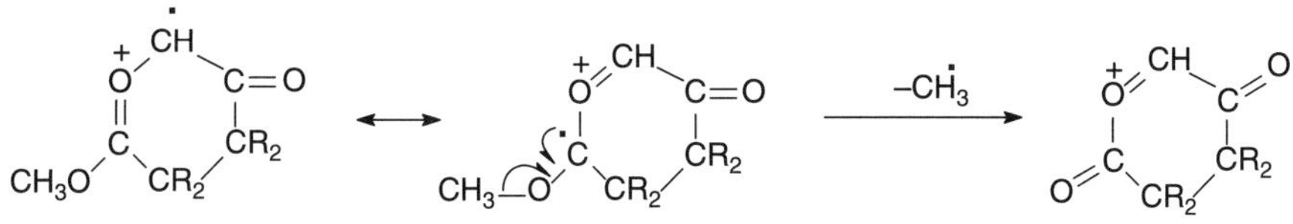

Scheme 5.22.

5.5.10. Fragmentation Scheme

The final stage of mass spectrum processing involves development of a fragmentation scheme. The scheme should reflect the most characteristic pathways of fragmentation of the molecular ion, composition of the fragment ions (and if possible their structures), the relationship of the fragment ions with one another, and sometimes the relative abundances of their peaks (Schemes 5.23 and 5.24).

It is impossible to fix into the scheme format all the ions forming during fragmentation. Many of them do not possess any notable structural information, while representation of all the possible fragmentation pathways camouflages really significant directions. There is no sense in representing trivial fragments. For example, peaks of m/z 77 and 76 ions due to the benzene ring are registered in the spectra of any aromatic compound. These peaks may be rather intense; however, the corresponding ions do not reflect peculiarities of the initial molecule. They may arise from any fragment of higher mass. Thus, the fragmentation scheme must include first of all the most important ions for the characterization of a certain group or of an individual compound. It is important not to miss the most intense ion peaks. However it is not less significant to mention the primary fragment and rearrangement ions, even if their peaks are of low intensity.

To be reliable the scheme should represent ions of known elemental composition, while the mentioned fragmentation pathways should be confirmed by metastable ion

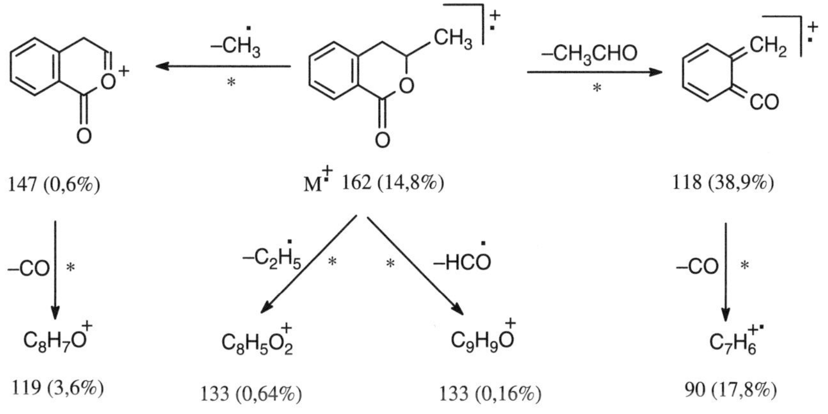

Scheme 5.23.

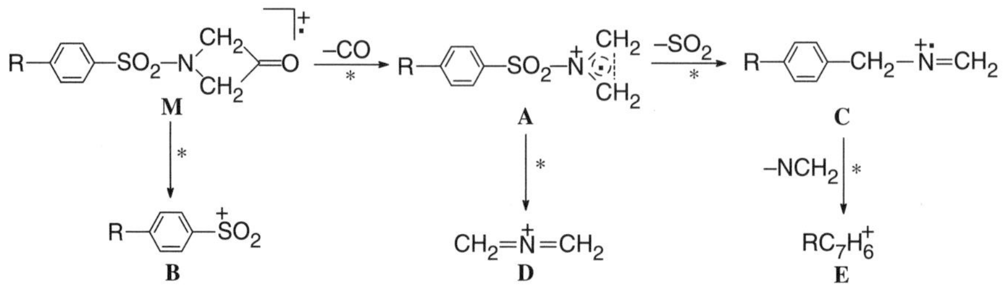

Scheme 5.24.

or collision activation spectra. The latter fact is represented in Schemes 5.23 and 5.24 as an asterisk near the arrow linking two ions.

Strictly speaking representation of the ion structures at the fragmentation schemes generally is not correct. Usually the principle of minor structural changes is applied. A researcher tries to retain the structure of the initial molecule intact, consequently taking away certain structural fragments. However, representation of an ion with its general formula is more justified. Sometimes the aim of the authors is the elucidation of the real structure of an ion, even if it differs notably from that of the authentic molecule. Then the structures are discussed specially in the text, taking into account the results of additional experiments and all the arguments pro and con.

If a scheme represents fragmentation pathways of an individual compound it is convenient to accompany m/z values with the relative intensity of the corresponding peak directly at the scheme (Scheme 5.23). However, more often a scheme represents fragmentation pathways for a group of structurally close compounds. In this case the most appropriate variant involves the labeling of an ion with a letter, while the quantitative parameters (relative abundances of the common fragment ions) are summarized in the table that accompanies the scheme (Scheme 5.24, Table 5.9). Upper case (A, B, C, ...) or lower case (a, b, c, ...) Latin letters are used to label fragment ions. In the earlier papers one finds labels F_1, F_2, F_3, etc. from the word "fragment."

To summarize the material of this chapter it is possible to propose an algorithm for spectral interpretation.

1. Do a library search on the spectrum.
2. Collect all available information (history) about the sample, including other types of spectra of the analyte, the method of synthesis and (or) isolation, possible class of organic compounds, possible impurities. Mark all the instrumental parameters that were used to record the mass spectrum.
3. Try to identify the molecular ion peak. Is it present in the spectrum? Use four necessary conditions to consider a peak as representing $M^{+\bullet}$. Do not forget about the nitrogen rule.
4. Using the intensities of the isotope peaks try to identify the elemental composition of all the ions (where it is possible). Calculate the unsaturation degree of these ions.

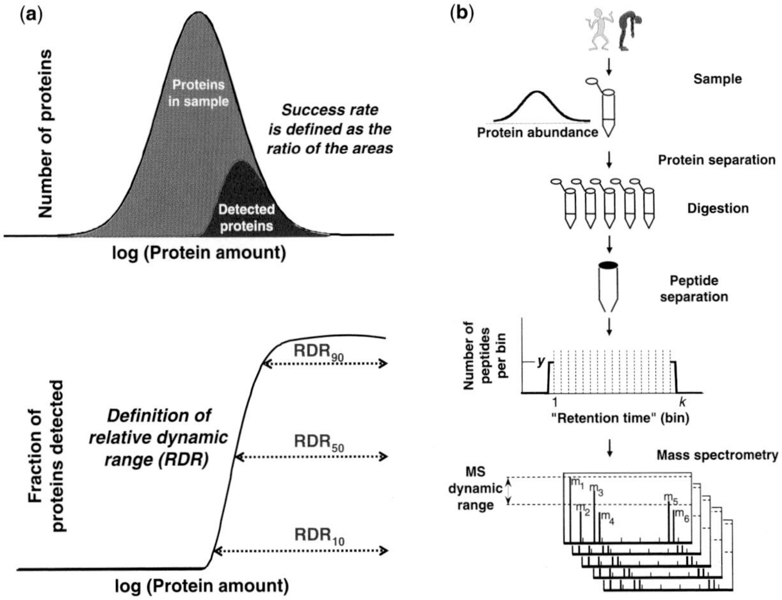

Figure 7.6. (a) Definitions of success rate and relative dynamic range. (b) Model of a proteomics experiment. (See page 218 for text discussion.)

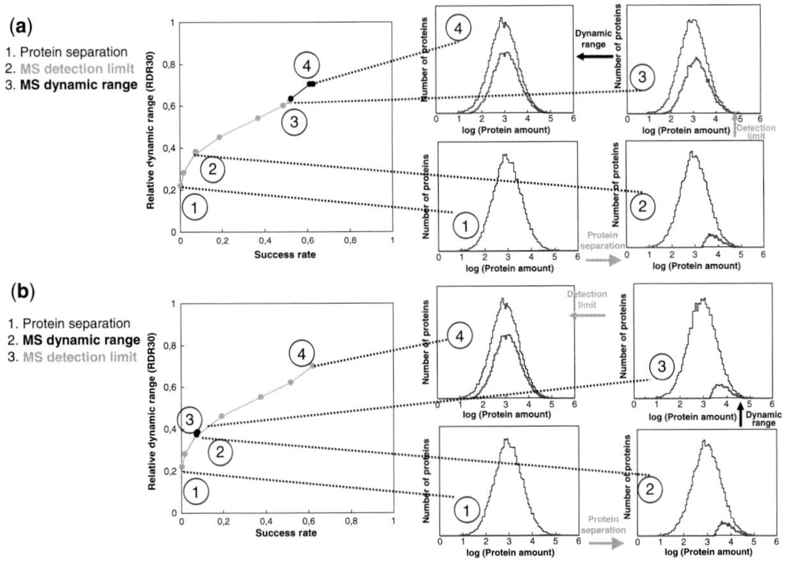

Figure 7.7. Results from model simulations showing the effect of protein separation and the effect of MS detection limit and MS dynamic range on the success rate and the relative dynamic range (RDR) for detection of proteins from *H. sapiens* tissue samples. (See page 219 for text discussion.)

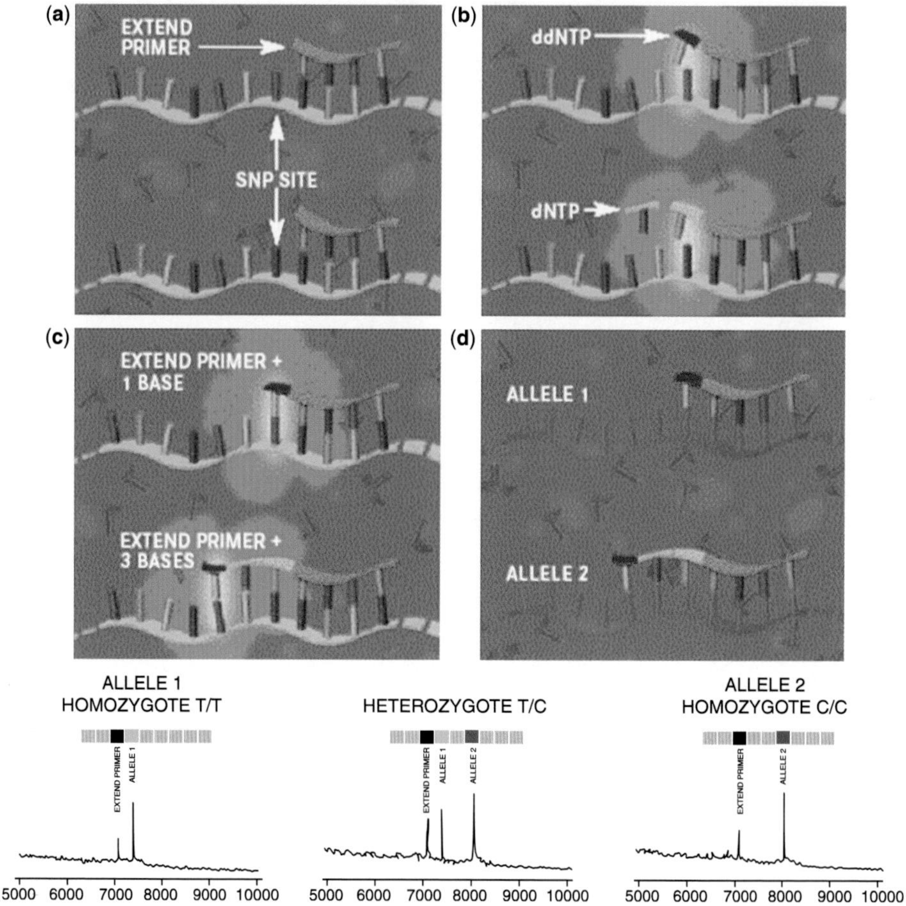

Figure 10.3. Mass array. (a) Primer binding; (b) primer extension; enzyme, ddATP and dCTP/dGTP/dTTP addition; (c) primer terminates; (d) primer extension products ready for MALDI-MS; (e) MS spectrum of primer extension products. Each addition of a nucleotide to the primer extension product increases the mass by 289 to 329 Da, depending on the nucleotide added. The mass difference is easily resolved by MALDI-TOF, which has the ability to detect differences as small as 3 Da. Printed by kind permission of Sequenom. (See pages 246–247 for text discussion.)

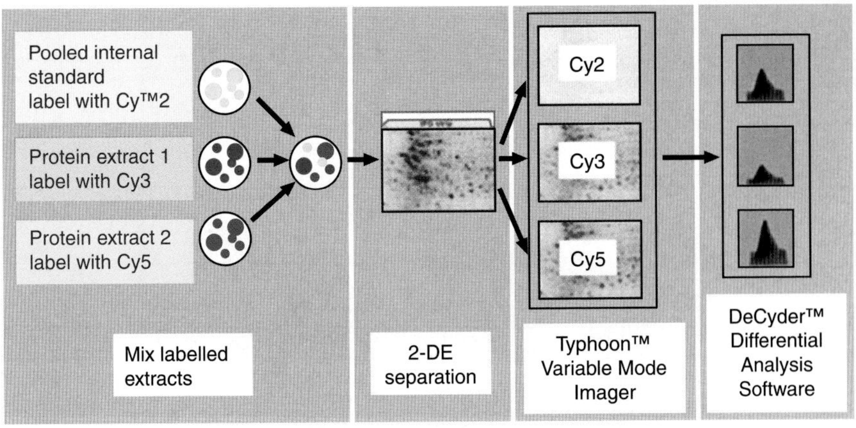

Figure 10.6. Principle of DIGE analysis: separation of control and treated sample on one gel and statistical validation using more than three repeated experiments. Printed by kind permission of GE Healthcare (formerly Amersham Biosciences). (See pages 249–250 for text discussion.)

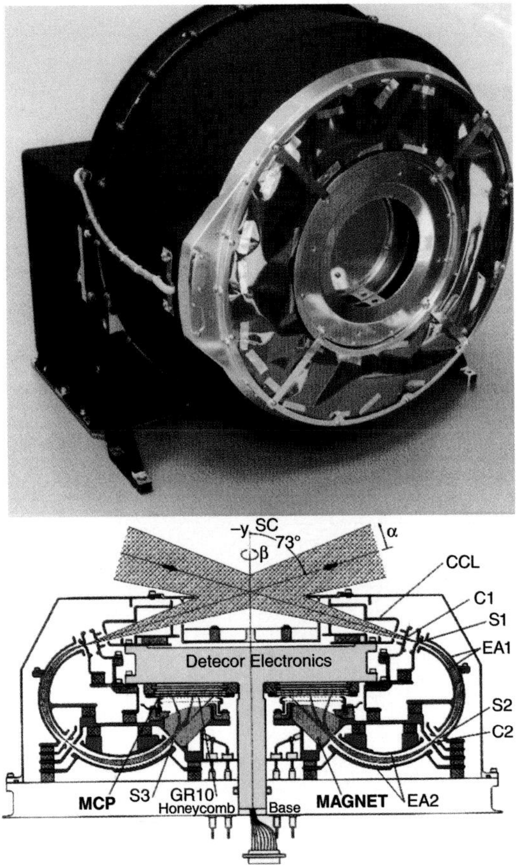

Figure 11.1. POLAR/TIMAS photo and schematic. (Courtesy W. Petersen, NASA.) (See page 257 for text discussion.)

Figure 11.2. ULYSSES/SWICS photo/schematic. (Courtesy G. Gloeckler, NASA.) (See page 260 for text discussion.)

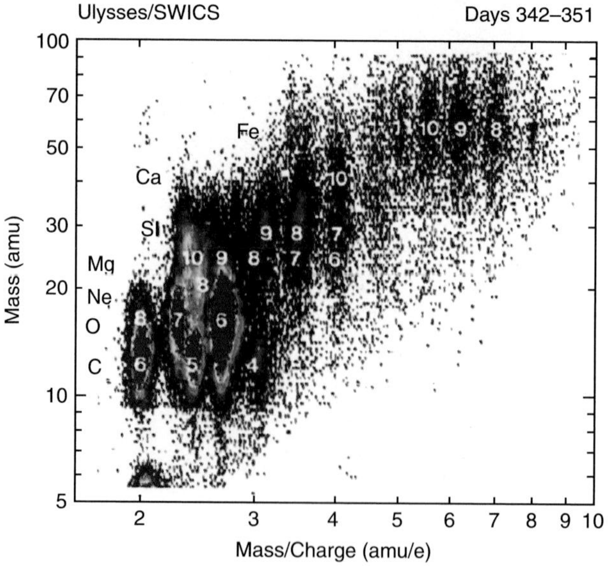

Figure 11.3. SWICS M vs M/Q analysis to enhance R. (Courtesy G. Gloeckler, NASA.) (See pages 261–262 for text discussion.)

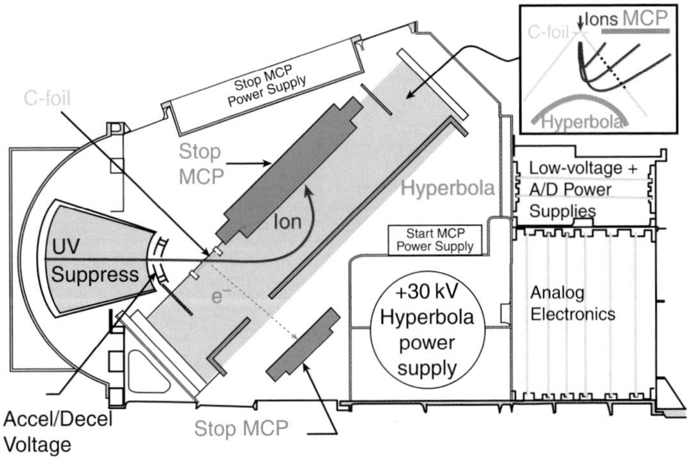

Figure 11.4. WIND/MASS schematic. (Courtesy R. Sheldon, NASA.) (See page 263 for text discussion.)

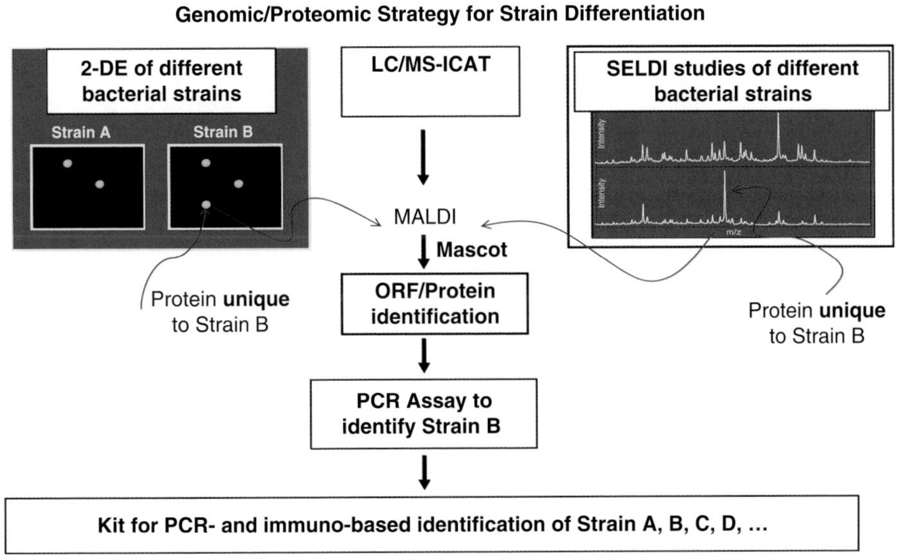

Figure 12.1. Different MS strategies for identification of biomarker proteins. (See page 270 for text discussion.)

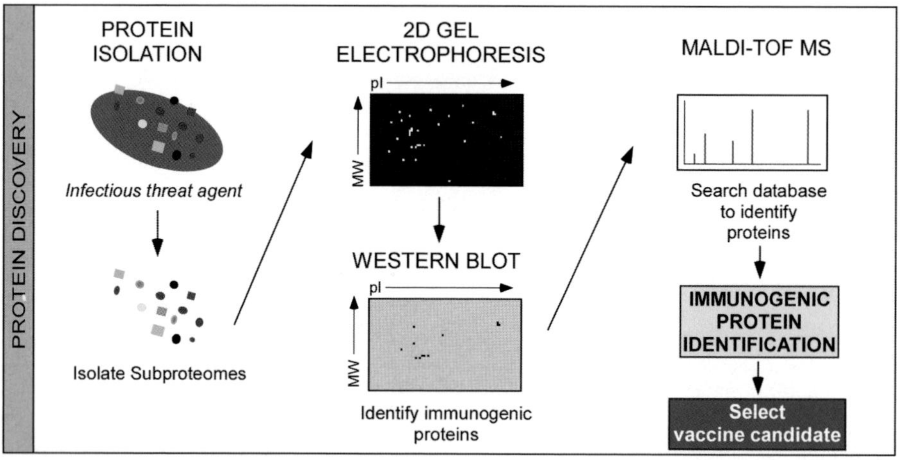

Figure 12.2. Identification of immunogenic proteins by 2DE/MALDI TOF-MS and Western blots probed with antisera from immunized animals or humans. (Reprinted from Khan, A. S. et al., 2006. Proteomics and Bioinformatics Strategies to Design Countermeasures Against Infectious Threat Agents. *J. Chem. Inf. Model.* 46: 111–115. With permission.) (See page 272 for text discussion.)

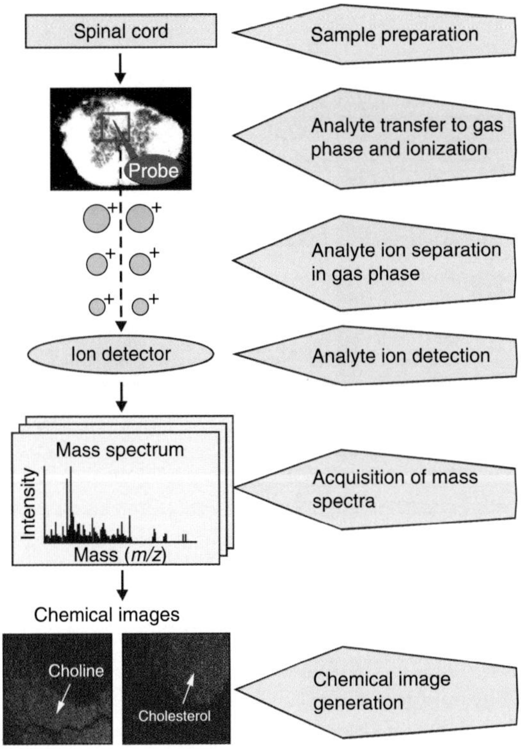

Figure 13.1. Imaging mass spectrometry. The acquisition of chemical images with mass spectrometry involves multiple, often elaborate, steps, beginning with sample preparation and ending with generation of the chemical image. The figure illustrates the process of imaging of a spinal cord section using a TOF-SIMS spectrometer equipped with a gold ion source, where the images of distribution of choline and cholesterol are obtained. The sample preparation consists of spinal cord dissection, freezing, cryostat sectioning, single section deposition on a wafer, and drying. (Reprinted from Rubakhin, S. S. et al., 2005. *DDT,* 10, no 12: 823–837. With permission from Elsevier.) (See page 276 for text discussion.)

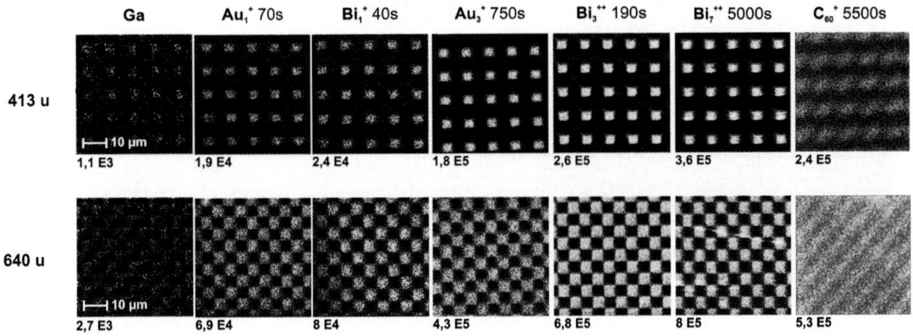

Figure 13.2. TOF-SIMS images of blue ($m = 413$ u) and green ($m = 641$ u) pigments of color filter array. Above each image the primary ion gun and the measurement time is displayed. Corresponding signal intensity of emitted secondary ions from an analyzed surface is given under suitable image. (Reprinted from Kollmer, F. 2004. *Appl. Surf. Sci.*, 231–232: 153–158. With permission from Elsevier.) (See page 278 for text discussion.)

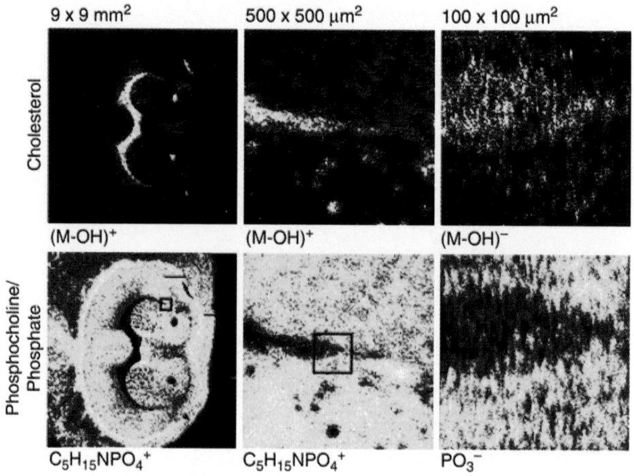

Figure 13.3. TOF-SIMS images showing the spatial signal intensity distribution from cholesterol and phospholine/phosphate of a mouse brain section at successively increased magnifications. The $9\times9\,mm^2$ and $500\times500\,\mu m^2$ images were obtained from measurements of positive secondary ions, showing the spatial intensity distribution of the $(M\text{-}OH)^+$ peak for cholesterol (369 u) and of the phosphocholine peak ($C_5H_{15}NPO_4^+$) with maximum image resolution 3 to 5 μm. The $100\times100\,\mu m^2$ images were obtained from measurements of negative ions, showing the spatial intensity distribution of the $(M\text{-}H)^-$ peak for cholesterol and the phosphate peak (PO_3^-) with maximum image resolution 0.2 to 0.3 μm. (Reprinted from P. Sjovall et al., 2004. *Anal. Chem.*, 76: 4271–4278. With permission from the American Chemical Society.) (See page 279 for text discussion.)

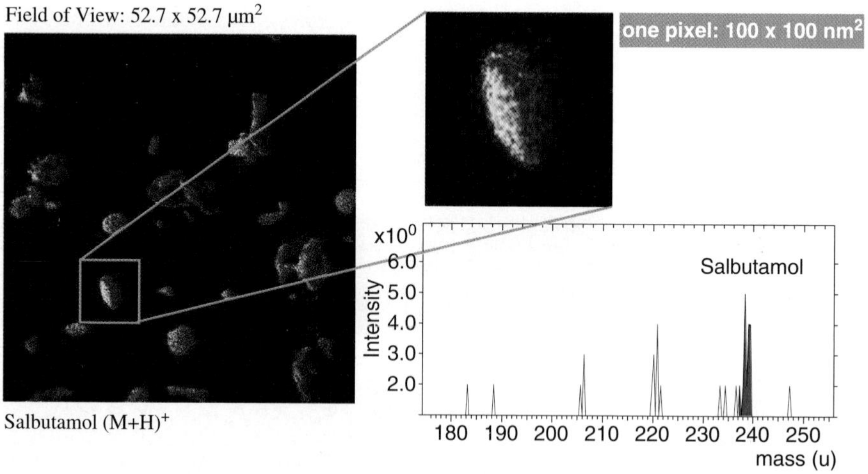

Figure 13.4. TOF-SIMS image and mass spectrum of molecular ion of salbutamol. A large area containing many beads (illustrated left) and the image of one bead enlarged (right). The pixel size in the image is $100\times100\,nm$, and the mass spectrum from the pixel selected shows significant intensity for the salbutamol molecular ion peak. Calculation shows that the amount of salbutamol in this pixel area was in the range of 2×10^{-20}mol. (Reprinted from Kollmer, F. 2004. *Appl. Surf. Sci.*, 231–232: 153–158. With permission from Elsevier.) (See page 280 for text discussion.)

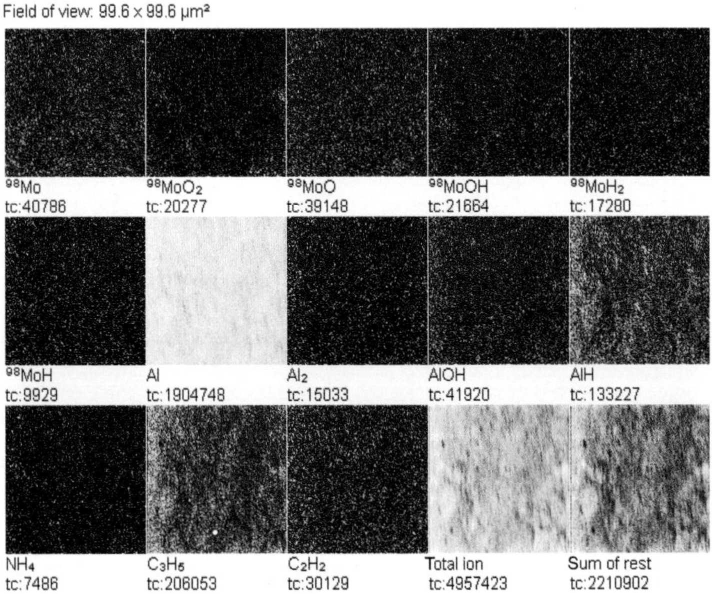

Figure 13.5. TOF-SIMS surface image of catalyst 10% Mo/Al_2O_3 calcinated at 500°C in air atmosphere (size 99.6 × 99.6 μm^2). (See page 281 for text discussion.)

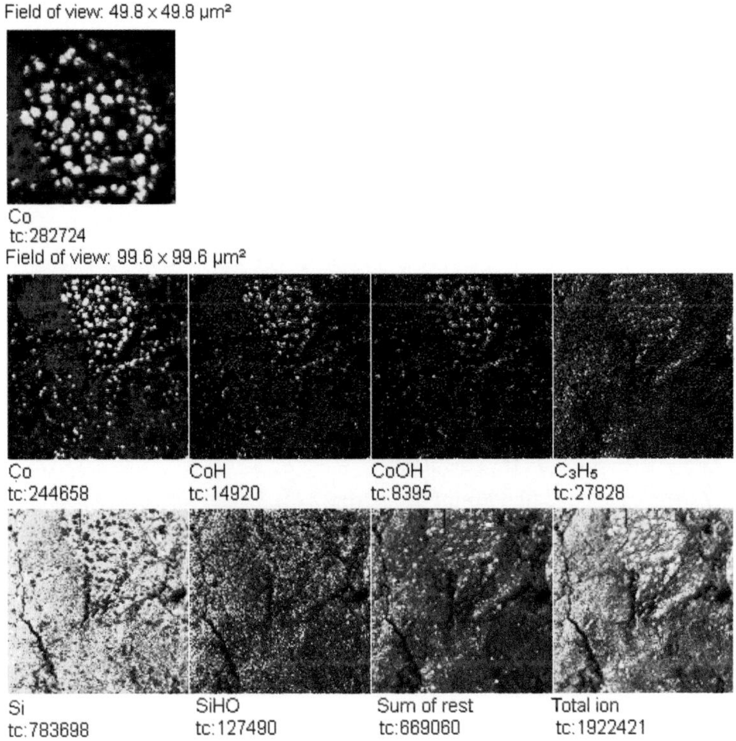

Figure 13.6. TOF-SIMS surface image of catalyst 15%Co/SiO_2 calcinated at 500°C in air atmosphere (size 99.6 × 99.6 μm^2) and Co image zoom (49.8 × 49.8 μm^2). (See page 281 for text discussion.)

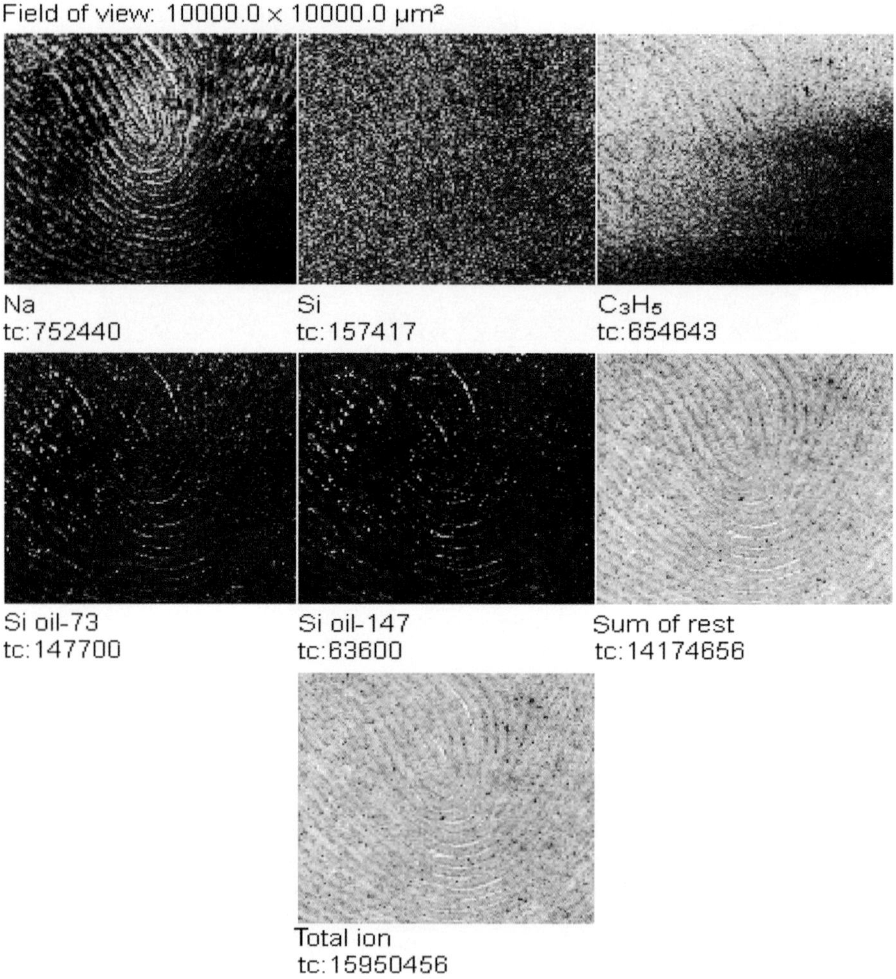

Figure 13.7. The image of a fingerprint taken from a glass sheet surface (10,000 $\times$ 10,000 μm). (Reprinted from Szynkowska, M. I. et al., 2007. Preliminary Studies Using Scanning Mass Spectrometry (TOF-SIMS) in the Visualisation and Analysis of Fingerprints, *Imaging Sci. J.*, 55: 180–187. With permission from Maney Publishing.) (See page 281 for text discussion.)

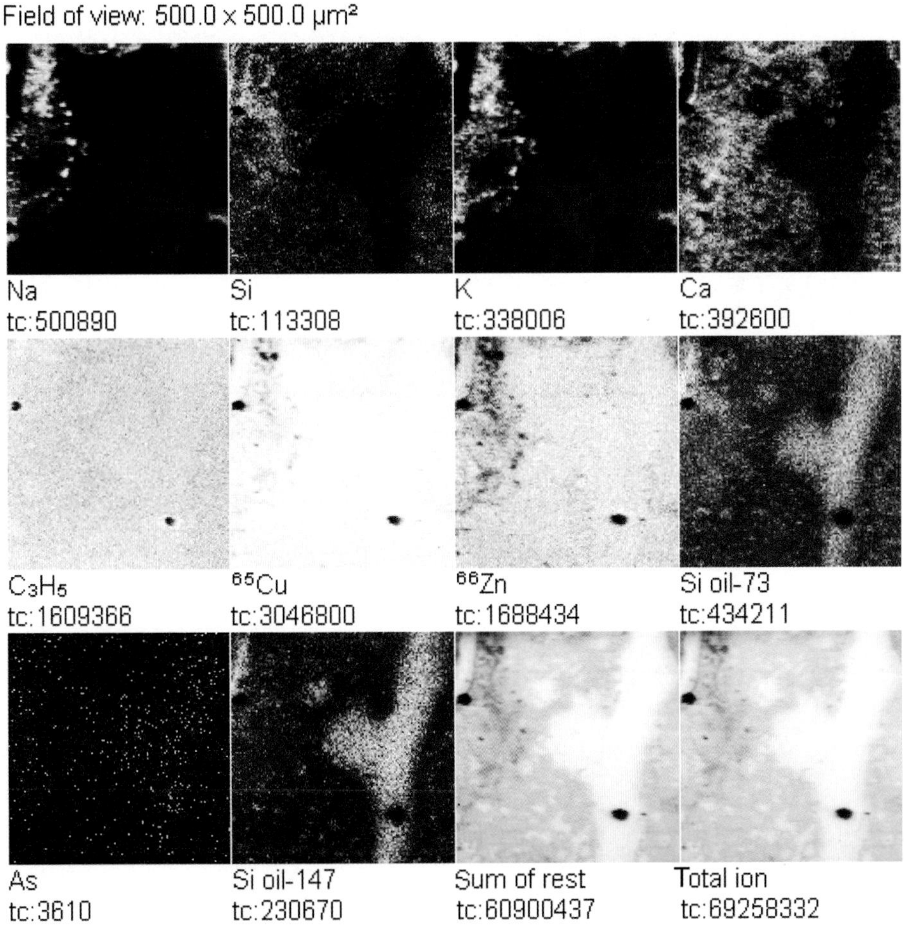

Figure 13.8. The image of a fingerprint after contact with As_2O_3 taken from a brass sheet surface (500 × 500 μm). (Reprinted from Szynkowska, M. I. et al., 2007. Preliminary Studies Using Scanning Mass Spectrometry (TOF-SIMS) in the Visualisation and Analysis of Fingerprints, *Imaging Sci. J.*, 55: 180–187. With permission from Maney Publishing.) (See pages 281–282 for text discussion.)

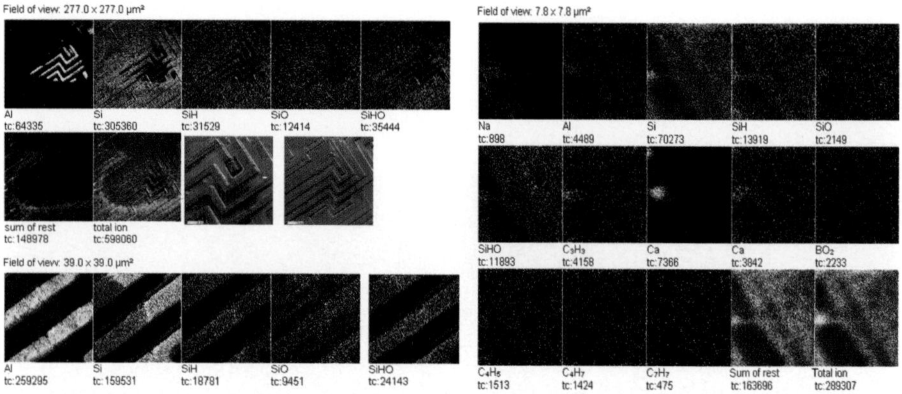

Figure 13.9. TOF-SIMS images of a silicon wafer taken from the fields 277 × 277 μm^2, 155 × 155 μm^2, 39 × 39 μm^2, and 7.8 × 7.8 μm^2. On the highest magnification image the existence and distribution of trace impurities of ions such as Na, Ca, BO_2, and C_xH_y are observed. (See page 283 for text discussion.)

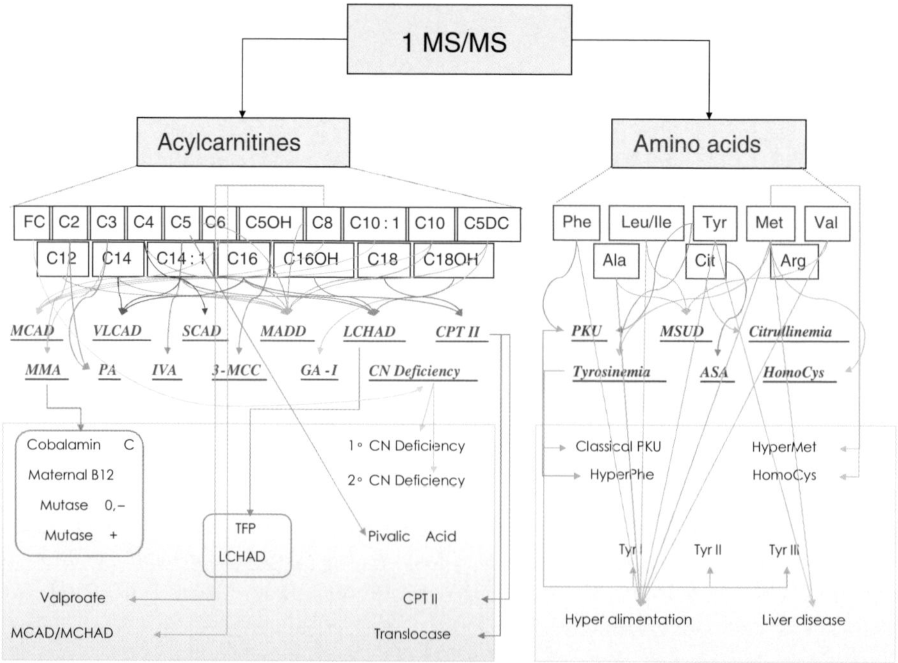

Figure 14.3. Schematic showing the relationship between the Tandem Mass Spectrometer, the metabolites it measures and the diseases or conditions detected. (See page 293 for text discussion.)

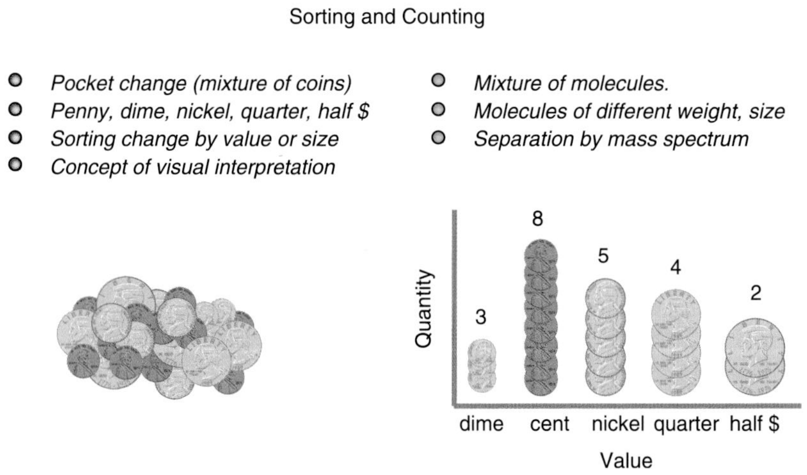

Sorting and Counting
Pocket change (mixture of coins)
Penny, dime, nickel, quarter, half $
Sorting change by value or size
Concept of visual interpretation
Mixture of molecules.
Molecules of different weight, size
Separation by mass spectrum
Quantity
3
8
5
4
2
dime
cent
nickel
quarter
half $
Value

Figure 14.4. Illustration of how a mass spectrometer can be compared to counting and sorting coins. (See pages 294–295 for text discussion.)

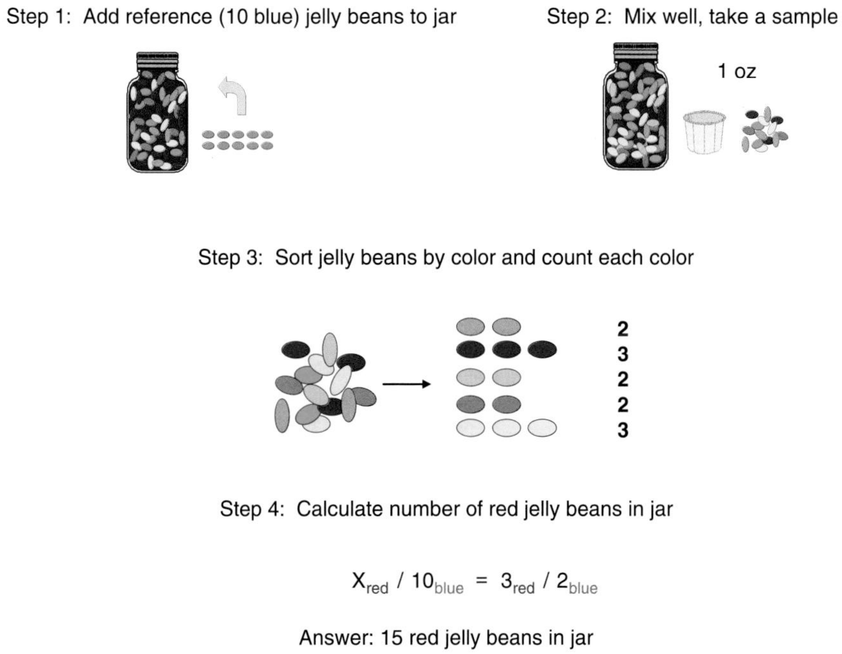

Step 1: Add reference (10 blue) jelly beans to jar
Step 2: Mix well, take a sample
1 oz
Step 3: Sort jelly beans by color and count each color
2
3
2
2
3
Step 4: Calculate number of red jelly beans in jar
$X_{red} / 10_{blue} = 3_{red} / 2_{blue}$
Answer: 15 red jelly beans in jar

Figure 14.5. An illustration of how a mass spectrometer measures compounds in blood by using reference standards (Stable Isotopes) using the example of jelly beans. (See pages 295–296 for text discussion.)

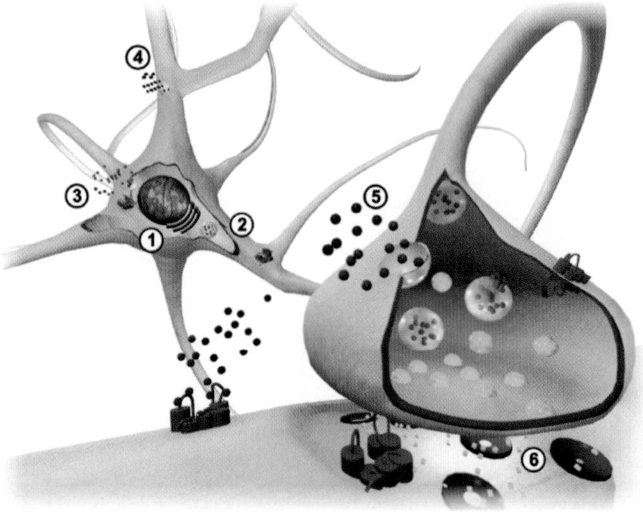

Figure 17.1. Neurotransmission (specific case of peptidergic cells). Production of the peptides in the cell body **(1)**. Packing of the peptides into large dense core vesicles **(blue)** for further transport to the axons **(2)**. Release of neuropeptides **(black)** from the cell soma **(3)** dendrites **(4)** and outside of the synapse **(5)**. Release of classic neurotransmitters **(green)** in the synaptic cleft **(6)**. G-protein-coupled type receptors **(red)**, which act as peptide receptors. (See page 322 for text discussion.)

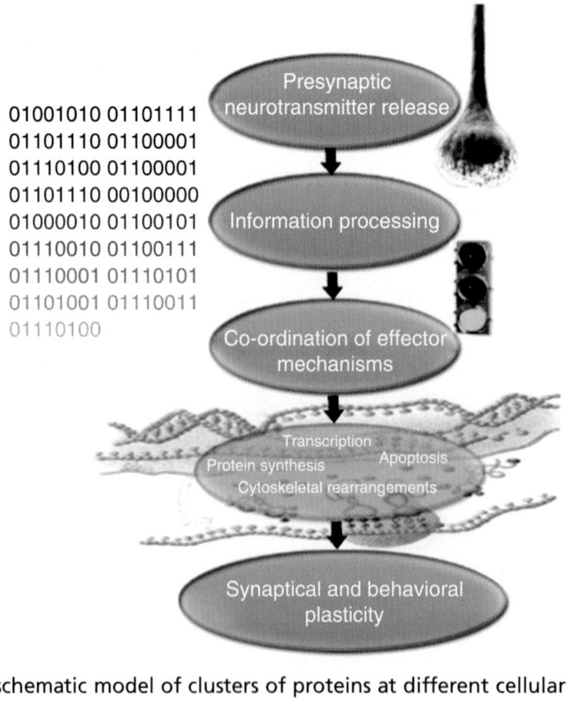

Figure 17.2. A schematic model of clusters of proteins at different cellular levels involved in learning and memory. (See page 324 for text discussion.)

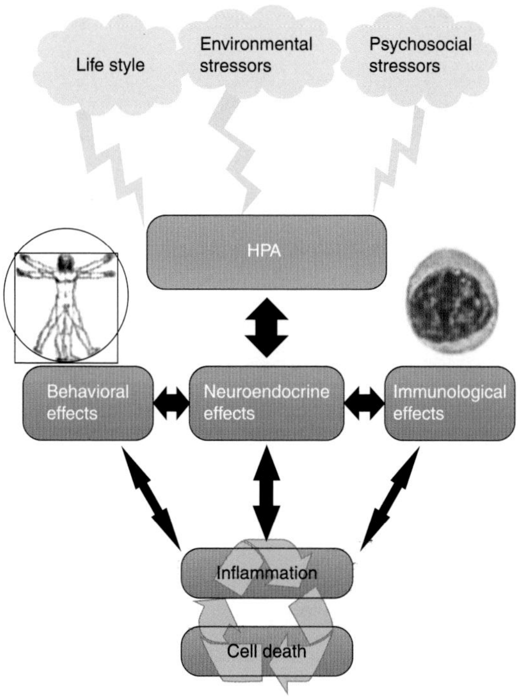

Figure 17.3. The brain in stress (HPA, hypothalamus-pituitary-adrenal axes). Schematic representation of some of the inflammatory response reactions that may even result in cell death as a response to prolonged inflammatory reactions. (See page 326 for text discussion.)

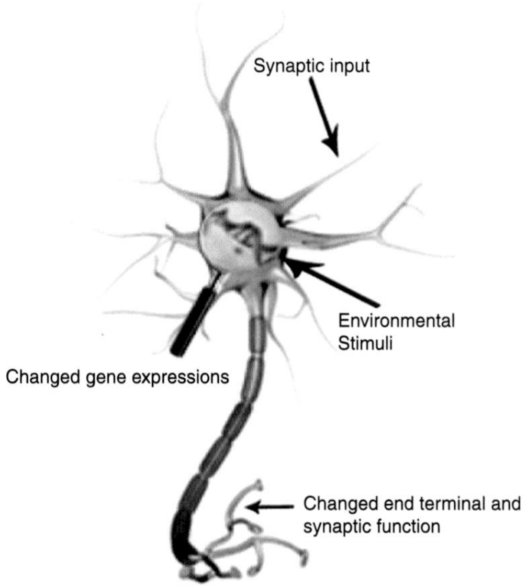

Figure 17.4. Epigenetics in the nervous system. Regulation occurs in response to synaptic inputs and/or other psychosocial-environmental stimuli. The external stimuli result in changes in the transcriptional profile of the neuron and eventually affects neural function(s). Many disorders of human cognition might involve dysfunction of epigenetic tagging. (See page 328 for text discussion.)

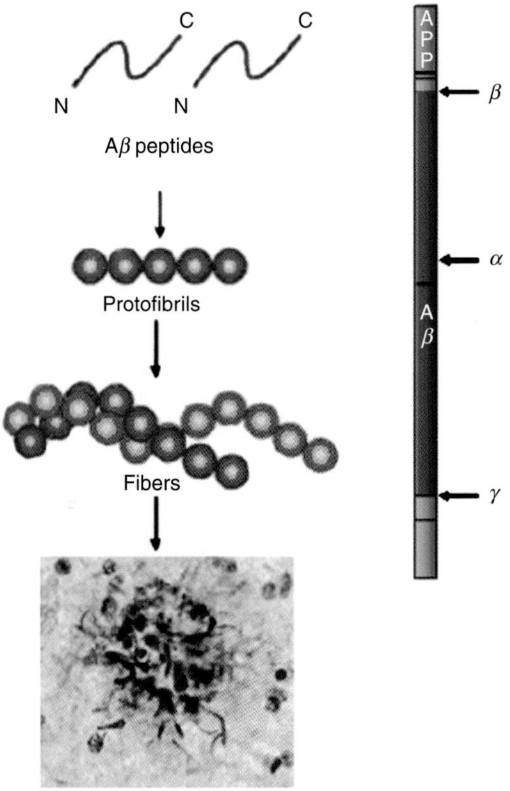

Figure 17.5. The precursor molecule APP and the three different proteases α, β, γ secretase that are involved in the processing of APP to β-amyloid peptide. The aberrant processing of the amyloid precursor protein (APP) leads to accumulation of beta-amyloid fragments, first as protofibrils and then as fibers that aggregate in the senile plaque structures. (See pages 332–333 for text discussion.)

TABLE 5.9. Relative Intensities of the Characteristic Fragment Ions in the Electron Ionization Mass Spectra of *N*-Arylsulfonylazetidin-3-ones

	Relative Abundances of the Ions (% in the Total Ion Current)					
R	M	A	B	C	D	E
H	–	0.5	0.3	0.1	75.0	1.1
CH_3	–	0.4	0.2	0.1	77.5	0.7
OCH_3	0.3	0.6	0.9	0.8	73.7	0.7
Cl	–	0.5	0.4	–	77.7	0.2
NO_2	–	0.1	0.1	0.1	65.7	0.5

5. Study the overall appearance of the spectrum.
6. Find all the peaks due to rearrangement ions.
7. Find all the homologous series of ions.
8. Find the most important primary fragment ions.
9. Try to estimate the nature of the most intense peaks in the spectrum.
10. Eliminate unacceptable choices done by computer library search.
11. Try to elaborate a possible structure of the molecule. Check whether it similar to any of the structures proposed by computer. In selecting a structure try to propose a fragmentation scheme, rationalizing all the most important peaks in the spectrum.
12. Maximum reliability may be achieved by recording a mass spectrum under the same conditions of the identified compound taken from another source or specially obtained.

REFERENCES

1. K. Biemann, *Mass Spectrometry, Organic Chemical Applications.* New York: McGraw-Hill, 1962.
2. W. Benz, *Massenspectrometrie Organischer Verbindungen.* Leipzig: Akademische Verlagsgeseschaft, 1969. R. C. Burnier, G. D. Byrd, and B. S. Freiser, *Anal. Chem.*, **52**(1980): 1641.
3. G. S. Hammond, *J. Am. Chem. Soc.*, **77**(1955): 334.
4. U. I. Zahorszky, *Org. Mass Spectrom.*, **14**(1979): 66.
5. D. P. Stevenson, *Disc. Faraday Soc.*, **10**(1951): 35.
6. H. E. Audier, *Org. Mass Spectrom.*, **2**(1969): 280.
7. A. G. Harrison, C. D. Finney, and J. A. Sherk, *Org. Mass Spectrom.*, **5**(1971): 1313.
8. J. A. Gilpin and F. W. McLafferty, *Anal. Chem.*, **29**(1957): 990.
9. A. G. Harrison, *Org. Mass Spectrom.*, **3**(1970): 549.
10. D. G. I. Kingston, J. T. Bursey, and M. M. Bursey, *Chem. Rev.*, **74**(1974): 215.
11. M. Karni and A. Mandelbaum, *Org. Mass Spectrom.*, **15**(1980): 53.

12. F. H. Field, in *Mass Spectrometry*, A. Maccoll (ed.) *MTP Int. Rev. Sci.* Butterworths, 1972, p. 133.
13. R. D. Bowen, B. J. Stapleton, and D. H. Williams, *Chem. Commun.*, **24**(1978).
14. H. Budzikievicz, C. Djerassi, and D. H. Williams, *Mass Spectrometry of Organic Compounds*. San Francisco: Holden Day, 1967.
15. M. H. Green, R. J. Cooks, J. M. Schwab, and R. B. Roy, *J. Am. Chem. Soc.*, **92**(1970): 3076.
16. J. M. Pechine, *Org. Mass Spectrom.*, **5**(1971): 705.
17. H. Suming, C. Yaozu, J. Longfrei, and X. Shuman, *Org. Mass Spectrom.*, **21**(1986): 7.
18. M. H. Green, in *Topics in Stereochemistry*, N. L. Allinger and E. L. Eliel (eds.) New York: Wiley Interscience, 1976, 35.
19. A. Mandelbaum, in *Stereochemistry*, Vol. 1, N. B. Kagan (ed.) Stuttgart: Thieme, 1977, 137.
20. A. Mandelbaum, *Mass Spectrom. Rev.*, **2**(1983): 223.
21. J. S. Splitter and F. Turecek (eds.) *Application of Mass Spectrometry to Organic Stereochemistry*. New York: VCH Publishers, 1994.
22. H. Schwarz, *Top. Curr. Chem.*, **73**(1978): 232.
23. A. T. Lebedev, T. N. Alekseeva, T. G. Kutateladze, S. S. Mochalov, Yu. S. Shabarov, and V. S. Petrosyan, *Org. Mass Spectrom.*, **24**(1989): 149.
24. A. T. Lebedev, I. V. Dianova, S. S. Mochalov, V. V. Lobodin, T. Yu. Samguina, R. A. Gazzaeva, and T. Blumental, *J. Am. Soc. Mass Spectrom.*, **12(8)**(2001): 956–963.
25. V. V. Lobodin, A. N. Fedotov, P. I. Dem'yanov, V. Ovcharenko, K. Pihlaja, T. Blumenthal, and A. T. Lebedev, *J. Am. Soc. Mass Spectrom.*, **16**(2005): 1739.
26. F. W. McLafferty and F. Turecek, *Interpretation of Mass Spectra*. Mill Valley, CA: University Science Books, 1993.
27. J. B. Lambert, H. F. Shurvell, D. A. Lightner, and R. G. Cooks, *Organic Structural Spectroscopy*. Upper Saddle River, NJ: Prentice Hall, 1998.
28. T. Clark, *J. Am. Chem. Soc.*, **109**(1987): 6838.
29. N. J. Jensen, K. B. Tomer, M. L. Gross, and P. A. Lyon, in *Desorption Mass Spectrometry*, P. A. Lyon (ed.) Washington, D.C.: American Chemical Society, 1985, 194.
30. R. L. Cerny, K. B. Tomer, and M. L. Gross, *Org. Mass Spectrom.*, **21**(1986): 655.
31. K. B. Tomer, N. J. Jensen, and M. L. Gross, *Anal. Chem.*, **58**(1986): 2429.
32. J. Adams and M. L. Gross, *J. Am. Chem. Soc.*, **111**(1989): 435.
33. J. Adams, *Mass Spectrom. Rev.*, **9**(1990): 141.
34. J. B. Lambert, H. F. Shurvell, D. A. Lightner, and R. G. Cooks, *Organic Structural Spectroscopy*. Upper Saddle River, NJ: Prentice Hall, 1998.

6

SEQUENCING OF PEPTIDES AND PROTEINS

Marek Noga, Tomasz Dylag, and Jerzy Silberring

6.1. BASIC CONCEPTS

Proteins and peptides are most often seen in the mass spectra as pseudomolecular ions, that is, molecules with attached charge-carrying protons (in the negative-ion mode, proteins and peptides lose protons and thus acquire a negative net charge). This additional proton has to be taken into consideration in order to predict correctly the m/z value at which the peptide of interest will be seen in a mass spectrum. For example, a peptide whose molecular weight (MW) (or molar mass) is equal to 2000 Da, when singly ionized, will be detected at 2001 m/z (for simplification, we assume the mass of proton as equal to 1):

$$\mathrm{MW} = 2000\,\mathrm{Da}$$
$$m/z = (2000 + 1)/1 = 2001 \quad (z = +1)$$

It may also happen that a peptide molecule will attach two protons. The m value is increased, and also the z (charge) is changed, which changes the ultimate m/z value (compare also Fig. 6.1):

$$\text{peptide mass} = 2000\ \mathrm{Da}$$
$$m/z = (2000 + 2)/2 = 1001 \quad (z = +2)$$

Mass Spectrometry. Edited by Ekman, Silberring, Westman-Brinkmalm, and Kraj

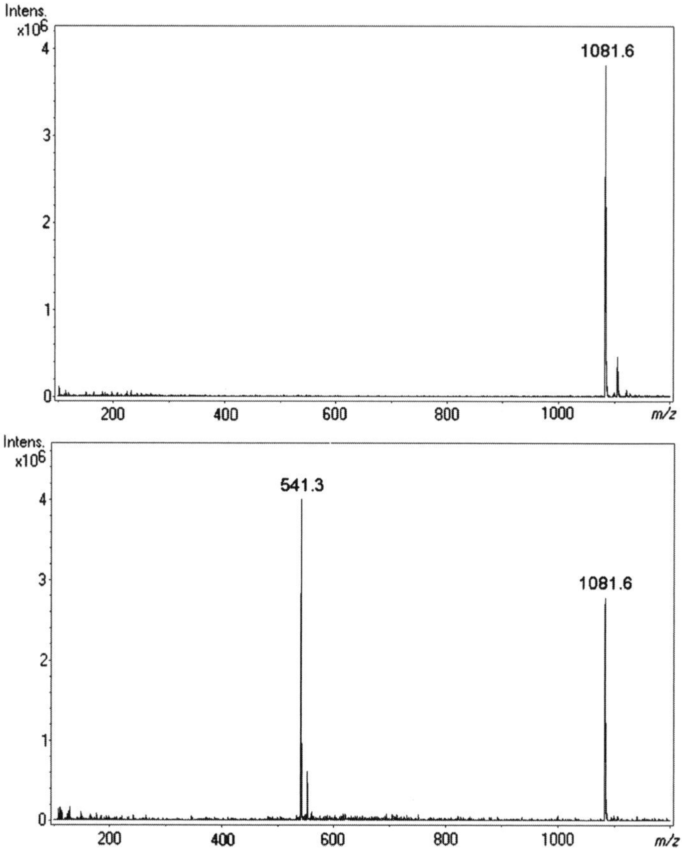

Figure 6.1. Mass spectra of synthetic peptide, FLFQPQRF-NH_2. Both spectra were recorded on an ESI ion trap mass spectrometer at different instrument settings (modifications of ion optics and ion-trap parameters).

The intensity ratios of singly, doubly, or higher-charged ions can vary for the same compound at different conditions. While sometimes only singly charged peptide ions might be seen, in other measurements only doubly or triply charged species are observed (Fig. 6.1). In general, the charge state depends on:

- The method of ionization, as electrospray ionization, ESI (see Chapter 2, Section 2.1.15) tends to promote multiple ionization, which is not as frequent in matrix-assisted laser desorption/ionization, MALDI (see Section 2.1.22) method.
- Peptide length, as longer peptides have more groups where additional protons can be attached (basic residues).
- Peptide sequence, as some amino acids (e.g., Arg or Lys) are more susceptible to ionization than others.
- Instrument settings and solvent pH and composition.

For studies of peptides and proteins, two major ionization techniques are used, MALDI and ESI (see Sections 2.1.15 and 2.1.22). MALDI is especially popular for peptide mass fingerprinting (see below) due to its high accuracy and sensitivity. It covers a broad mass range, typically from several hundred to several hundred thousand m/z and, for low molecular weight compounds, yields predominantly singly charged ions. However, direct hyphenation of MALDI with a liquid chromatography system is not so straightforward and thus, sophisticated fraction collectors have to be used. On the other hand, one of the main features of ESI is the possibility of direct coupling to LC. Its main advantages include the ease of conducting advanced fragmentation experiments, especially with triple quadrupole analyzers. Unlike MALDI, ESI often produces multiply charged species. Neither MALDI nor ESI induce strong fragmentation on their own and, therefore, enable determination of molecular weight. In proteomics, MALDI and ESI are used almost exclusively nowadays.

6.2. TANDEM MASS SPECTROMETRY OF PEPTIDES AND PROTEINS

Tandem mass spectrometry (see Chapter 3) can be applied to structural studies of various types of compounds, provided their molecular weights do not exceed approximately 2 to 4 kDa (there are also ways to analyze the sequence of larger molecules (proteins) by MS in the "top-down" strategy—see Figs. 6.2 and 6.3). In general, the larger

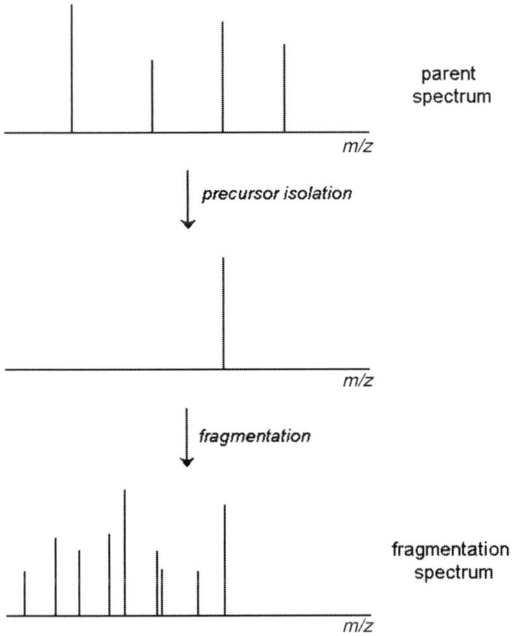

Figure 6.2. Stages of the tandem MS experiment.

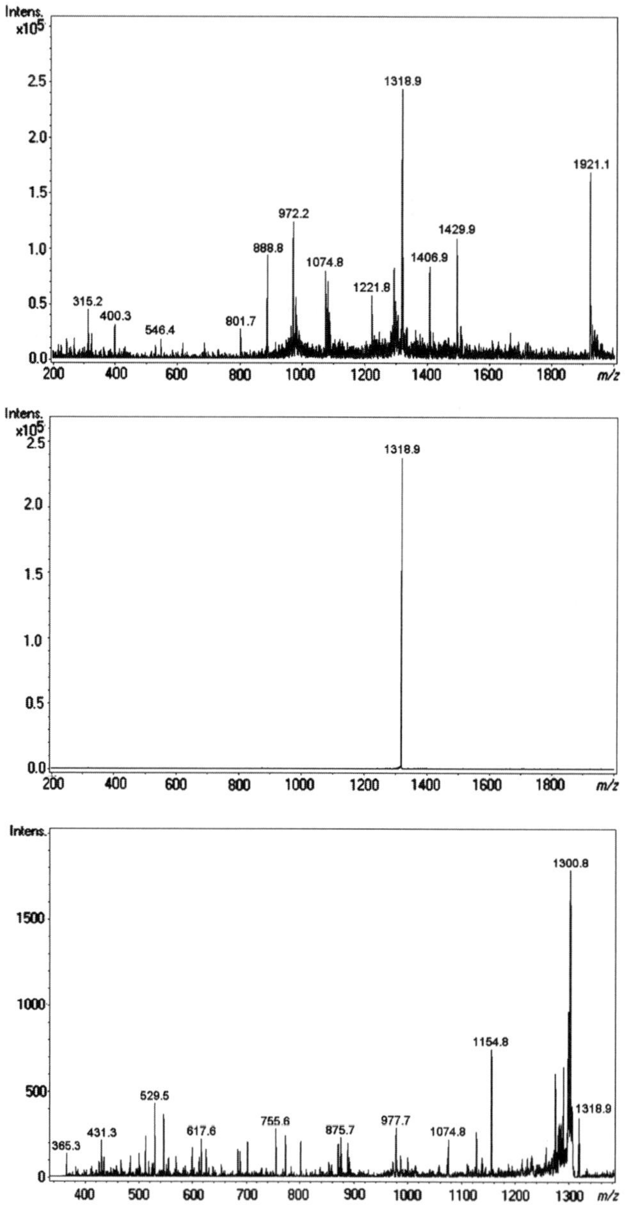

Figure 6.3. Real-life example of a tandem MS experiment in an electrospray ion trap instrument. Top panel: a complex peptide mixture. Middle panel: ion at 1318.9 m/z was isolated from other sample components. Note the lack of any other peaks and a very low background. Bottom panel: fragmentation spectrum of the selected parent ion (1318.9 m/z), note the different scale of the m/z axis. All peaks seen in this mass spectrum are product ions that were formed due to the controlled fragmentation of the parent ion. The main peak at 1300.8 m/z corresponds to the loss of water molecule, a lower intensity parent ion at 1318.9 m/z is also seen.

the compound, the more difficult are fragmentation and interpretation of the data obtained. The case of peptides is exceptional, since their molecules are relatively large but contain many bonds of similar chemical stability. Most important is that their fragmentation patterns contain information on the sequence of amino acids, which is of crucial value for their identification. In an ideal situation, MS/MS spectrum of a peptide will provide a ladder of peaks, and the distances between them reflect the sequence of amino acid residues (Fig. 6.4).

For example, in the mass spectrum shown in Fig. 6.4, the mass difference between ions at 387.2 and 500.3 m/z is equal to 113.1, which suggests that they differ by Leu or Ile. Going to the right, $572.5 - 500.3 = 72.2$, which corresponds to Ala, etc. It is often convenient to start such calculations from the losses of the precursor ion, that is, to analyze the spectrum from right to left. It should be noted here that only the mass differences between two adjacent ions of the same series are informative, for example, y_7-y_6, b_5-b_4 etc, but not y_7-b_6. This strategy will be described in detail in Section 6.6 on de novo sequencing.

6.3. PEPTIDE FRAGMENTATION NOMENCLATURE

Peptide sequencing by MS is not easy because various covalent bonds can get broken during fragmentation. The breaking bonds can encompass both backbone linkages and side chain groups. In order to precisely describe these events, special nomenclature is used. Unless stated otherwise, all information in this chapter relates to positive ions, because they are typically analyzed during peptide sequencing.

6.3.1. Roepstorff's Nomenclature

A simple convention for description of peptide fragments formed in the mass spectrometer was proposed by Roepstorff and Fohlman in 1984 [1] and further modified by Johnson in 1987 [2]. Fragment ions are described by single lower-case letters with additional indexes (Fig. 6.5). If we consider fragmentation of a peptide backbone only, six ion series can be formed due to fragmentation at:

—CHR-|-CO—NH— for a and x type ions
—CHR—CO-|-NH— for b and y type ions
—CO—NH-|-CHR— for c or z type ions

Ions derived from the N-terminus of the original peptide are termed a, b, or c (in other words, the charge is retained on the N-terminus), while those originating from the C-terminus are named x, y, or z. Numerical subscripts contain information on the number of amino acid residues present in a given ion. It must be remembered that we are interested only in charged species since the neutral ones are not detected and do not yield peaks in the spectrum. The main idea behind this system is shown in Fig. 6.5, and the structures of selected ion types are depicted in Fig. 6.6.

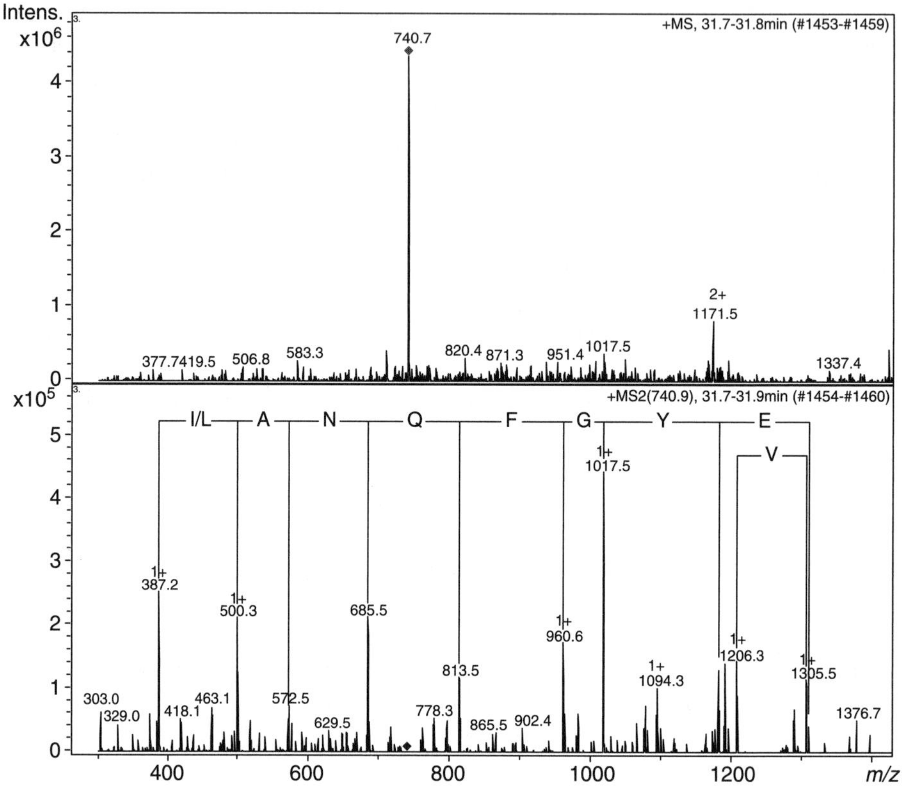

Figure 6.4. Fragmentation spectrum of a tryptic peptide obtained from bovine serum albumin. Peptide sequence: LGEYGFQNALIVR, monoisotopic $[M+H]^+ = 1479.796$, monoisotopic $[M+2H]^{2+} = 740.402$. Upper panel: full scan MS spectrum. Lower panel: MS/MS spectrum of a doubly-charged ion at 740.7 m/z with a ladder of y ions, the distances between which correspond to amino acid residues (upper row of letters). A shorter series of b ions is also seen (lower row of letters). See Fig. 6.5 for description of nomenclature. Note the often observed phenomenon where multiply-charged ions lose the charge during fragmentation process and, therefore, have higher m/z values than the original parent ion.

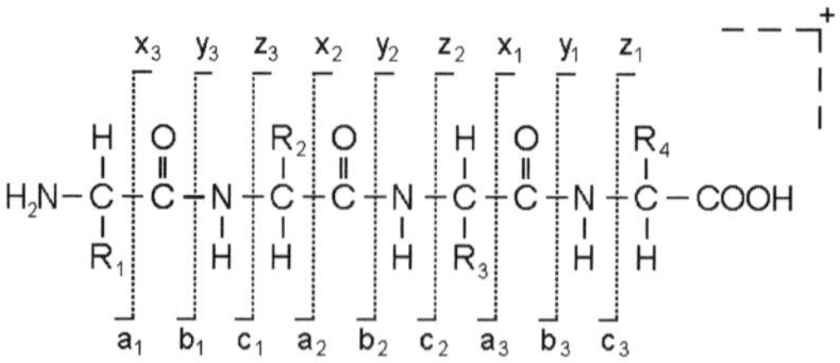

Figure 6.5. Nomenclature of peptide ions resulting from backbone fragmentation.

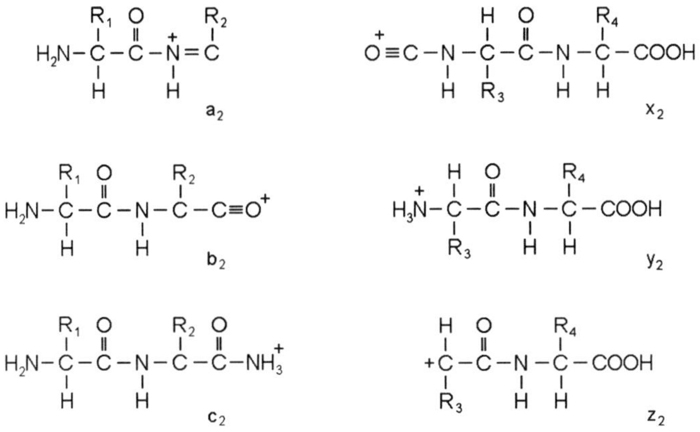

Figure 6.6. Chemical structures of selected product ions. Note they already carry a single charge.

6.3.2. Biemann's Nomenclature

Biemann further improved the nomenclature scheme by description of immonium and satellite ions [3].

Immonium ions are fragment ions of the single amino acid residues (Fig. 6.7). They result from breakage of the peptide backbone on both sides of the same residue and have important diagnostic information thus helping to confirm the presence (but not location in a certain position in the peptide chain) of several amino acids. For example, a peak at m/z 120 is characteristic for phenylalanine, while ion at m/z 70 is typically observed for proline-containing peptides (Table 6.1). Only the pairs of Leu with Ile and Lys and Gln do not yield discriminating immonium ions as they have identical molecular weights. Immonium ions are abbreviated with single capital letters, with respect to the original amino acid. Although the presence of an immonium ion usually suggests the presence of a given amino acid in the analyzed peptide, the lack of a certain immonium ion does not exclude anything.

At higher energies, fragmentation of the peptide molecule can also occur outside its backbone, that is, in the side chains. Ions formed in this way are termed *satellite ions* (Fig. 6.8).

In some cases they may serve to differentiate between selected amino acids. In particular, they allow one to distinguish between leucine and isoleucine, which have identical molecular weights but yield different satellite ions (Fig. 6.9).

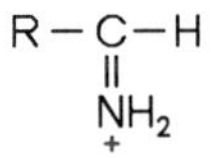

Figure 6.7. A general chemical structure of immonium ions.

TABLE 6.1. List of the Most Important Immonium Ions

Amino Acid	Immonium Ion (*m*/*z*)	Other Ions (*m*/*z*)
Ala A	44	
Arg R	*129*	59, 70, *73*, 87, 100, 112
Asn N	*87*	*70*
Cys C	*76*	
Gln Q	101	56, 84, 129
Gly G	30	
His H	**110**	*82, 121, 123, 138*, 166
Ile I	**86**	44, *72*
Asp D	*88*	*70*
Glu E	*102*	
Leu L	**86**	44, *72*
Lys K	*101*	*70*, 84, 112, 129
Met M	*104*	61
Phe F	**120**	*91*
Pro P	**70**	
Ser S	60	
Thr T	74	
Trp W	**159**	77, 117, **130**, 132, **170**, **171**
Tyr Y	**136**	*91, 107*
Val V	72	41, 55, 69

Note: Numbers in bold represent the most abundant peaks, those in italics peaks of lower intensity.

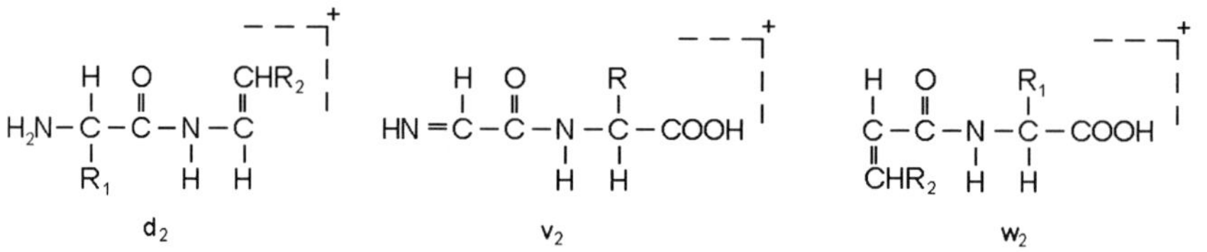

Figure 6.8. Satellite ions.

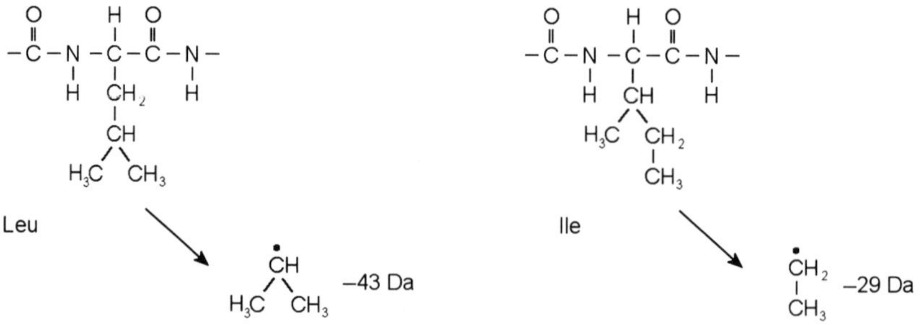

Figure 6.9. The analysis of satellite ions may help to differentiate between leucine and isoleucine, which have identical molecular weights and cannot be distinguished in most MS/MS experiments. In high energy CID, Leu loses a 43 Da radical, while Ile loses a 29 Da fragment.

TABLE 6.2. Data for Calculation of Molecular Weights of Various Types of Fragment Ions

Ion Type	Molecular Weight of Neutral Species
a	[N]+[M]-CHO
a*	a-NH_3
a°	a-H_2O
b	[N]+[M]-H
b*	b-NH_3
b°	b-H_2O
c	[N]+[M]+NH_2
d	a—partial side chain
v	y—complete side chain
w	z—partial side chain
x	[C]+[M]+CO-H
y	[C]+[M]+H
y*	y-NH_3
y°	y-H_2O
z	[C]+[M]-NH_2

Source: Taken from http://www.matrixscience.com/help/fragmentation_help.html, with permission.
Note: N is the molecular weight of the neutral N-terminal group, C is the molecular weight of the neutral C-terminal group, and M is the molecular weight of the neutral amino acid residues. In order to obtain m/z values for positive ions, add the appropriate number of protons to obtain the required charge and divide by the number of charges.

It may happen that an ion formed in the analyzer loses a water molecule, formed by the OH^- anion from the carboxy terminus and the H^+ ion from the amino terminus. Such ions, deprived of water, are denoted with the ° symbol, for example, a°. If an ion loses the $-NH_2$ group, it is described with *, for example a*. Formation of multiply charged fragments is also possible, especially if the precursor ion was originally endowed with a multiple charge. If the charge is different from +1, it is shown in the top right-hand side of the symbol, for example, $b_2{}^{2+}$.

It is useful to be able to predict m/z values of different ions that may be formed from the peptide of interest. They can be easily calculated based on Table 6.2.

6.3.3. Cyclic Peptides

Cyclic peptides are analyzed less frequently then the linear ones but their sequencing and nomenclature are much more complicated. There is no generally accepted nomenclature scheme but we describe here the one proposed by Ngoka and Gross [4]. It takes account of Roepstorff-Fohlmann's and Biemann's conventions but introduces amendments to address cyclic peptide fragmentation.

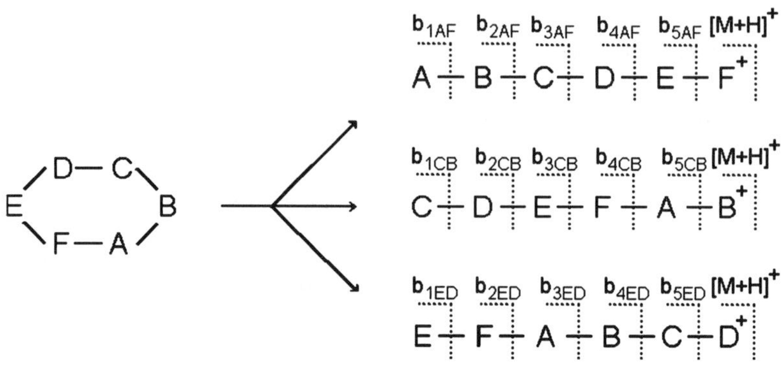

Figure 6.10. Nomenclature of product ions of cyclic peptides.

In most cases, protonation of a cyclic peptide results in ring opening. The newly formed ions have their apparent N- and C-termini if the peptide bond was cleaved, and they still can undergo further fragmentation. As a result, various linear fragment ions can be formed from the cyclic precursor. They are described by a letter indicating the ion type (e.g., a or b), and subscripts carrying information on the number of residues present in the fragment ion and peptide sequence at which ring opening occurred. For example, the 3XZ subscript informs that the newly formed fragment ion contains three amino acid residues, and that the fragment was formed as a result of protonation at the Z amino acid and ring opening at the X-Z peptide bond (Fig. 6.10).

It is worth remembering that the same cyclic peptide can give rise to various linear fragment ions of the same molecular weights, if the bonds in different parts of the backbone were opened. Linear fragment ions can undergo further rearrangements, for example, after a b series ion loses the CO group, it becomes a corresponding a series ion ($b_{3AF} \rightarrow a_{3AF}$). Linear ions formed from cyclic peptides can also lose a water or ammonium molecule, as happens with typical linear peptides subjected to fragmentation.

6.4. TECHNICAL ASPECTS AND FRAGMENTATION RULES

If we take a look at the technical aspect of fragmentation, various types of spectrometers can be used that can be grouped according to the type of analyzer(s). Depending on this criterion, either "tandem in space" or "tandem in time" experiments can be conducted (see Chapter 3, Section 3.1). From the point of view of fragment ion formation, there are also different types of fragmentations (see Section 3.2). The list of peptide fragment ions typically formed in different types of instruments is shown in Table 6.3. It should be noted that, depending on the spectrometer type, MS/MS spectra of the same substance may provide different levels of information. It is not always convenient if too many ion series are present since this can render the spectrum too complicated or even obscure for analysis. On the other hand, the wealth of

TABLE 6.3. Types of Ions Formed in Different Types of Instruments

	ESI Q-TOF	MALDI TOF PSD	ESI IT	ESI Q	ESI FTICR	MALDI TOF TOF	ESI 4 sector	FTMS ECD	MALDI Q-TOF	MALDI Q-IT-TOF
1^+ fragments	+	+	+	+	+	+	+	+	+	+
2^+ fragments if precursor 2^+ or higher	+		+	+	+		+	+	+	
Immonium ions		+				+	+		+	+
a series ions		+				+	+			+
a-NH_3 if fragment includes RKNQ		+				+				+
a-H_2O if fragment includes STED		+				+				+
b series ions	+	+	+	+	+	+	+		+	+
b-NH_3 if fragment includes RKNQ	+	+	+	+	+	+	+		+	+
b-H_2O if fragment includes STED	+	+	+	+	+	+	+		+	+
c series ions								+		
x series ions										
y series ions	+	+	+	+	+	+	+	+	+	+
y-NH_3 if fragment includes RKNQ	+		+	+	+	+			+	+
y-H_2O if fragment includes STED	+		+	+	+	+			+	+
z series ions							+	+		
internal yb <700 Da						+	+		+	+
internal ya <700 Da						+	+		+	+
z+H series ions								+		
d or d' series ions						+				
v series ions						+				
w or w' series ions						+				

Source: From http://www.matrixscience.com/help/search_field_help.html, with permission.
Abbreviations: Q—quadrupole, IT—ion trap, ECD—electron capture dissociation.

TABLE 6.4. Amino Acids or their Combinations of Similar Molecular Weights, Difficult to Distinguish in Low Resolution Spectra

Amino Acid		Monoisotopic Mass
Asp	D	115.02694
Asn	N	114.04293
Gly-Gly	GG	114.04292
Glu	E	129.04259
Gln	Q	128.05858
Lys	K	128.09496
Gly-Ala	GA	128.05857
Val-Val	VV	198.13682
Pro-Thr	PT	198.10044

information can be useful for unambiguous confirmation of a compound identity in comparison with reference spectra.

Not every fragment of a given peptide will be seen in a fragment mass spectrum. This spectrum will contain only peaks corresponding to ions, that is, charged molecules (some fragment ions can carry a double charge and they can also be subjected to fragmentation). Fragments with zero net charge are not detected. Moreover, the fragmentation pattern is dependent on several factors, including peptide sequence, charge state, the type of spectrometer, and the fragmentation energy. For most effective elucidation of the peptide sequence, the series of fragment ions should be complete and the mass differences between ions should be accurate.

High accuracy of the measurements is of utmost importance. There are 20 natural amino acids that, except for the Leu/Ile pair, have different molecular weights. However, the some of them or of some of their combinations have very similar molecular weights. The case of combinations of amino acids with similar molecular weights is especially important when the spectrum lacks a full string of ion peaks (Table 6.4).

6.5. WHY PEPTIDE SEQUENCING?

Until a decade ago, the main tool for protein sequencing was chemical Edman degradation. This technique is powerful and robust but it is also time consuming and fails if a chemically modified protein is to be analyzed. Mass spectrometry, on the other hand, enables sensitive, automated, high-throughput analysis, therefore it is the method of choice for protein sequencing in proteomics where a large number of samples have to be dealt with. On the other hand, MS-based peptide sequencing may give errors for amino acids of identical or similar molecular weights, such as leucine and isoleucine or lysine and glutamine. For this reason, mass spectrometry does not allow for unambiguous sequencing of just any protein. Nevertheless, as will be shown later, MS can efficiently be used for protein identification based on sequence correlation with database records. Although the most straightforward application of mass spectrometers is determination of molecular weight, such knowledge is not sufficient for identification because a huge number of proteins of similar molecular weights exist in nature and their possible posttranslational modifications make such analysis uncertain. Fortunately, there are ways to alleviate this problem, thanks to the improvements in peptide analysis and peptide sequencing technology. The techniques for protein identification by mass spectrometry are based either on measured peptide masses or peptide fragmentation patterns. But why is peptide analysis of great interest for proteomics that predominantly deals with proteins? It is just because proteins can efficiently be analyzed (identified), provided they are first subjected to proteolytic cleavage to generate shorter peptides. Such an approach, where the protein of interest is digested prior to MS analysis, is termed the "bottom-up" strategy.

Peptide mass fingerprinting (PMF) is a basic identification strategy, in which the protein of interest is digested by a proteolytic enzyme, for example, by trypsin, to obtain a so-called peptide map (i.e., a mass spectrum of the digested fragments), which is unique for each protein. If trypsin is used for proteolysis, the generated fragments are called tryptic peptides. Based on a simple mass spectrum (not a MS/MS experiment) of this peptide mixture, a list of molecular weights of the peptide map components is generated. The masses are then searched against a database for protein identification. For reliable results, the sequence coverage (i.e., the ratio: number of peptides matching a protein, relative to the full protein sequence) should reach approximately 20 to 30%. For PMF, the MALDI mass spectrometers are usually used due to their accuracy, sensitivity, and ability to simultaneously detect a larger number of components along the mass spectrum. The PMF method fails if several proteins are present in one sample or if the analyzed protein is not deposited in databases.

More advanced approaches rely on recording the fragmentation spectra of peptides obtained after enzymatic digestion of a protein. This is due to the fact that only peptides of up to about 2 to 3 kDa can be directly subjected to fragmentation. To analyze a larger protein, it needs to be cleaved by a proteolytic enzyme to generate shorter fragments. In the peptide sequence tag approach, a selected peptide from the peptide map is fragmented to reveal its sequence of amino acids (i.e., sequence tag). This information, accompanied by additional data (e.g., peptide molecular mass), is used for database search and protein identification. In another approach, known as uninterpreted MS/MS search, the database search is based on a complete list of fragment ion masses.

It is worth noting that although various proteinases can be used for protein digestion, trypsin is most commonly used in proteomics due to its activity, selectivity, and stability. It cleaves peptide bonds at the C-terminal side of Arg and Lys residues, provided that the adjacent C-terminal amino acid is not Pro. This substrate specificity is well suited for mass spectrometry as all generated peptides contain a basic amino acid at the C-terminal that readily incorporates a proton, which improves ionization yield. Additionally, trypsin generates peptides of a length that is suitable for fragmentation, neither too long nor too short.

For identification purposes, the higher the number of peptides and the longer the sequence of known fragments, the higher the reliability of protein identification.

For a protein of about 20 kDa, identification can be based on as few as three to five peptide fragments. If a bigger protein is to be identified, even 10 to 50 peptides should be (partly) sequenced and used for database search input. Unfortunately, peptide fragmentation is a complex process and complete peptide sequencing is not always possible. What is more, the assignment of a protein to a peptide sequence may fail if the protein is not deposited in a database or if a database contains mistakes. It should be clearly stated that such identification, based on the sequencing of several fragments only, provides information on the presence of a given protein but does not allow for its full (complete) characterization, including, for example, mutations or posttranslational modifications.

In recent years, a novel approach to protein identification emerged, called top-down sequencing. Here the entire nondigested protein is analyzed. Apart from accurate MW measurement, the protein ion is fragmented by the electron capture dissociation (ECD) method (see Chapter 3). This provides in-depth information on the sequence of protein. Such analysis can be performed only with FTICR instruments (see Section 2.2.6) that ensure high resolution and accuracy but, at the same time, they are exceptionally expensive. However, as very large ions are analyzed, even the high accuracy of FTICR is sometimes not sufficient, and it is recommended that such analyses are accompanied by more traditional bottom-up approaches.

Another method for performing the top-down sequencing of an intact protein is application of the in-source dissociation with the MALDI ion source (MALDI-ISD). In this approach, the use of increased laser energy in the MALDI source causes fragmentation of the protein molecule. The resulting spectrum contains fragmentions with mass differences corresponding to the amino acid sequence, thus allowing for direct sequence readout. Unfortunately, this method has some important drawbacks:

- It requires very high concentration of protein in the sample to be effective
- The protein must be pure (i.e., not in a mixture with other components), as the fragmentation process takes place in the ion source and it is not possible to perform ion selection prior to fragmentation.

6.6. DE NOVO SEQUENCING

De novo sequencing is a difficult and time-consuming procedure for determination of the amino acid sequence of a peptide without relying on the information gained from other sources. Due to the existence of isobaric residues (of the same molecular weights) and possible groups of amino acids with very similar masses (discussed earlier in this chapter), this procedure seldom allows for a complete and unambiguous structure elucidation. However, thanks to excellent sensitivity of mass spectrometry, this technique can provide vital information that is often beyond the capabilities of conventional sequencing methods.

Most problems in de novo sequencing are caused by the high number of possible bonds that may be cleaved and because not all recorded fragment ions are sequence-specific. In almost every MS/MS spectrum there is a high number of ions that themselves do not provide information about backbone sequence. Such ions include loss of water and ammonia from amino acid side chains.

6.6.1. Data Acquisition

It is often neglected that the first step of de novo sequencing is data acquisition. The quality of the spectrum or spectra used for sequencing is the most critical parameter of the entire procedure. First of all, the mass spectrometer should be well calibrated and tuned. If it can operate in different modes, the one with the highest possible mass accuracy and resolution should be applied. If the experimenter has more spectrometers to choose from, the one with the highest mass accuracy and resolution should be used, provided it shows good fragmentation efficiency.

In the initial attempt at sequencing, collision energy should be kept as low as possible, thus providing complete fragmentation of the precursor ion. Typically, the lower the energy, the less fragmentation is observed, and thus the MS/MS spectra contain fewer peaks and are simpler to annotate. Later, if the information extracted from the spectra is not sufficient, the experiment can be repeated with higher fragmentation energy. For example, in the initial sequence readout, it is not crucial to differentiate Leu and Ile by high energy side-chain breakdown. On the contrary, ions resulting from such cleavages will just make the initial steps much more troublesome. Modern instruments often contain an option called ramping. This feature provides automatic adjustment of the collision energy by the software and is very useful during online LC MS/MS experiments.

If sequencing is performed with electrospray ionization, often several multiple-charged ions will be available for fragmentation. Which one to use? The best answer is "all." Typically, the fragmentation pattern changes with the charge state, the number of amino acids, and their chemical properties. Fragment ion spectra resulting from fragmentation of singly and doubly charged precursors will often provide different but complementary fragmentation. Even if the analysis of one MS/MS spectrum does not reveal the whole sequence, there is still a chance that the fragment ion spectrum from different precursor charge state(s) will provide the missing information. Good practice suggests starting the analysis with a spectrum showing more high intensity peaks, preferably equally abundant over the whole mass range.

If the instrument allows for multistage fragmentation, the MS^3 spectra of the most intense MS/MS fragment ions should also be acquired. There is never too much data in a de novo sequencing experiment. Even if the MS^2 spectrum provides the complete sequence, data obtained from additional stages can still be used for sequence validation. In case of three-dimensional ion-traps that suffer from low mass cut-off, MS^n spectra are indispensable to cover low mass fragments. For further explanation of this phenomenon, please refer to Chapter 3.

It should be kept in mind that, most probably, we will not be able to annotate all peaks on the spectrum. Fragmentation patterns of peptides are not fully understood yet. Not all peptides will break down into fragments described here, as many additional cleavages and rearrangements are possible.

6.6.2. Sequencing Procedure Examples

The main principles of de novo sequencing will be discussed here using a few d spectra that were acquired using an ion trap mass spectrometer equipped with an electrospray ion source, under low energy collision induced dissociation (CID).

Ions present on the fragment ion spectrum can either be sequence specific (they can provide information about the sequence) or non-seqence-specific (they cannot provide direct information on the sequence). For example, all ions belonging to six basic ion series (a, b, c, x, y, z) are sequence specific, whereas immonium ions are not (but they still are valuable diagnostic ions).

Generally speaking, the sequencing procedure leads to identification of sequence-specific ions and assigning them to the correct ion series. Mass differences between sequence-specific ions will correspond to masses of amino acid biradicals (also called amino acid residue, amino acid molecule without hydrogen from the α-amino group and hydroxyl group from α-carboxyl group), resulting in a sequence string. Assignment of peaks to the correct ion series allows determination of whether such a string is a C- or N-terminal string. An effective approach to de novo sequencing solves both problems simultaneously and is also applied for double-checking of the sequence revealed from both sides.

In the examples below, we will use the terms N- or C-terminal fragments to mean fragments that contain the N- or C-termini, respectively.

Example 1. Let us start with a very simple example: fragmentation of a short peptide using a singly charged precursor. The precursor mass is 574.3. The unannotated fragment ion spectrum is shown in Fig. 6.11. This spectrum contains a small number of intense peaks equally distributed over the m/z range from around 200 Th up to the m/z of the precursor. A number of low intensity peaks emerge from the background. This is a good example of a rich fragment ion spectrum with good fragmentation efficiency.

A good way to start sequencing is to begin near the precursor mass. Mass differences between precursor and first ions in C- and N-terminal series are specific to the ion series. Using this approach, we may be able to identify first N- and C-terminal amino acids, respectively.

Fragmentation near the C-terminus typically starts with the loss of water (−18 Da) to produce first b-ion, b_n for n-AA long peptide. In our example spectrum, there is indeed a peak with m/z 556.1 Th, corresponding to an 18 Da loss from the precursor. Although this peak is not sequence specific itself, it is a specific starting point for the entire b-ion series. Starting with this value, we should be able to find other peaks belonging to the b series, just looking for mass differences equal to amino acid biradicals. In our spectrum (Fig. 6.11), we immediately notice that the peak at 425.1 is shifted by 131.1 Da from the peak at m/z 556.1, which corresponds to methionine (M). Next, we find that peaks at 278.0 Th and 221.0 Th fit for mass shifts of 147.1 (phenylalanine, F) and 57.0 (glycine, G). We cannot pursue this process any further as there are no lower mass fragments visible. This effect is caused by the low-mass cut-off of the ion trap analyzer applied during spectra acquisition. However, thanks to multistage fragmentation offered by the ion trap, we can "extend" the effective mass range for fragmentation

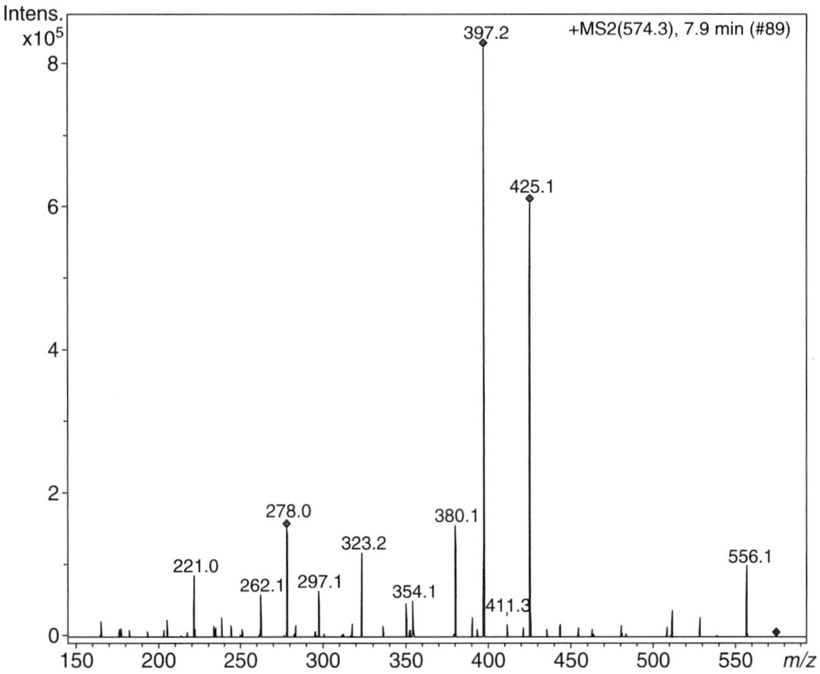

Figure 6.11. Product ion spectrum used in Example 1. The spectrum was obtained on the quadrupole ion trap mass spectrometer with ESI source and low energy fragmentation. Precursor mass is 574.3 Da.

by acquiring the MS^3 spectrum by fragmenting one of the lower mass peaks selected from the MS^2 spectrum.

B-ions are often accompanied by a-ions resulting from different backbone cleavage. The mass of an a-ion is always lower by 28 Da from that of the corresponding b-ion (loss of CO group). Pairs of a- and b-ions can be used as a high confidence marker of correct sequence assignment, especially if they cover multiple adjacent AA residues. A quick look at our example spectrum shows that the peak at 397.2 Th is an a-ion accompanying 425.1 Th b-ion. This observation makes us more confident about previous assignments of b-ions. Unfortunately, there are no more a-ions visible. This is not very surprising as a-ions are typically less abundant than b-ions.

After a short survey, we managed to identify a partial sequence of the analyzed peptide GFM-OH. The total mass of the identified fragments is 353.1 Da [57.02 (G) + 147.06 (F) + 131.04 (M) + 18.01 (H_2O)], leaving 221.3 Da for remaining amino acids and suggesting that there are only 2 AA missing.

As we cannot obtain any further information from the N-terminal fragments, we should try to supplement our sequence coverage with the information obtained from C-terminal fragments. First the N-terminal amino acid can be identified through a mass difference between the mass of the precursor ion and the heaviest of the y-ion series. The first fragment of y-series, y_{n-1}, has a mass difference of the residue biradical mass relative to the full precursor mass. If a peak with such a mass shift is found, more peaks from the y-series can be found by searching for mass differences equal to the masses of AA biradicals. Unlike the N-terminal series, in low energy fragmentation, y-ions are very rarely (if ever) accompanied by x- or z-ions. The reason for this is that a-ions are believed to emerge from fragmentation of b-ions rather than from direct backbone cleavage between α-carbon and carbonyl carbon. This would result in formation of a- and x-ions.

In our example spectrum (Fig. 6.11), the majority of most intense peaks belong to N-terminal fragments, leaving only low intensity peaks for potential C-terminal ions. In such cases, care should be taken on peak annotation as such low intensity peaks can also result from detector noise. However, they can be used for sequencing as long we can see at least a two to three AA sequence string, and the information obtained this way is consistent with other results.

From all unannotated peaks in a high-mass region of the spectrum shown in Fig. 6.11, only one—at 411.3 Th—fits to AA-specific mass shift, 163.3 Da, that is, Tyr. Starting from that peak we can find two more specific signals, corresponding to two glycine residues (peaks at 354.1 and 297.1), suggesting N-terminal sequence of NH_2-YGG. The second glycine identified here was already identifed in the C-terminal sequence string obtained previously. This way we can assume that the peptide was successfully sequenced.

The last task to be done is verification of the revealed sequence. The first, most obvious, task is to check that the molecular weight of the obtained sequence fits the mass of the precursor ion. In our case, it does. The second task that can be done using an ion trap instrument is acquisition of the MS^3 mass spectrum of identified fragments to confirm their identities. For this purpose, it will be sufficient to check fragmentation of one of the lower mass b-ions to see if we will be able to confirm the sequence string obtained from the low intensity y-ions. In the MS^3 spectrum of 278.2 Th (Fig. 6.12) we can see the previously observed b_2 ion at 221.0 Th, but also few a-ions. Ion a_2 is observed at 193.0 Th and an ion a_1 at 136.1 has a mass shift of 56.9 Th, corresponding to glycine. These results very strongly support our previous sequence assignment procedure with annotation of all identified peaks and mass differences. Fig. 6.13b shows an MS^3 spectrum used for sequence verification.

Example 2. The second example describes fragmentation of a slightly larger peptide using both singly and doubly charged precursors. Spectra shown in Figs. 6.14 and 6.15 present fragmentation of the singly and doubly charged peptide. Learning from the experience obtained in the previous case, it appears that starting with the singly charged precursor might be a good idea. The MS^2 spectrum obtained from a doubly charged precursor looks highly complex, as it contains more peaks distributed over the entire m/z range. This is a common situation in electrospray ionization, where multiply charged peptides yield better fragmentation efficiency and provide more sequence-specific fragment ions than singly charged peptides. For that reason it is wise to begin the sequencing from a doubly charged precursor. This will be the approach used in this example.

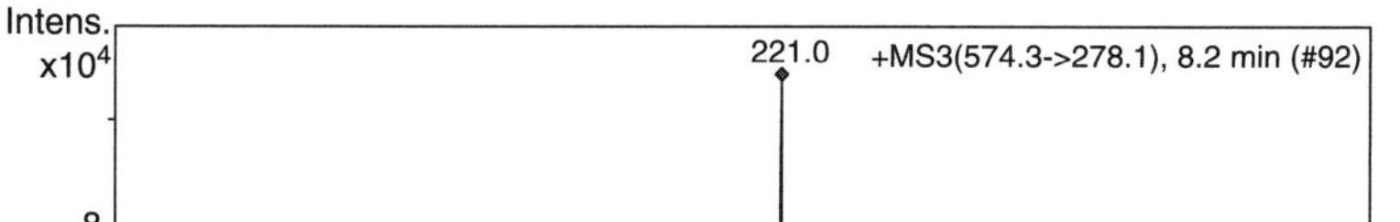

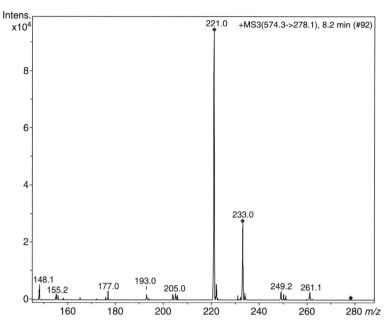

Figure 6.12. MS^3 of the peptide used in Example 1 from Fig. 6.11. Precursor mass in the first stage was 574.3 Th, and 278.1 Th in the second stage.

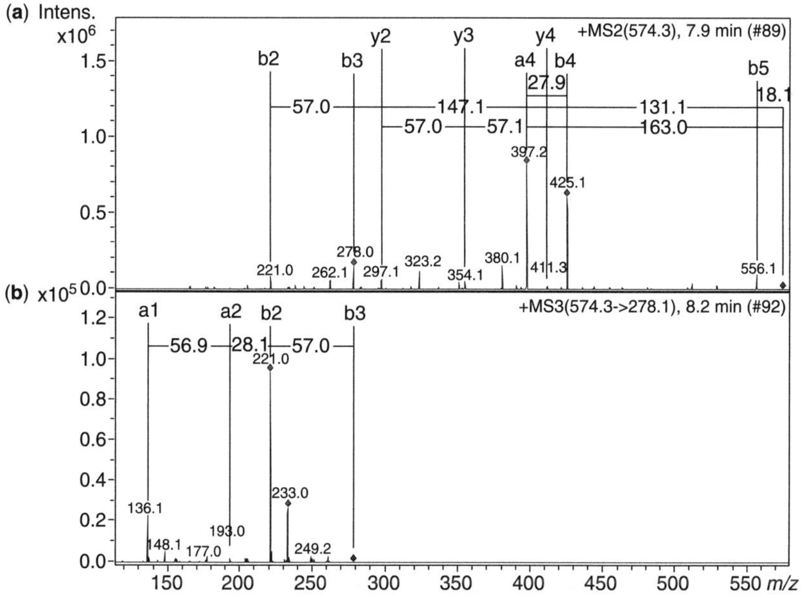

Figure 6.13. Annotated spectra used in Example 1. Upper panel (a) presents MS^2 spectrum of precursor at 574.1 Th, lower panel (b)—MS^3 spectrum of secondary precursor at 278.0 Th.

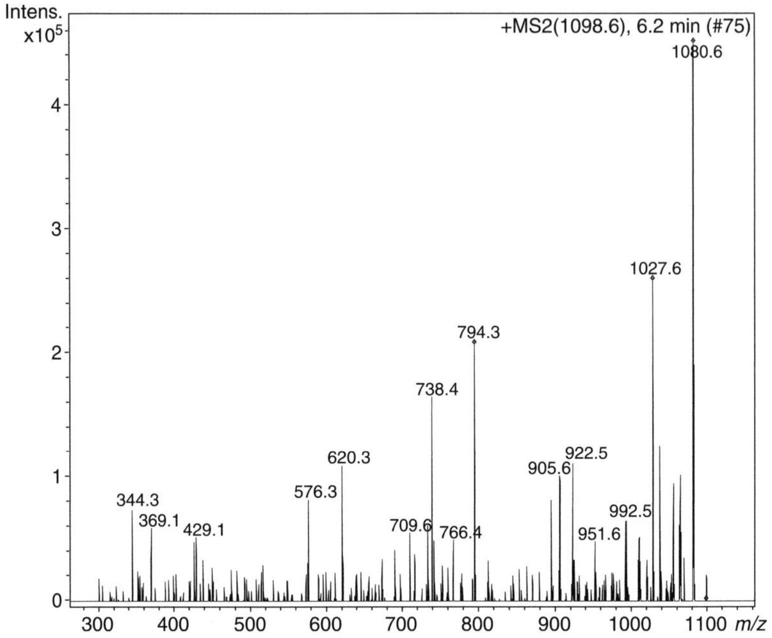

Figure 6.14. Example 2: fragmentation spectrum of singly charged peptide, precursor mass of 1098.6.

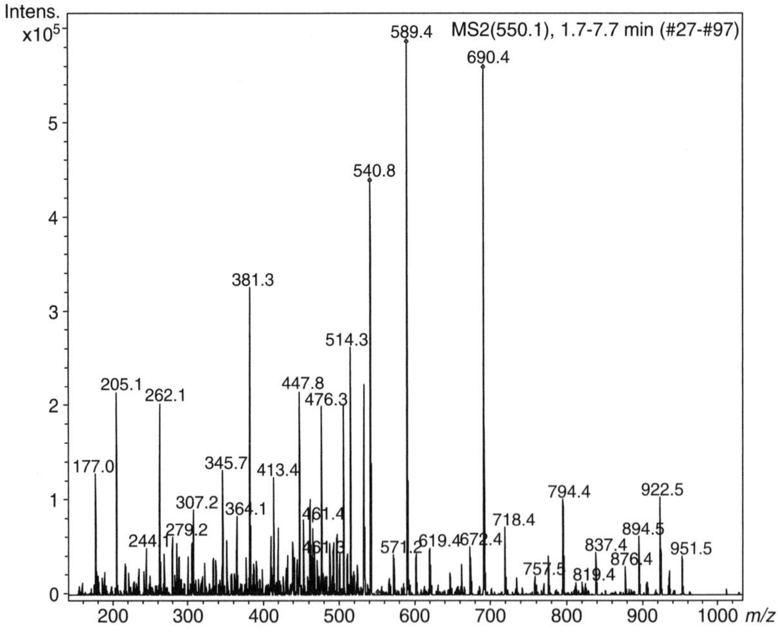

Figure 6.15. Product ion spectrum of the doubly-charged precursor (550.1 Th) of the same peptide as shown in Fig. 6.14.

Because our precursor was doubly charged, the fragment ions can either be singly or doubly charged. All doubly charged fragment ions will have an m/z lower than the precursor, but it may happen that a singly charged ion will have its m/z higher than the m/z of the precursor. Keeping this in mind, we can be sure that all peaks with m/z over 550 Th are singly charged. In the region of m/z lower than 550 Th, the spectrum contains a mixture of singly and doubly charged peaks, and each isotopic cluster should be analyzed to determine the exact charge state. Sometimes, differently charged fragment ions have very similar m/z values, which makes reliable charge determination impossible. In such cases, all sequence assignments with peaks of uncertain charge state should be verified by other fragment series.

Let us try to start sequencing in a manner similar to the previous example, with fragment ions close to the precursor. We can search for the mass differences starting from 2+ precursor mass (550.1) or from its 1+ equivalent (1098.6). In the first case, we will look for mass shifts corresponding to masses divided by two. Just as in our previous example, the b-ion series starts with a loss of water from the original molecule (−18 Da, or −9 Th as doubly charged). Indeed, there is a peak at 540.8 Th that most likely is the one we are looking for. Starting from here, we can look for a mass shift corresponding to the mass of the first C-terminal amino acid. A short survey shows that there is a peak shifted by 35.3 Th (i.e., 71 Da, alanine) at 505.3 Th. Now we can look for the next amino acid's mass shift, and we have the peak at 461.8 Th with a shift of 43.5 corresponding to 87 Da (serine). Looking further, we find a mass shift of 64.1 Th that can correspond to lysine (128.095 Da or glutamine 128.059). In this case, the mass accuracy is not sufficient to make unambiguous identification. The search for another mass shifts yields no success as there are no significant peaks that fit to any of the amino acid residues.

After analysis of the available b-ion series, we can try to find some C-terminal ions, namely y-ions. Starting from a 2+ precursor mass, there is a mass shift of 73.9 Th (147 Da, phenylalanine) followed by 28.5, 28.4, and 73.6 Th (two glycines at 57 Da, and another phenylalanine with 146 Da). No more peaks can be assigned to this ion series, but there are still singly charged fragment ions available that may supplement our results. After the survey for the doubly charged ions, the sequence already found is H-FGGF-...-(Q/K)SA-OH.

When starting with singly charged series, we should first validate sequence strings obtained from doubly charged ions. Unfortunately, there are no singly charged fragment ions covering the first C-terminal amino acids. However, there is a peak at 951.5 that corresponds to phenylalanine loss from the N-terminus. Moreover, the previously obtained entire sequence string FGGF can be verified, as it is characterized by a very intense peak at 690.4 Th. Fortunately, this ion series can be extended; the peak at 589.4 Th is equal to 101 Da mass shift, which corresponds to threonine and, further on, the peak at 532.3 is shifted 57 Da, revealing glycine. Processing further towards the low mass region is troublesome as there is a mixture of numerous singly and doubly charged peaks.

A search for the singly charged N-terminal fragments shows that a peak at 922.5 corresponds to a loss of serine and alanine from the C-terminus. A shift of 128 Da specific for glutamine/lysine is also present (peak at 794.4 Th), but unfortunately, this sequence string cannot be extended any further. All that can be done is to find the singly charged b-ions corresponding to the sequence string identified by using y-ion series. No new sequence information was obtained during this step, and no other ion series exists.

Here we encountered a typical situation in the de novo sequencing—there is a part of the sequence that is not covered by any ion series. Not all bonds between amino acids are of equal strength, and some of them might be particularly resistant to collisions, which in turn results in the missing mass-shifts (and missing residues).

Short examination of the fragment ion spectrum from a singly charged precursor is not particularly helpful. Some previously found mass shifts can be verified but no peaks with m/z lower than 800 Th can be assigned. Detailed peak annotation is presented in Fig. 6.16.

How to proceed in such a situation? One may try to examine most of the peaks on the spectrum, including really low-intensity ones, in an attempt to find the missing sequence string. This method will be time consuming because of the large number of peaks to examine, and error prone, as some mass shifts may be fitted by pure coincidence. The other way is to try to calculate the possible amino acid composition of the missing sequence, and check the shifts corresponding to a limited group of amino acids. We will try the second approach.

The mass of the identified sequence strings with terminal groups equals 870.4 Da. The mass of the peptide is 1097.6 Da (the precursor is a singly charged pseudo molecular ion $[M+H]^+$), so the mass of the remaining amino acids is 227.2 Da. This mass is specific to three amino acid pairs: AlaArg, (Ile/Leu)Asn, (Gln/Lys)Val and two triplets: AlaGlyVal and GlyGly(Leu/Ile). Unfortunately, none of these combinations might easily be proven. However, if we take a closer look at the peak at 461.8 Th (doubly charged b-series ion, Fig. 6.17), its isotopic cluster suggests that there are two overlapping fragments. The low-mass "shoulder" at 461.3 Th might belong to a different fragment. Moreover, this m/z fits to the mass shift of 156.0 Da (arginine) relative to 305.3 Th and 71.0 Da (alanine) shift towards 532.3 Th, fitting into the singly charged y-ion series.

If we include this peak, the complete sequence of the peptide will be: H-FGGFTGAR(K/Q)SA. This identification is still not complete and, unfortunately, not the most reliable. This example shows the common situation that de novo sequencing often cannot provide the entire sequence. There are only indirect clues supporting the

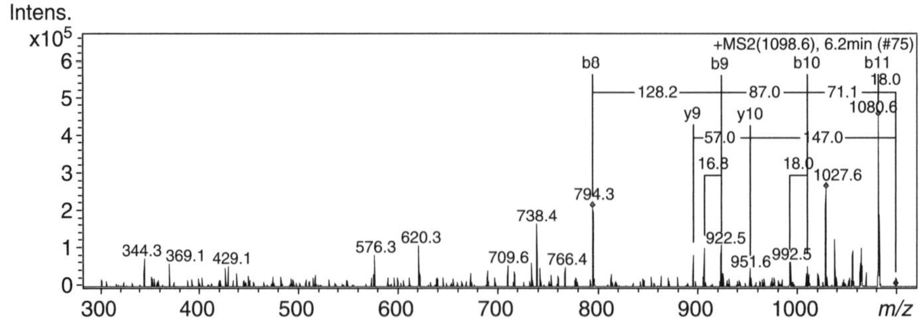

Figure 6.16. Annotated product ion spectra from Fig. 6.14—singly-charged precursor.

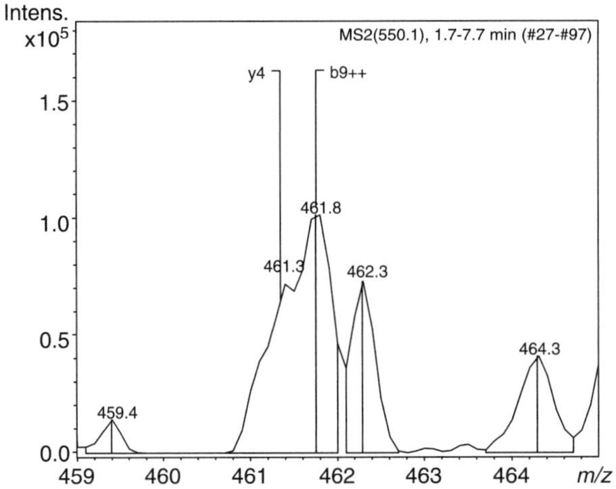

Figure 6.17. Overlapping product ions. Close-up y_4 (461.3 Th) and b_9^{++} ions (461.8 Th).

above proposal. One of the missing amino acids is a highly basic arginine. After a breakdown of the peptide's backbone, the fragments containing this amino acid preferably hold charges. Hence, most of the fragment ions along the spectrum contain basic amino acid and therefore, its mass shift cannot be found. If the next amino acid in the sequence is a basic lysine (and not the glutamine) we obtain a pair of basic amino acids that emphasizes this effect even more. This phenomenon also explains why any b-ions below b_8 (after loss of lysine) cannot be found in the case of a singly charged precursor fragmentation.

In such cases, when the peptide is derived from natural sources (endogenous origin), it is advisable to search for possible sequences in the sequence databases. In our case, this approach allows anticipating that the analyzed peptide is the Nociceptin/OrphaninQ (1–11) fragment with a sequence of H-FGGFTGARKSA-OH. Figures 6.16 and 6.18 cover the annotated fragment ion spectra.

Example 3. The last example covers the biggest peptide obtained from a tryptic digestion of an unknown protein. Although there are better ways to identify such peptides than de novo sequencing, this example allows us to present some additional hints that can be used in sequencing strategies.

The example spectrum is shown in Fig. 6.19. The precursor m/z is 741.0 Th and the charge is 2+. Brief examination of the spectrum shows that there are almost no abundant doubly charged ions. Also, the spectrum does not cover the mass range near the precursor m/z region. Therefore, we cannot begin with the same procedure as previously described. On the other hand, there are a number of very intense peaks with mass differences specific for particular amino acids.

Starting with the peak at 274.2 Th, a mass shifts of 113.2 (Leu/Ile), 113.3 (Leu/Ile), 70.8 (Ala), 114.2 (Asn), 128.0 (Lys/Gln), 147.1 (Phe), 57.1 (Gly), 163.1 (Tyr), and 128.8 (Glu) can be found. The entire series ends up with signal at 1309.6 Th and

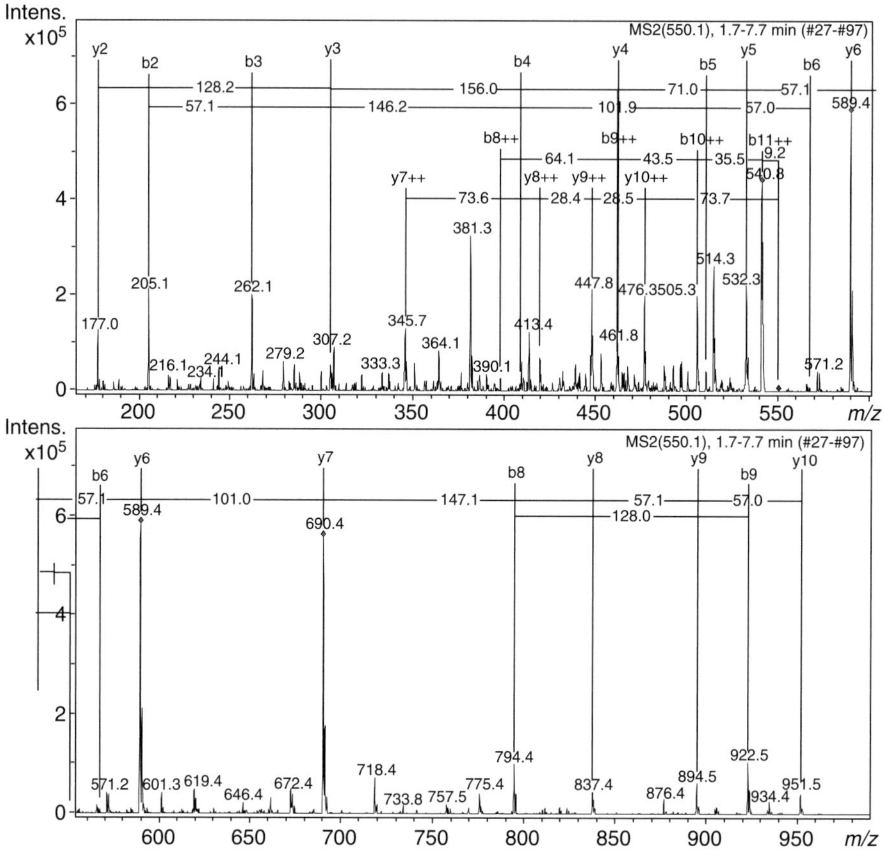

Figure 6.18. Annotated product ion spectra, doubly charged precursor.

covers nine amino acid residues, which reveals a major part of the sequence. Still, it is unknown whether these ions are N- or C-terminal fragments.

There are three basic approaches to identify the ion series. The most straightforward approach is based on a-ions linked to b-ions. If there is at least one or two mass shifts of 28 Da associated with any of the already identified peaks, there is a high probability that we found a b-ion series. However, this can only be considered speculation.

The second approach is more time consuming but can provide more reliable results. As the sequence string covers most of the peptide's sequence, the unknown terminal fragments are no longer than two to three amino acids, so their masses may be specific enough to identify them.

The third idea is focused on finding the reverse ion series that might cover some additional amino acids. Masses of b-ions and y-ions emerging from the cleavage of the same bond add to the mass of the whole precursor increased by 1 Da. If the peptide is 10 AA long, the sum of masses of, for example, b_8 and y_2 ions is equal to the mass of the singly charged precursor +1 Da. This additional dalton is because

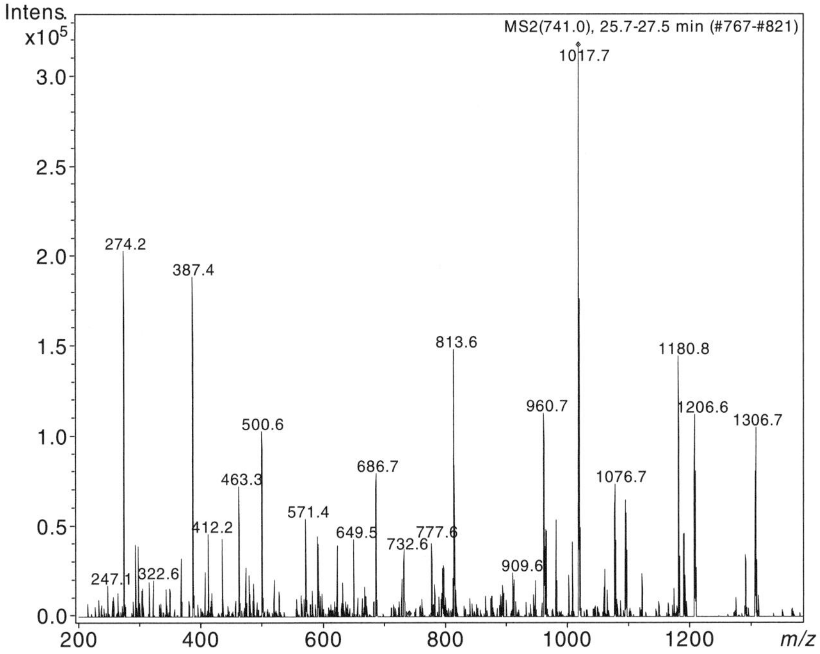

Figure 6.19. Fragmentation spectrum of doubly charged peptide precursor at 741.0 Th, used in Example 3.

each of these fragments contains a charge, which is typically a proton adduct, so the sum of their masses contains two additional protons, while the singly charged precursor contains just one. Therefore, we must add 1 Da extra to the precursor signal.

In this example, we will use the third approach—as one ion series is already available, finding the reverse ion series should be easy. Indeed, starting from the peak at 300.1 Th (which comes from the cleavage of the same bond as the 1080.8 peak) we see the mass differences of 163.2 and 56.9 Th, corresponding to tyrosine and glycine. Using the same procedure, the entire previously assigned sequence can be proven correct, but also—extend the already revealed sequence by one more residue—the mass shift between 1206.6 and 1305.8 equal to 99.2 Th corresponds to valine.

The last identified peak at 1305.8 nicely fits to the mass shift of 174.0 Th, which corresponds to a loss of arginine and C-terminal water from the precursor (156.3 + 18 Da). These results fit very well into our expectations, as the sequenced peptide is derived after tryptic digestion. Trypsin is a proteolytic enzyme cleaving selectively at the C-terminal side of basic amino acids—arginine and lysine, unless the next residue is proline. Bearing this in mind, we may assume that the previously found mass shift of 128 Th in the middle of the sequence is most probably glutamine and not lysine. Otherwise, the peptide would be cleaved at the C-terminal side of this amino acid. Of course, the missed cleavages also may occur and, again, this information can be only used as speculative and must be proven by other data.

Based on these findings the revealed sequence of the analyzed peptide is: . . .EYGFQNALIVR-OH. The mass of the still unidentified sequence string is 171.1 Da, which corresponds to one of two possible pairs: glycine with leucine/isoleucine or alanine with valine. It is not possible to prove either of these combinations, based on current data, or to identify the order of these amino acids.

Complete sequencing of this peptide requires acquisition of additional tandem mass spectra, preferably MS^3 fragmentation, of one of the low-mass y-ions. Because the peptide of interest is derived from a biological source, yet another possibility might be the use of sequence databases, similarly to the previous example. Actually, this approach works very well in this case, allowing identification of the peptide of interest as H-LGEYGFQNALIVR-OH, the 421–433 fragment of bovine serum albumin.

Figure 6.20 covers the annotated spectrum from Fig. 6.19. Only one a-ion was identified along the spectrum (435.3 Th) showing that relying on the a-b ion pairs for sequence assignment has often limited importance. However, sometimes there are other non-sequence-specific ions that might prove useful for sequencing. Note that all of the high-mass b-ions (b_7-b_{11}) are accompanied by the peaks with m/z 17 Th lower. This mass shift corresponds to the mass of ammonia and shows typical neutral loss of a molecule containing either amide (N or Q) or arginine residue. In the example spectrum, such ions accompanied all fragments containing both N and Q residues, and were no longer visible after these residues were detached from the shorter fragments

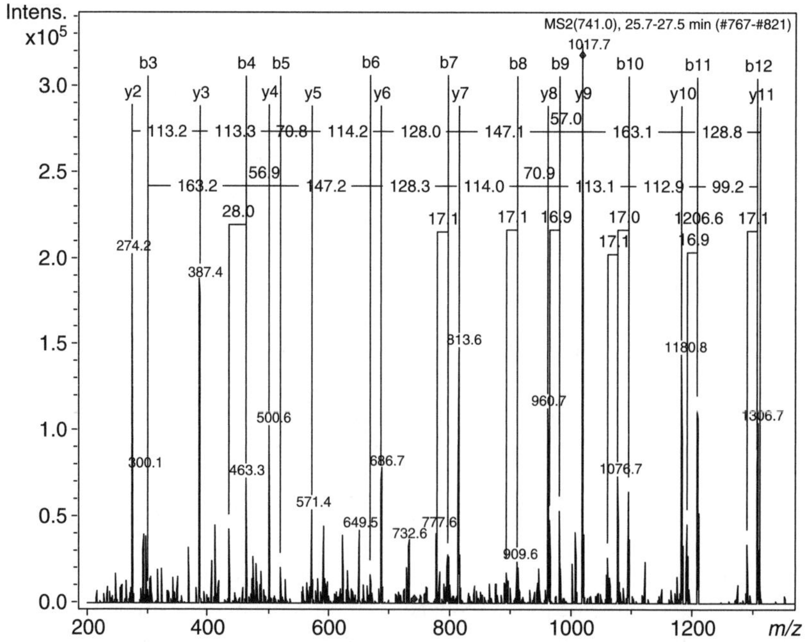

Figure 6.20. Annotated spectrum used in Example 3. The 17 Th mass difference corresponds to ammonia loss from the amide amino acids side-chains. Such peaks being non-sequence-specific themselves, can be very useful during sequencing.

(with mass lower than b_7). This observation can be used as a clue to distinguish between glutamine (Q) and lysine (K) residue when the mass resolution is not sufficient. Also, if two or more consecutive peaks forming a sequence string are accompanied by -17 Th peaks, there is a high probability that this sequence assignment is correct.

Similarly to 17 Th neutral loss, 18 Th mass shifts are also very common, corresponding to the loss of water. Such shifts are (in most cases) caused by elimination of hydroxyl group from serine and threonine side chains.

Conclusions. Three examples discussed in this chapter show three simple peptides with increasing sequence length and difficulty, but also posessing different properties. These three examples show a high impact of basic amino acids on the character of peptide fragmentation. The first peptide contained neutral amino acids and the only basic group, which preferably holds the charge in the positive-ion mode, was the N-terminal amino group. Therefore, the most intense peaks belonged to N-terminal fragments, corresponding to the a- and b-ion series.

The second example contained two basic amino acids in the middle of the sequence, but closer to the C-terminus. The fragmentation spectrum contained a huge number of both C- and N-terminal peaks, with a high number of doubly charged ions, including both basic amino acids. The presence of basic residues caused a more balanced number of b- and y-ions but also made it very difficult to obtain fragments with bond cleavage near basic residues.

The third example contained one basic amino acid, arginine, located at the C-terminus. Both C-terminal arginine and N-terminal amino group were capable of retaining a charge, therefore almost complete series of b- and y-ions were visible along the spectrum. Typically, both arginine and lysine have slightly higher affinity to protons than α-amino group; therefore, typically the C-terminal ions were more intense. This example showed an additional advantage of using trypsin for protein digestion prior to sequencing, as peptides released by this method provide fragment ion spectra with good coverage of both N- and C-terminal ion series.

Examples shown in this chapter were chosen to show basic principles and ideas for de novo sequencing. In order to get acquainted with sequencing procedures, the reader is encouraged to practice on his or her own using peptides with increasing length and difficulty. Please note that it will not always be possible to obtain the entire sequence using the approaches described here. For such cases it will be necessary to use one (or several) of the supporting techniques.

6.6.3. Tips and Tricks

6.6.3.1. Low Mass Region. All spectra shown in the examples were acquired using the quadrupole ion trap mass spectrometer. As noted previously, this widely used and relatively cheap mass analyzer suffers the low-mass cut-off phenomena. In addition to techniques used in the examples shown above, other mass analyzers applied for tandem mass spectrometers may cover the low mass region of the fragmentation spectrum that can be information rich.

First of all, the shortest fragments in both b- and y-ion series should be available. Such ions can be used to assign the sequence strings to either N- or C-terminal ion series, just in the same way as the heaviest ions, with masses close to the precursor. The mass of the y_1 ion is equal to the mass of C-terminal amino acid +18 Da, and mass of the b_1 ion equals the mass of N-terminal amino acid enlarged by 1 Da (for N-terminal hydrogen). In such cases, the peak assignment procedure can be started either with heavy ions like in examples discussed above or with light ions using the information from the low mass region.

The second advantage of the low mass region, which can be utilized in conjunction with higher energy fragmentation, is the presence of immonium ions. Such ions correspond to an amino acid side chain with α-carbon and result from the cleavage of both backbone bonds of the α-carbon. Though these ions do not provide any information about the order of amino acids in the sequence, they can be used as highly specific markers to determine whether a given amino acid exists in the sequence.

6.6.3.2. High Mass Accuracy Analyzers. Certain mass analyzers, such as Q-TOF or FTICR, can provide much more accurate spectra than can be provided by an ion trap or quadrupole. Better ion resolution minimizes the problems with overlapping peaks, such us those encountered in Example 2, or allows distinguishing between lysine and glutamine based on the exact mass of the amino acids. Because of that, it is possible to sequence larger peptides. For the ion trap experiments, the largest peptide that can be sequenced is around 15 amino acids long. This can be extended even twice with the use of an FTICR mass analyzer.

6.6.3.3. Database Search Tools. If the peptide investigated is obtained from biological material, it is always wise to use the sequence databases. Certain bioinformatic tools also provide the possibility of searching the uninterpreted MS-MS spectra against all possible sequences in the database(s). Tools such as Mascot, Protein Prospector, or Sequest work best when the specificity of an enzyme releasing the peptide from protein sequence is known. Peptide from Example 3 can be identified in about 10 seconds by the Mascot search engine provided that the user knows it is a tryptic peptide. If the specificity of the releasing enzyme is unknown, the identification procedure is more complicated. Submitting the spectrum from Example 2 to the Mascot without any enzyme specificity does not allow peptide identification, based on the software's scoring algorithm, but the best matching sequence is still correct. Very often this procedure might be used for fast screening of tandem MS spectra for potential matching sequences that will be proven manually later on.

6.6.3.4. Enzymatic Digestion. If the peptide to be sequenced is in a pure form or can be isolated and purified, the more "classic" methods can be used to support de novo sequencing. However, they are unlikely to provide good results when working with mixtures of peptides. Below, several additional techniques are listed.

1. Edman Degradation. This technique requires more material than MS-based sequencing and its sensitivity decreases with the number of amino acids detected. The use of Edman degradation sometimes allows determination of those N-terminal amino acids that were not detected during MS sequencing.

2. Caboxypeptidase Y. This enzyme cleaves consecutively C-terminal amino acids from the peptide. Used with proper activity and incubation time, it can produce a mixture containing a ladder consisting of the original peptide and peptides with several amino acids removed from the C-terminus. Mass differences between these peptides correspond to the masses of consecutive C-terminal amino acids. An additional advantage of this approach is that shortened peptides may still be analyzed using tandem mass spectrometry and their fragment ion spectra may provide more sequence information than the original peptide.
3. Aminopeptidase N. This enzyme works in a manner similar to carboxypeptidase Y, but cleaves amino acids from the N-terminus. It can be used in the same way as carboxypeptidase Y.
4. Use of Sequence-Specific Enzymes. Enzymes such as trypsin (cleaves C-terminal site of Lys and Arg, unless the next amino acid is Pro) or protease V8 (cleaves C-terminal side of Asp and Glu) can be used to digest the peptide of interest into smaller fragments and to sequence them one by one. The main difficulty of this approach is that even after complete sequencing of all fragments, there is still a need to assign the order of those fragments within protein. In order to achieve the complete sequence, it is necessary to perform two independent experiments using two enzymes with different specificities and/or to obtain at least partial sequence information from direct de novo sequencing of the original peptide.

6.6.3.5. Peptide Derivatization. The sequence information obtained from a peptide can be significantly improved by chemical modification. These approaches are discussed in more detail in the following section.

6.7. PEPTIDE DERIVATIZATION PRIOR TO FRAGMENTATION

The straightforward approach to de novo sequencing sometimes fails, for example, due to the low quality tandem mass spectra. Often it is not caused by the equipment settings or operator's capabilities, but just by unfavorable fragmentation pattern of a given peptide. Among possible approaches to solve such issues is chemical derivatization of peptides.

There are two basic aims of chemical derivatization: simplification of the fragmentation pattern by either promoting or demoting formation of a chosen ion series, and stable-isotopic labelling for simple assignment of ions to proper ion series (Fig. 6.21). Both methods may provide good results and their advantages and limitations will be discussed.

It should be noted that the majority of the derivatization techniques modify the peptide's N-terminus. The reason is that the N-terminal amine group is easier to modify than the C-terminal carboxyl group. Also, due to differences in pKa value in the ε-amino group of lysine, there are possible reaction that modify the N-terminus only, while the lysine side chains remain intact. Modifications of carboxyl groups

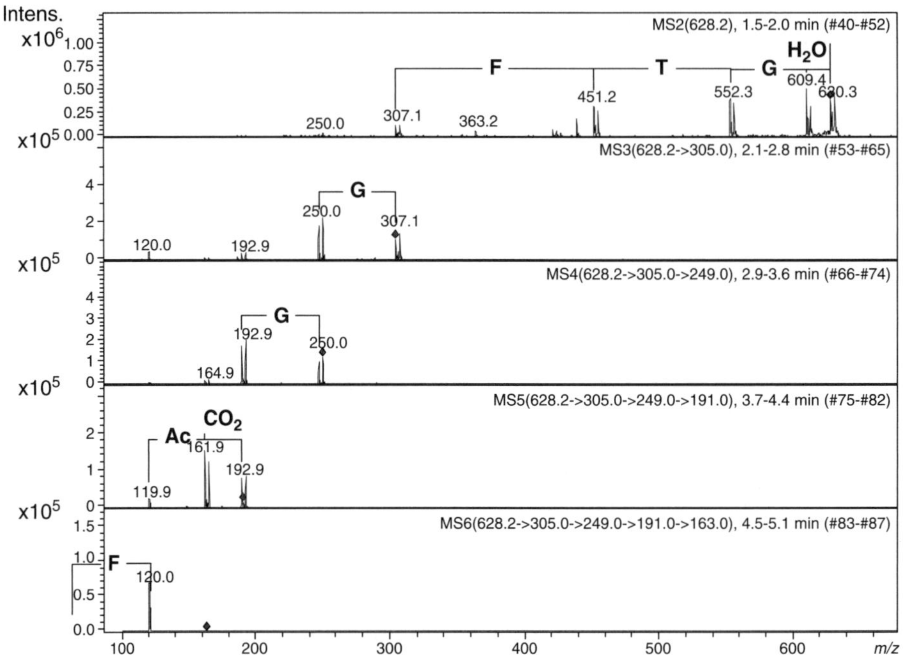

Figure 6.21. MS^6 fragmentation of FGGFTG peptide with H_3/D_3-acetyl N-terminal isotopic label. The labels reveal labeled b-ion series, acetylated a_1-ion in MS^5 spectrum, and a_1-ion without a labeling group on MS^6 spectrum.

(such as esterification) tend to affect both C-terminus and acidic side-chain groups to a similar extent.

6.7.1. Simplification of Fragmentation Patterns

Examples presented in a previous part of this chapter showed a profound impact of chemical properties of amino acids on peptide fragmentation patterns. In particular, the number and location of basic groups, retaining charges in the most commonly used positive-ion mode, strongly affect the abundance and intensity of the ion series along the fragmentation spectrum. For the same reason, introduction of highly basic groups into the peptide can promote formation of certain ions, whereas highly acidic groups can cause the opposite effect.

Most of the research performed in this field is based on tryptic peptides. As discussed earlier, such peptides contain basic amino acid residues on their C-terminus, which causes formation of the high intensity C-terminal ion series, mostly y-ions. In such peptides the N-terminal ions have lower intensity and do not provide important sequence information. Introduction of a highly basic group, such as dimethylalkyl-ammonium acetyl (DMAA) or *tris*(2,4,6-trimethoxyphenyl)phosphonium acetate into the peptide's N-terminus significantly increases the intensity of the N-terminal ion series (b- and a-ions). On the other hand, attachment of the highly negatively charged sulfonic acid into the N-terminus completely removes the N-terminal ion series from the fragment ion spectrum, thus promoting formation of the C-terminal ion series.

Such modifications can be very helpful when the fragmentation efficiency of a peptide is poor or the number of sequence-specific peptides is limited.

6.7.2. Stable Isotopes Labeling

The idea behind the use of chemical labeling is a straightforward approach to simplify peak assignment into the proper ion series. If either peptide's C- or N-terminus is labeled by specific isotopic cluster, the labeled ions are easily recognizable on the spectrum and determination of the sequence string is no longer a problem.

One of the exemplary methods for stable isotopic labeling is derivatization of the peptide by the 1 : 1 mixture of H_6- or D_6-acetic anhydride. After such treatment, a half of the peptide molecules contain "light" acetyl groups on their N-terminus, while the other half has the "heavy" acetyl group. Both populations of the molecules differ in mass by 3 Da, and the mass spectrum of such peptide mixtures will show characteristic isotopic pattern with "doublets" of the equally intense peaks. When the fragment ion spectrum (MS^2) is obtained, the operator can enlarge the isolation/fragmentation window to allow both populations to fragment simultaneously. Fragment ion spectra with fragments containing N-terminus, will derive from a mixture of "light" and "heavy" populations, and thus retain the specific isotopic "doublets," whereas C-terminal ions will not contain the isotopic label and will appear as simple ions. This way, both ion series will be recognized immediately even by an inexperienced operator.

The general concept behind this technique can be utilized using many different approaches with their individual benefits and disadvantages:

- N-terminal acetylation, such as described above, can be specific to the N-terminal amino group only, not affecting lysine side chains. Its main disadvantage is that the acetyl group blocks the N-terminus from retaining the charge during the fragmentation process and may cause the N-terminal ion series to be less intense if the peptide contains basic amino acid on its C-terminus, which is the case of tryptic peptides.
- The N-terminal amino group can be modified by different reagents, including succinic anhydride, propionic anhydride, and *N*-acetoxysuccinimide. Different approaches allow the use of different mass shifts between "light" and "heavy" populations and usually differ in fragmentation patterns.
- If the peptides to be labeled emerge from an enzymatic digestion, the labeling might be performed during the digestion step. If the digestion buffer contains 50% $H_2{}^{16}O$ and 50% $H_2{}^{18}O$, then the peptide's C-terminus will be labeled with a 2 Da-split doublet. The main disadvantage of this technique is that the oxygen atoms of the carboxyl group are labile and ^{18}O can be exchanged with ^{16}O from the solution, and thus the label can be damaged.

- The peptide's C-terminal carboxyl group might be esterified with a 1:1 mixture of methanol/D_3-methanol, resulting in a 3 Da-split doublet, which labels all C-terminal ions. Whereas this approach acetylates the peptide's N-terminus only, it will always modify the carboxyl group of glutamic and aspartic acid residues too. Therefore, if the peptide contains such amino acids, the isotopic patterns of ions will become very complex. Although these patterns can still be used to identify acidic residues, the strategy is neither easy nor recommended.
- Instead of increasing the fragmentation/isolation window to cover the entire isotopic cluster, the operator may apply another procedure. Both "light" and "heavy" precursors are fragmented separately, with very narrow isolation window. Two fragment ion spectra are acquired and the labeled ions are recognized by comparing them—labeled peaks will have their position shifted by the label mass difference between the spectra, whereas nonlabeled ones will remain at the same place.

ACKNOWLEDGMENTS

Tomasz Dylag was supported by a scholarship from the Foundation for Polish Science. Marek Noga was supported by a research grant from the Polish Ministry of Science and Higher Education, grant no. N204 136 32/3396. The work was partially supported by grant from the International Center for Genetic Engineering and Biotechnology (ICGEB), Trieste, Italy, CRP/POL05/02.

REFERENCES

1. P. Roepstorff and J. Fohlman, Proposal for a Common Nomenclature for Sequence Ions in Mass Spectra of Peptides. *Biomed. Mass Spectrom.*, **11**(1984): 601.
2. R. S. Johnson, S. A. Martin, K. Biemann, J. T. Stults, and J. T. Watson, Novel Fragmentation Process of Peptides by Collision-Induced Decomposition in a Tandem Mass Spectrometer: Differentiation of Leucine and Isoleucine. *Anal. Chem.*, **59**(1987): 2621–2625.
3. K. Biemann, Nomenclature for Peptide Fragment Ions. In *Methods in Enzymology*, Vol. 193, J. A. McCloskey (ed.), Elsevier Science and Technology, Orlando, FL, 1990, 886–888.
4. L. C. M. Ngoka and M. S. Gross, A Nomenclature System for Labeling Cyclic Peptide Fragments. *J. Am. Soc. Mass Spectrom.*, **10**(1999): 360–363.

ONLINE TUTORIALS

http://www.ionsource.com/tutorial/spectut/spec1.htm for electrospray mass spectra interpretation.

http://www.abrf.org/ResearchGroups/MassSpectrometry/EPosters/ms97quiz/SequencingTutorial.html for low energy CID sequencing, a very clear description.

http://www.ionsource.com/tutorial/protID/idtoc.htm for a protein identification tutorial.

http://www.ionsource.com/tutorial/DeNovo/DeNovoTOC.htm for a de novo sequencing tutorial.

7

OPTIMIZING SENSITIVITY AND SPECIFICITY IN MASS SPECTROMETRIC PROTEOME ANALYSIS

Jan Eriksson and David Fenyö

Proteomics studies aim at answering questions about a biological system by characterizing all its proteins (see also Chapter 10). The proteins are typically characterized by analyzing carefully chosen samples from the biological system by mass spectrometry [1]. The mass spectrometric information should ideally be sufficient to answer two questions about each sample: what does it contain and how much? In order to answer these questions appropriately a researcher has to face the three central problems of mass spectrometry based proteomic research: (i) the design of the experiment to allow for detection of proteins that are present in low abundance in the biological system [2]; (ii) the optimal use of the experimental information to allow for statistically significant identification [3] and quantitation [4] of the proteins detected; and (iii) the accurate assignment of the significance levels of the results [5].

The success in solving these three central problems will depend on many factors in a given experiment. We will here use different terms to describe how the approach in a given experiment can handle the central problems: *Success rate* and *relative dynamic range* [2], which are specific to proteomics experiments and will be defined stringently below, and the terms *sensitivity* and *selectivity*, which originate from mathematical statistics. The *sensitivity* is a measure of how good the method employed is at identifying a protein that is actually present in the sample. The *specificity* is a measure of how good the method is at not reporting a result when a protein is absent from the sample. The focus in this chapter is on the question of what is in the sample, that is, identification of proteins, but

Mass Spectrometry. Edited by Ekman, Silberring, Westman-Brinkmalm, and Kraj

we will first describe briefly what should be considered with respect to the information obtained from experiments aiming at answering the question of how much there is of different components in the sample using MS-based quantitation.

7.1. QUANTITATION

In proteomics, quantitation is typically a *comparison of two biological systems*, for example, cells in a normal state versus cells in a transformed state. MS-based quantitation utilizes analyses of digested extracted proteins (Fig. 7.1a). The comparison

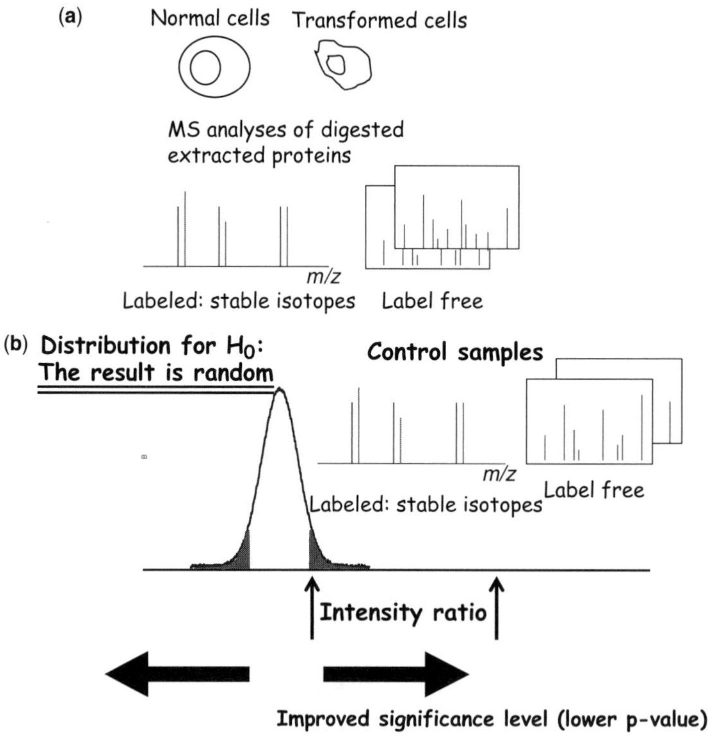

Figure 7.1. (a) Quantitative comparison of MS-signals from two different cell systems can be done by either of two basic principles: (i) stable isotope labeling of one system followed by mixing the systems prior to the MS-analysis and comparison of the intensities of pairs of signals from labeled and unlabeled (or differently labeled) ions; (ii) by label-free analysis where mass spectra are acquired separately from the systems and comparisons of signal intensities of specific m/z values are done between the spectra: (b) The statistical significance of ratios between signal intensities from the system can be judged once the distribution of intensity ratios for control samples with no known systematic difference have been obtained. The significance level of a quantitation measurement is better the smaller the overlap with the distribution for the control samples. The significance level is given by the red area under the distribution curve.

between the two systems is done either using stable isotope labeling of one of the systems and mixing of the proteolytic peptides from the two systems prior to MS analysis, or by so-called label-free analysis in which spectra are acquired separately from each system [6]. It is an advantage to introduce the stable isotope label and mix the samples as early in the experimental protocol as possible, because experimental variation is then minimized. In all of these approaches the intensity ratios between MS signals from individual peptides must be determined and are employed as a measure. It is important to determine whether it is plausible that the ratio represents a true difference between the two systems, that is, if the result can be discerned from a result corresponding with no difference between the systems. In order to answer that question control samples with no biological difference between the systems must be analyzed and all intensity ratios computed. This set of intensity ratios yields a distribution that represents the hypothesis that a given result is random. Hence, from this distribution the p-value (significance level) of a result (intensity ratio) from a real quantitation experiment can be computed (Fig. 7.1b).

7.2. PEPTIDE AND PROTEIN IDENTIFICATION

The identification of peptides and proteins using MS information can be done in three different fashions: (i) de novo sequencing, (ii) library searching, and (iii) sequence collection searching.

De novo sequencing utilizes the information from an MS-MS spectrum of a peptide isolated and fragmented in the mass spectrometer (see also Chapter 2). The spectra are analyzed with respect to mass differences that correspond to mass values of individual amino acids or stretches of peptide sequences. This information is employed for proposing the most likely sequence of the peptide analyzed [7, 8]. Advantages of this approach include no need for a sequence collection, allowing the sequencing of proteins from organisms that have not yet been sequenced. The main disadvantage is the need for excellent data quality.

Library searching compares MS-MS spectra of a peptide isolated and fragmented in the mass spectrometer to a library of peptide MS-MS spectra [9–11]. The analysis aims at identifying a peptide by finding the best similarity between an MS-MS spectrum and a member of the spectrum library. This approach is very fast and sensitive, since the comparisons involve real spectra of observed peptides using the intensity information in the comparison. It is, however, important that only high-quality spectra are included in the libraries. This approach does not work for analysis of peptides not already detected, but there is a rapidly growing number of peptide MS-MS spectra in the public domain [12–14], allowing the construction of spectrum libraries with good coverage for many mammals, fungi, and bacteria.

Sequence collection searching aims at identifying proteins or peptides from mass spectrometric information and information from protein sequence collections generated from genome sequencing. Sequence collection searching is the major approach for protein identification and exists in two different versions: (i) peptide mass fingerprinting, which utilizes a mass spectrum of proteolytic peptides from an individual protein digested with a specific enzyme and assumes that the proteolytic peptide masses yield a fingerprint of the protein [15, 16]. It is also assumed that the fingerprint can be recognized when searching a set of theoretical mass fingerprints derived by computing the mass values resulting from in silico digestion of each sequence in a sequence collection. (ii) Sequence searching using MS-MS information, where a set of mass values detected from a proteolytic peptide ion isolated and fragmented in the mass spectrometer is compared with theoretical proteolytic peptide fragment mass values generated in silico for each proteolytic peptide in a protein sequence collection (Fig. 7.2) [17–20].

Significance testing is important for minimizing false results. Identification using any of the methods mentioned above involves the scoring of each comparison between the experimental data and the model, followed by ranking of the models. Unfortunately, there is a risk of obtaining false results, since mass values measured are not unique for an individual peptide or peptide mass fragment (Fig. 7.3) [21]. In analogy with what is described above for quantitation there is a need to evaluate an identification result with respect to its statistical significance. The significance level (p-value) of a result can be determined once the distribution of scores for false (random) results is known (Fig. 7.4). Such distributions are specific for each algorithm employed in the scoring procedure. There are three different ways of generating the score distribution for false results: simulation [5], collecting statistics during the search [22–27], and direct computation [3].

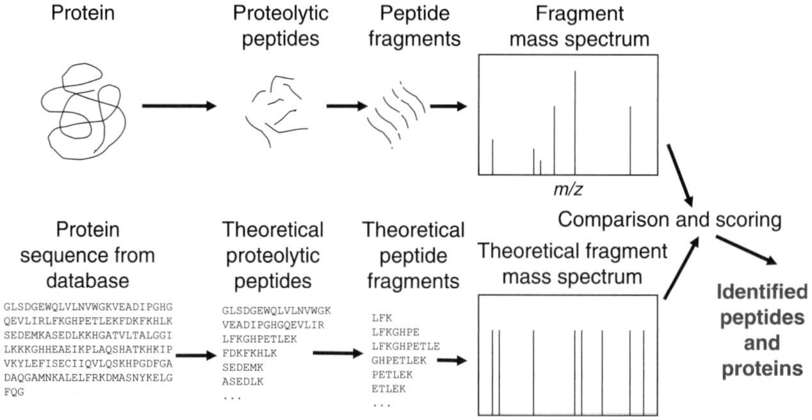

Figure 7.2. Protein identification using sequence collection searching and MS/MS data of proteolytic peptides fragmented in the mass spectrometer. Search conditions such as fragmentation pathways and mass accuracy are specified prior to the search and in the search procedure a computational algorithm compares the mass information from the experiment with theoretical mass information obtained by in silico fragmentation of each proteolytic peptide in a sequence collection. The peptides in the sequence collection are given a score that measures the degree of matching with the experimental MS/MS information and the peptide in the sequence collection that displays the best score is given the highest rank and is assumed as the identification result.

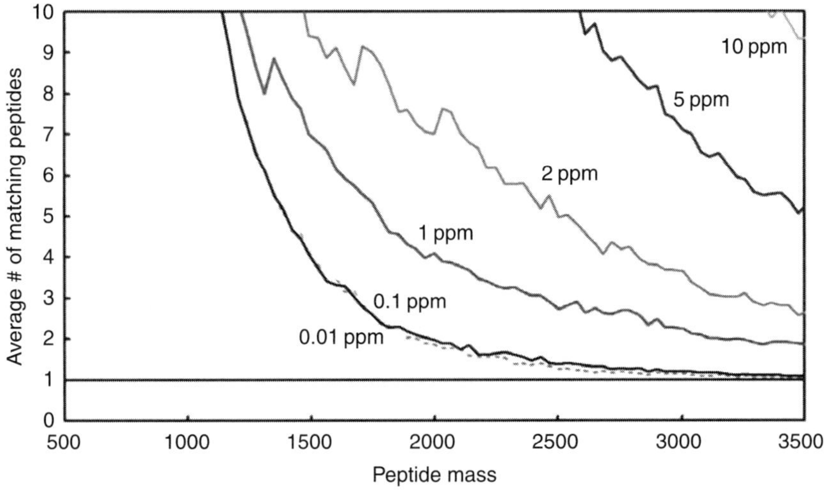

Figure 7.3. The average number of peptides matching within various mass windows (ppm) as a function of the peptide mass (Da) for proteins from *H. sapiens* completely digested with trypsin. Note that there is negligible increase in the information value (no reduction in the number of matches) below 0.1 ppm.

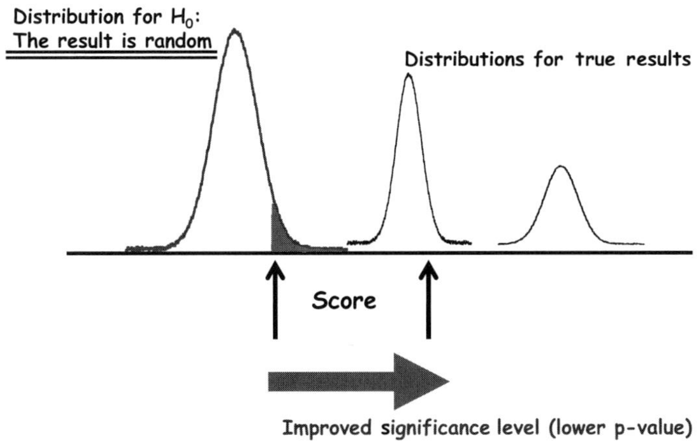

Figure 7.4. The significance level of an identification result can be determined once the distribution of scores for false identification results is known. Score distributions for true results can vary between experiments and are typically unknown, in contrast with the distribution of scores for false identification results, which can be derived by various methods (see text for details). A score that is in a region with little overlap with the distribution for false results yields a good significance level (the gray area is small).

It is critical to optimize the experimental design and data analysis to maximize the resulting information. Score distributions for true results vary between experiments and typically these distributions are unknown, since it is difficult to prove that a result is true unless the data used is synthetic or the data is from a control sample characterized with an independent and reliable method. It is desirable that the score distributions for true and false results are well separated so that the score itself can be employed as a means for minimizing the number of false results not rejected and to minimize the number of true results rejected (Fig. 7.4). An indirect view of the separation between these distributions is given by a so-called ROC-curve. In the simplest form a ROC-curve is plotted with the frequency of true results as a function of the frequency of false results and with the data points organized so that the score becomes worse with increasing distance from the origin of the graph. This can be slightly modified into plotting the *sensitivity* versus 1 − *selectivity*, where

$$\text{Sensitivity} = \frac{\#\text{ of true results not rejected}}{\text{total}\,\#\text{ of true results}}$$

and

$$1 - \text{Selectivity} = \frac{\#\text{ of false results not rejected}}{\text{total}\,\#\text{ of false results}}$$

The sensitivity and the selectivity depend on the choice of algorithm. A simple way to examine what influences the sensitivity and the selectivity is to employ synthetic data and simulate protein or peptide identification. Figure 7.5a displays a ROC-curve comparing peptide mass fingerprint-based identification of *Saccharomyces cerevisiae* proteins using a set of PMFs generated in silico where in each PMF four mass values were correlated with a single protein and 16 mass values were chosen randomly. The same data set was employed for searching the *S. cerevisiae* sequence collection using two different search algorithms: algorithm 1, Probity and algorithm 2, which ranks simply based on the number of matching mass values in each PMF. It is seen in Fig. 7.5a that for algorithm 1 there is a region along the y-axis where good scores yield only true results, whereas for algorithm 2, the score more or less arbitrarily indicates a true or a false result also for the best scores. From this simulation example we learn that the sensitivity and the selectivity depend on the choice of algorithm.

The sensitivity and the specificity depend on the search conditions. Fig. 7.5b indicates results from a simulation using synthetic MS-MS spectra generated in silico from the *S. cerevisiae* sequence collection. Each spectrum contained 25 peptide fragment mass values, but only seven mass values corresponding to an individual peptide. These spectra were employed for searching the *S. cerevisiae* sequence collection using the algorithm X! Tandem in two sessions employing different search conditions. In one session the windows for accepted mass errors of both the peptide itself and its fragments were ten times larger than in the other session. Based on the distinct different in the outcome from the two sessions displayed in Fig. 7.5b we conclude that the sensitivity and the specificity depend on the search conditions.

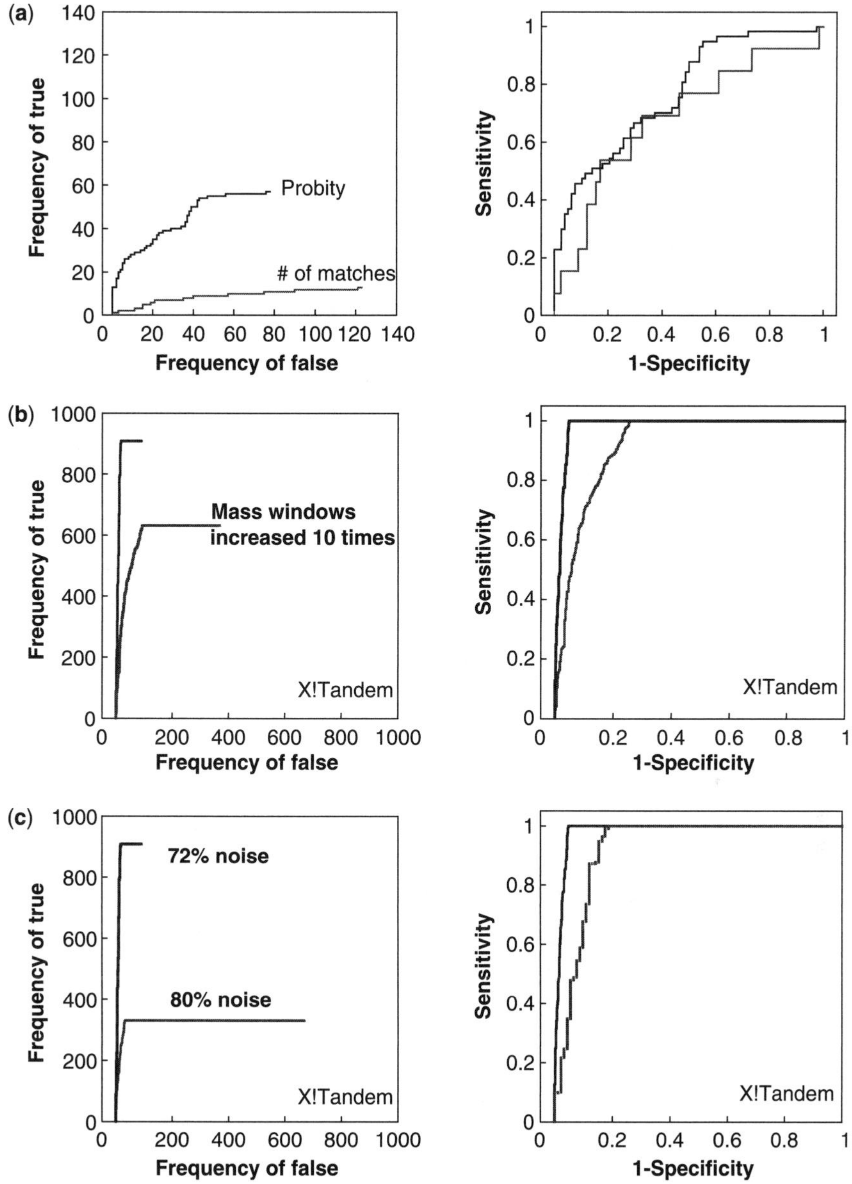

Figure 7.5. Simulation results that elucidate how the sensitivity and the selectivity of a proteomics experiment depend on various features: (a) The choice of algorithm. The probity algorithm displays better sensitivity and selectivity than an algorithm that ranks strictly based on the number of matches. (b) The search conditions. Increasing the mass window of a search 10 times when searching with data that display small mass errors yields worse sensitivity and selectivitry. (c) The quality of the data. Data with less noise yields better sensitivity and selectivity.

The sensitivity and the selectivity depend on the data quality. Figure 7.5c displays results from a simulation employing the same data set as was used in Fig. 7.5b, together with a simulation in which the MS-MS spectra have only five mass values corresponding to an individual peptide (out of 25). Hence we see that the sensitivity and the selectivity depend on the data quality.

7.3. SUCCESS RATE AND RELATIVE DYNAMIC RANGE

We have already concluded that the data quality is an issue for the sensitivity and the selectivity for protein identification in MS-based proteomics experiments. A related issue is that we do not acquire data for all the proteins actually present in the sample. The reason for this is that there is a discrepancy between the experimental dynamic range and the range of protein abundances in the proteome. The bell-shaped curve shown in Fig. 7.6a is an approximation of the protein amount distribution measured for yeast (*S. cerevisiae*) using immunodetection methods [28]. The range of protein abundances in yeast spans six orders of magnitude. It is believed that for human body fluids the range of protein abundances is at least 10^{10}. The dynamic range of a mass spectrometer can be as low as 10^2 (for generating signals from two substances present in the sample at a given point in time). Proteomics researchers have realized that the complexity and the range of protein abundance of a proteome make it necessary to

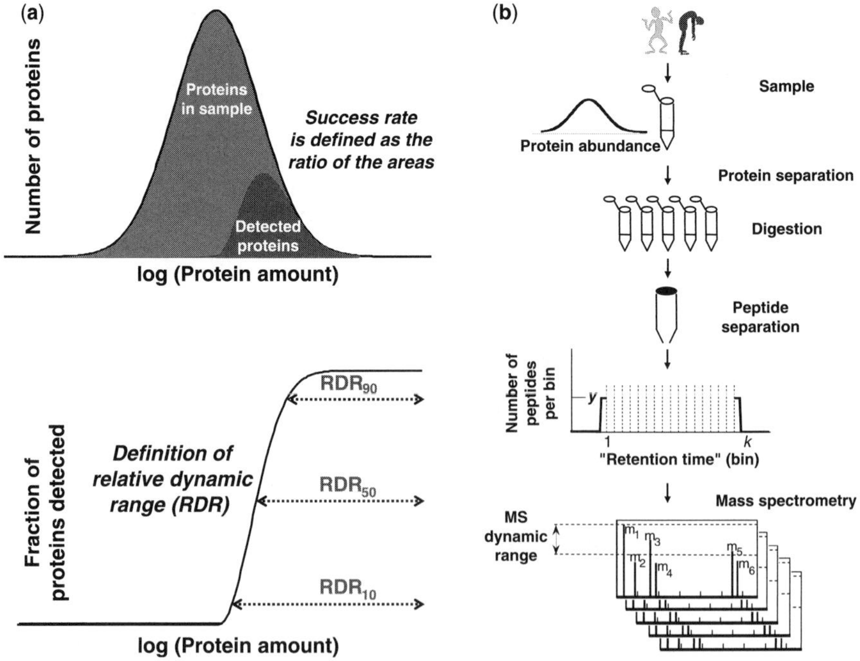

Figure 7.6. (a) Definitions of success rate and relative dynamic range. (b) Model of a proteomics experiment. (See color insert.)

apply various separation protocols prior to the MS analysis. There are many options to choose from in this respect and it is impossible to examine the merits of all combinations experimentally. By constructing a model of a proteomics experiment and a model of the protein abundance distribution of a proteome it is possible to use computer simulations to examine how good a particular experimental design would be for detecting the proteins of that proteome. In such simulations the quantities studied are the success rate and the relative dynamic range (RDR), where the success rate indicates how many proteins are detected divided by the total number of proteins in the proteome and the RDR indicates how deep down into the low abundance proteins an experimental design can manage to detect proteins (see Fig. 7.6a). The experimental design can be described by a set of parameters (Fig. 7.6b) and we will here give an example of how one feature of the sample preparation and two features of the mass spectrometer influence the success rate and theRDR: the degree of protein separation, the MS detection limit,

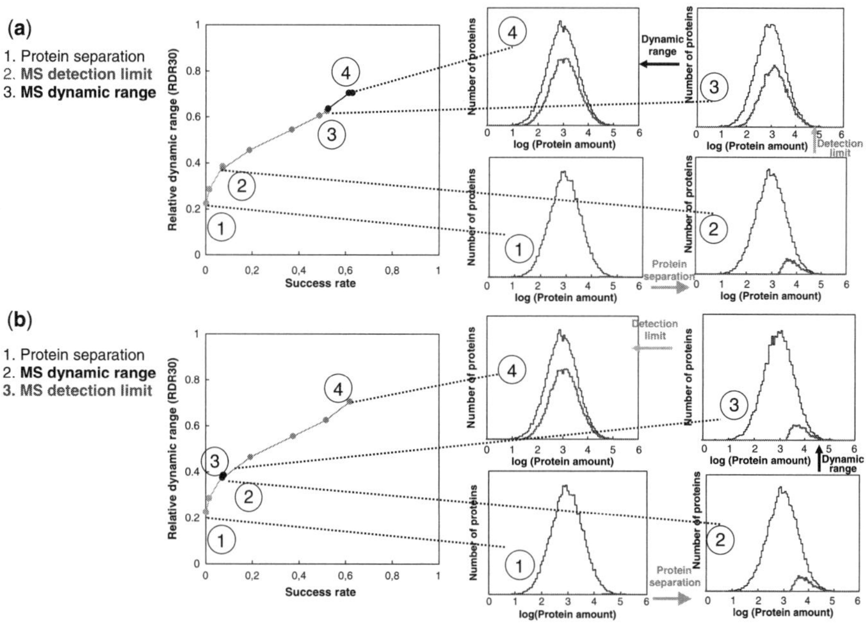

Figure 7.7. Results from model simulations showing the effect of protein separation and the effect of MS detection limit and MS dynamic range on the success rate and the relative dynamic range (RDR) for detection of proteins from *H. sapiens* tissue samples. (a) *Left:* RDR as a function of success rate when first improving the protein separation going from 30,000 proteins (1) to 300 proteins (2), then enhancing the sensitivity of the mass spectrometer from 1 fmol to 1 amol (3), and finally improving the MS dynamic range from 10^2 to 10^4 (4). *Right:* The protein abundance distribution assumed for human tissue together with the distribution of the proteins detected for the experimental designs 1 to 4. (b) Same as in (a), but with the MS dynamic range improved prior to improving the MS detection sensitivity. Note that the effect of improving the dynamic range is negligible compared with the effect of improving the detection sensitivity. (See color insert.)

and the MS dynamic range. The top left panel of Fig. 7.7a indicates how the success rate and the RDR vary when first improving the protein separation, then improving the MS detection limit and finally improving the MS dynamic range. The right panel of Fig. 7.7a shows the protein abundance distribution model employed in the simulation together with the distribution of the proteins detected for the initial design (1), the design with improved protein separation (2), after improving the detection limit (3), and after enhancing the MS dynamic range (4). It is evident that all these three features of the experimental design can influence strongly the outcome of an experiment. The way in which design parameters are changed can, however, have a strong influence on the result. For example, if instead of improving the protein separation, the MS dynamic range is improved, there is no improvement of the success rate and the RDR until the MS detection limit also is improved (Fig. 7.7b, 1–4).

7.4. SUMMARY

Computations and simulations are important tools for examining the performance of mass spectrometry-based proteomic research. Computations are necessary for deriving distributions for results corresponding with "no difference between the systems" for quantitation experiments and for results corresponding with "a false result" for identification experiments. We have demonstrated using simulations that the sensitivity, that is, the ability to identify a protein present in the sample, and the selectivity, that is, the ability to not report proteins absent from the sample, depend on three factors: (i) the choice of protein identification algorithm (including search conditions), (ii) the data quality, and (iii) the experimental design.

REFERENCES

1. R. Aebersold and M. Mann. Mass Spectrometry-Based Proteomics. *Nature*, **422**(2003): 198–207.
2. J. Eriksson and D. Fenyo. Improving the Success Rate of Proteome Analysis by Modeling Protein-Abundance Distributions and Experimental Designs. *Nat. Biotechnol.*, **25**, no. 6 (2007): 651–655.
3. J. Eriksson and D. Fenyo. Probity: A Protein Identification Algorithm with Accurate Assignment of the Statistical Significance of the Results. *J. Proteome Res.*, **3**, no. 1 (2004): 32–36.
4. R. Aebersold. Quantitative Proteome Analysis: Methods and Applications. *J. Infect. Dis.*, **187**, Suppl 2(2003): S315–20.
5. J. Eriksson, B. T. Chait, and D. Fenyo. A Statistical Basis for Testing the Significance of Mass Spectrometric Protein Identification Results. *Anal. Chem.*, **72**, no. 5 (2000): 999–1005.
6. C. Fenselau. A Review of Quantitative Methods for Proteomic Studies. *J. Chromatogr. B, Analyt. Technol. Biomed. Life Sci.*, **855**, no. 1 (2007): 14–20.
7. J. A. Taylor and R. S. Johnson. Sequence Database Searches via de novo Peptide Sequencing by Tandem Mass Spectrometry. *Rapid Commun. Mass Spectrom.*, **11**, no. 9 (1997): 1067–1075.

8. A. Frank and P. Pevzner. PepNovo: De novo Peptide Sequencing via Probabilistic Network Modeling. *Anal. Chem.*, **77**, no. 4 (2005): 964–973.
9. R. Craig, et al., Using Annotated Peptide Mass Spectrum Libraries for Protein Identification. *J. Proteome Res.*, **5**, no. 8 (2006): 1843–1849.
10. B. E. Frewen, et al., Analysis of Peptide MS/MS Spectra from Large-Scale Proteomics Experiments Using Spectrum Libraries. *Anal. Chem.*, **78**, no. 16 (2006): 5678–5684.
11. H. Lam, et al., Development and Validation of a Spectral Library Searching Method for Peptide Identification from MS/MS. *Proteomics*, **7**, no. 5 (2007): 655–667.
12. R. Craig, J. P. Cortens, and R. C. Beavis. Open Source System for Analyzing, Validating, and Storing Protein Identification Data. *J. Proteome Res.*, **3**, no. 6 (2004): 1234–1242.
13. L. Martens, et al., PRIDE: The Proteomics Identifications Database. *Proteomics*, **5**, no. 13 (2005): 3537–3545.
14. F. Desiere, et al., The PeptideAtlas Project. *Nucleic Acids Res.*, **34**, Database issue (2006): D655–D658.
15. W. J. Henzel, et al., Identifying Proteins from Two-Dimensional Gels by Molecular Mass Searching of Peptide Fragments in Protein Sequence Databases. *Proc. Natl. Acad. Sci. U.S.A.*, **90**, no. 11 (1993): 5011–5015.
16. W. Zhang and B. T. Chait. ProFound: An Expert System for Protein Identification Using Mass Spectrometric Peptide Mapping Information. *Anal. Chem.*, **72**, no. 11 (2000): 2482–2489.
17. M. Mann and M. Wilm. Error-Tolerant Identification of Peptides in Sequence Databases by Peptide Sequence Tags. *Anal. Chem.*, **66**, no. 24 (1994): 4390–4399.
18. J. Eng, A. L. McCormack, and J. R. Yates, III, *J. Am. Soc. Mass Spectrom.*, **5**(1994): 976–989.
19. D. N. Perkins, et al., Probability-Based Protein Identification by Searching Sequence Databases Using Mass Spectrometry Data. *Electrophoresis*, **20**, no. 18 (1999): 3551–3567.
20. R. Craig and R. C. Beavis. TANDEM: Matching Proteins with Tandem Mass Spectra. *Bioinformatics*, **20**, no. 9 (2004): 1466–1467.
21. D. Fenyo, J. Qin, and B. T. Chait. Protein Identification Using Mass Spectrometric Information. *Electrophoresis*, **19**, no. 6 (1998): 998–1005.
22. H. I. Field, D. Fenyo, and R. C. Beavis. RADARS, a Bioinformatics Solution that Automates Proteome Mass Spectral Analysis, Optimises Protein Identification, and Archives Data in a Relational Database. *Proteomics*, **2**, no. 1 (2002): 36–47.
23. A. Keller, et al., Empirical Statistical Model to Estimate the Accuracy of Peptide Identifications made by MS/MS and Database Search. *Anal. Chem.*, **74**, no. 1 (2002): 5383–5392.
24. R. E. Moore, M. K. Young, and T. D. Lee. Qscore: An Algorithm for Evaluating SEQUEST Database Search Results. *J. Am. Soc. Mass Spectrom.*, **13**, no. 4 (2002): 378–386.
25. D. Fenyo and R. C. Beavis. A Method for Assessing the Statistical Significance of Mass Spectrometry-Based Protein Identifications using General Scoring Schemes. *Anal. Chem.*, **75**, no. 4 (2003): 768–774.
26. A. I. Nesvizhskii, et al., A Statistical Model for Identifying Proteins by Tandem Mass Spectrometry. *Anal. Chem.*, **75**, no. 17 (2003): 4646–4658.
27. J. Peng, et al., Evaluation of Multidimensional Chromatography Coupled with Tandem Mass Spectrometry (LC/LC-MS/MS) for Large-Scale Protein Analysis: The Yeast Proteome. *J. Proteome Res.*, **2**, no. 1 (2003): 43–50.
28. S. Ghaemmaghami, et al., Global Analysis of Protein Expression in Yeast. *Nature*, **425**, no. 6959 (2003): 737–741.

PART III

APPLICATIONS

INTRODUCTION

What do the Shroud of Turin, cerebrospinal fluid, soil, gasoline, the "Anatomy Lesson of Dr. Nicolaes Tulp" by Rembrandt van Rijn, apple cider, and solar wind have in common? One answer is that they all can be mass spectrometry samples. This point of view, though essentially correct, seems somewhat limited and biased, however. In this part of the book, we are turning the common mass spectrometry perspective upside down. Instead of listing different application areas for mass spectrometry, we let ten different researchers or research groups introduce their respective fields and describe how mass spectrometry can aid them in their work.

Mass Spectrometry. Edited by Ekman, Silberring, Westman-Brinkmalm, and Kraj

8

DOPING CONTROL

Graham Trout

Sport is a major part of the lives of many people both as active participants and as passive observers. Those who are actively involved enjoy social and health benefits. However, sport can be very competitive and the use of drugs to enhance performance is evident throughout the history of sport. In modern times, the rewards for success continue to increase and the competition is fierce. For some athletes the will to win is so strong that they will consider anything to enhance their performance. The vast sums of money associated with high profile sports are also powerful incentives to cheat. The problem was originally recognized and acted upon by the International Olympic Committee (IOC), which initiated systematic drug testing at the Olympic Games in 1972. In 1999 an international body called the World Anti-Doping Agency (WADA) was created, with the mandate of overseeing doping control in sport. WADA produced a World Anti-Doping Code in 2003, which has been accepted by all national Olympic committees and most international sporting federations. Each year WADA produces an updated list of prohibited substances that are not permitted to be used in sport [1]. The list contains nine groups of prohibited substances and three groups of prohibited methods. The main reasons for adding substances to the list are that they are performance enhancing or have the potential to be so, or are likely to be injurious to health. A summary of the 2008 list is shown in Table 8.1. In addition to setting the code and preparing the list WADA also accredits 33 laboratories around the world that are permitted to carry out international sports drug testing

Mass Spectrometry. Edited by Ekman, Silberring, Westman-Brinkmalm, and Kraj

TABLE 8.1. Summary of the WADA 2008 Prohibited List

	Prohibited Substances
S1	Anabolic Agents—includes anabolic androgenic steroids such as stanozolol and other anabolic agents such as clenbuterol.
S2	Hormones and Related Substances—includes erythropoietin (EPO) and growth hormone (hGH).
S3	Beta-2 Agonists—all are prohibited but some, such as salbutamol and terbutaline, are permitted by inhalation.
S4	Hormone Antagonists and modulators—includes aromatase inhibitors such as anastrazole and selective estrogen receptor modulators such as tamoxifen.
S5	Diuretics and other Masking Agents—includes diuretics such as amiloride and chlorothiazide and masking agents such as probenecid and plasma expanders.
S6	Stimulants—such as amphetamine, cocaine, mesocarb, and strychnine.
S7	Narcotics—such as buprenorphine and dextromoramide.
S8	Cannabinoids
S9	Glucocorticosteroids—all are prohibited when administered orally, rectally, intravenously or intramuscularly. Topical preparations are permitted.
	Prohibited Methods
M1	Enhancement of Oxygen Transfer—includes blood doping and the use of hemoglobin-based blood substitutes.
M2	Chemical and Physical Manipulation—includes tampering with samples.
M3	Gene Doping.

Note S1 to S5 are prohibited at all times, whereas S6 to S9 are only prohibited in competition. Alcohol and beta-blockers are prohibited in some sports in competition.

and sets the standards that these laboratories must follow. Laboratories are required to use assays that have been shown to detect only the substance of interest and can discriminate between compounds of closely related structures. Given that most substances must be detected at low nanogram per milliliter levels, this requirement effectively means that mass spectrometry coupled with some form of chromatography must be used whenever possible. In fact the IOC, which accredited sports drug testing laboratories prior to WADA, required that mass spectrometry be used for the confirmation of prohibited substances, other than for macromolecules such as the peptide hormones where the use of immunoassays was allowed.

Because of the penalties associated with a positive drug test (a two-year ban and loss of sponsorship) precautions must be taken to ensure that the sample being analyzed in the laboratory actually came from the athlete in question. Samples are either urine or blood, with most samples being urine. The sample is collected under supervision and placed by the athlete into two uniquely numbered bottles (so called A and B samples). The bottles are sealed with security lids having the same number and which, once closed, cannot be opened without breaking the lid. The bottles are transported to the testing laboratory along with paperwork which has no mention of the identity of the athlete. The A sample is opened for testing, while the B sample is retained unopened. If the A sample returns a positive result the athlete has the right to request the analysis of the B sample and to be present at its opening and analysis. Of course it is just as important to be sure that the analyte found in the sample has been correctly identified and this is where the skill of the chemist and the specificity and selectivity of hyphenated chromatography mass spectrometry techniques are required. A comprehensive review of the analytical methods used in sports drug testing has been published [2], and a more recent review concentrating on the application of mass spectrometry in sports drug testing has been published by Thevis and Schanzer [3].

The mass spectrometry techniques that are most commonly used are GC-MS (see Chapter 4) using typical bench-top quadrupole systems (see Chapter 2), such as the Agilent 6890/5973, and for lower detection levels GC-MS/MS with ion traps (Chapter 2) or GC-HRMS (high resolution mass spectrometry) using magnetic sector instruments (Chapter 2). In recent years liquid chromatography coupled with mass spectrometry has been used more often (see Chapter 4). Table 8.2 shows this trend by comparing the screening methods used by the National Measurement Institute, New South Wales, Australia, in the Sydney 2000 Olympic Games and in the Melbourne 2006 Commonwealth Games. Despite adding a whole new class of drugs (S9, the glucocorticosteroids), as well as new drugs in the other classes, the number of analytical methods needed has not increased, primarily due to the capabilities of LC-MS-MS. In addition the number of extractions required for each sample has decreased from four to three, with only one requiring derivatization compared to three in 2000.

The analytes are typically extracted from the biological matrix using solvent extraction or solid phase extraction (SPE). Most analytes require some form of chemical derivatization prior to analysis by GC-MS techniques, whereas with LC-MS-MS no further treatment of the extract is required. The extracts obtained from urine are relatively dirty because of the many endogenous compounds that are present. It is for this reason that the very selective techniques of GC-MS-MS, GC-HRMS, or LC-MS-MS are required to detect some of the prohibited substances that have low detection levels.

It can be seen that most of the substances prohibited in sport are screened for by hyphenated chromatographic mass spectrometric techniques. Confirmation is also done using such techniques. The reason for this is that mass spectrometry coupled with chromatography provides the sensitivity and selectivity required to detect and

TABLE 8.2. Chromatographic Screening Methods for Prohibited Substances

Methods Used in 2000	Methods Used in 2006
1. GC-MS method for S1 (anabolic agents), S3 and S4	1. GC-MS method for S1 (anabolic agents), S3, and S4.
2. GC-HRMS method for low level S1	2. GC-HRMS method for low level S1
3. GCNPD and GC-MS for S6 (stimulants)	3. GCNPD and GC-MS for S6 (stimulants)
4. GC-MS for S5 and S8	4. LC-MS/MS for S5, some S1, S8 and some S6
5. GC-MS for S7 and some S6	5. LC-MS/MS for S7 and S9

confirm the presence of nanogram per milliliter concentrations of drugs. A typical bench-top GC-MS system used for drug testing consists of a gas chromatograph (GC) with a 15 to 30 m capillary column connected to a quadrupole-based mass spectrometer (MS). The GC resolves the mixture of compounds injected into a multitude of peaks with typical widths of a few seconds each. The MS used to detect the eluted peaks can be used in two modes, either full scan or selected ion monitoring (SIM). In full scan mode the MS is a universal detector producing mass spectra at a rate of approximately one scan per second. The mass range of interest typically spans from 40 to 500 amu (u). The quadrupole is basically a mass filter with approximately unit resolution, which lets the ions produced in the ion source pass sequentially to the detector. For a mass range of 500 this means that each ion is only detected for approximately 2 ms of a one second scan. Thus, if one is looking for a particular ion of interest then it is only detectable for 0.2% of the time. This reduces the potential detection level. The alternative mode of operation of the MS is to use SIM where only selected ions are looked for. This markedly reduces the background noise and improves the signal-to-noise ratio. The MS is now acting as a highly selective detector looking for target compounds. If the MS were programmed to only look for four ions in SIM rather than operate in full scan the time spent on each ion would increase from 2 ms to 250 ms. This would result in an improvement in signal to noise of more than ten with a corresponding improvement in limit of detection. With prohibited substances such as the exogenous anabolic steroids we are looking for low concentrations of compounds in a profusion of chemically similar endogenous steroids and the full scan mode of operation lacks the sensitivity required. Thus, SIM is used to look for the known metabolites of the prohibited substances. Because the target compounds elute at different times it is possible to program the MS so that several different groups of ions are collected during the chromatographic run.

The use of GC-MS in SIM mode has ample sensitivity to detect even the low concentration steroids when presented to the instrument as a pure standard. However, the samples analyzed by the laboratory are not pure standards but dirty urine extracts. Even though the GC-MS in SIM mode is a selective detector there are unfortunately often many compounds present that have the same mass ions (at unit resolution) as the target analyte, resulting in chemical background noise. As a result of this high background a prohibited drug can be hidden in the noise. Prohibited substances are detected by printing out mass chromatograms around the known retention time of the analyte of interest. In the case of stanozolol it is detected by looking for silylated derivatives of hydroxylated metabolites for which two major ions are 545 and 560. A trideuterated internal standard of 3′-hydroxystanozolol is added to each sample giving a 563 ion. In some extracts the background in these ion traces is so high that the presence of the metabolite of stanozolol may not be suspected. It is for just such samples that GC-HRMS is used in our laboratory to screen all steroid extracts.

Why is GC-HRMS able to detect cases of doping that conventional quadrupole GC-MS may miss? It is not just because of the increased sensitivity that an HRMS has but mainly because of the enhanced selectivity. HRMS instruments can operate at a resolution that is capable of distinguishing masses that differ by less than 0.01 amu (u), whereas a quadrupole MS can only achieve 0.1 amu (u) resolution at best. In fact

HRMS can be used to calculate the empirical formula of a molecule by measuring its mass to a precision of 0.001 to 0.002 amu (u). Thus, the ion 86 which comes from $C_5H_{12}N$ (86.0970) can be distinguished from $C_5{}^{13}CH_{13}$ (86.1017). The mass 86.1017 arises frequently as it can come from any compound with a carbon chain longer than six carbon atoms. Thus, if mass 86 is the major ion in an analyte of interest due to $C_5H_{12}N$ then one can expect high background levels of ion 86 due to $C_5{}^{13}CH_{13}$ when operating at low resolution. However, with HRMS the hydrocarbon background is not seen and detection of the analyte is much easier. The effect of resolution on the separation of an ion with m/z of 331 and its P + 1 (due to ^{13}C) at 332 is shown in Fig. 8.1. The separation achieved at a resolution of approximately 500 (the lowest resolution at which the instrument can be set) is still significantly better than that achieved in a typical bench-top quadrupole system. At higher resolutions the large separation between 331 and 332 makes it clear how ions with masses differing by much less than 1 amu (u) can be distinguished. The effect of resolution in a real sample can be seen in Fig. 8.2 where the same urine sample extract has been analyzed for the presence of a metabolite of stanozolol at decreasing resolutions. The ions 545.3415 and 560.3650 come from 3′-hydroxystanozolol, whereas the 563.3838 ion comes from the trideuterated 3′-hydroxystanozolol which is added to every sample as an internal standard. At 4000 resolution there are clearly two aligned peaks at 10.3 minutes due to the presence of 3′-hydroxystanozolol, with the peak in the 560 ion chromatogram being clearly the largest in the window. When the resolution is reduced to 500 a peak at 10.4 minutes, which was barely observable at 4000 resolution, becomes dominant, despite the fact that the peak at 10.3 minutes has actually increased in intensity by

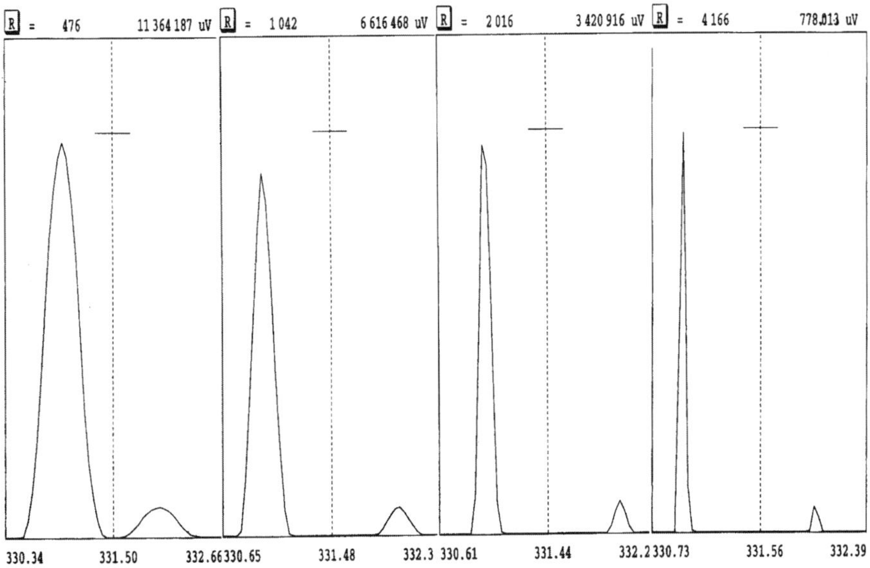

Figure 8.1. Effect of mass spectral resolution on the separation of a tuning ion of mass 331 and its P + 1 peak at mass 332.

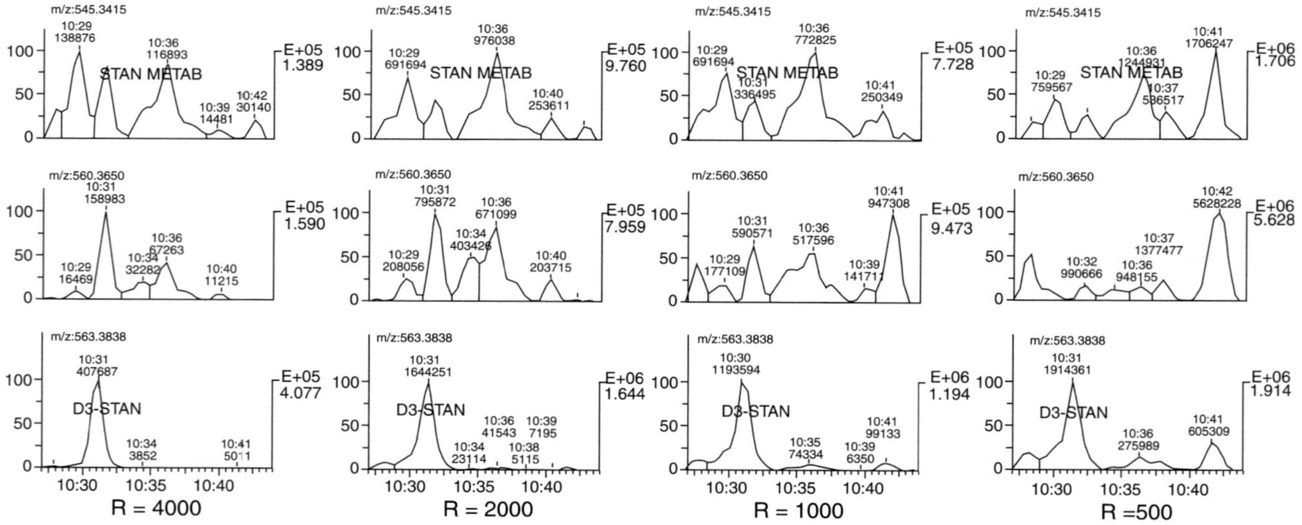

Figure 8.2. Effect of mass spectral resolution on the detection of a metabolite of stanozolol in a real sample.

more than a factor of five. The peak in the 560 ion trace at 10.3 minutes due to the stanozolol metabolite is now part of the noise. The ion chromatograms shown at resolution 500 are similar to those that would be obtained from a bench-top quadrupole system. Comparison of the ion chromatograms at 4000 and 500 resolutions demonstrates how HRMS makes detection of low level steroids relatively easy. A confirmation procedure using a more selective extraction method confirmed the presence of 3′-hydroxystanozolol at a concentration of approximately 3 ng/mL in the athlete's urine.

Prior to 2000 all WADA accredited laboratories would have had several bench-top GC-MS systems but relatively few would have had LC-MS or LC-MS-MS systems and these would not have been used for high volume drug screening. However, now LC-MS and in particular LC-MS-MS is being used more frequently for drug screening. With the requirement to analyze for glucocorticosteroids in all competition samples being introduced in 2004 many laboratories turned to the use of LC-MS-MS as corticosteroids are difficult to derivatize and analyze using GC-MS. Our laboratory has changed from GC-MS to LC-MS-MS for the detection of diuretics and narcotics as derivatization is not needed for LC-MS-MS.

The question could be asked "why is MS-MS needed for most LC based drug screens whereas MS is adequate for most GC based drug screens?" The answer has much to do with the mode of ionization used for each method and, to a lesser extent, the comparative resolving power of the two chromatographic methods. In GC by far the most common method of ionization used is electron impact (or electron ionization (EI)) (see Section 2.1.6) where a beam of electrons typically at 70 eV is used to bombard the analytes eluting from the GC column into a vacuum. As a result the compounds fragment and produce a characteristic pattern of ions referred to as a mass spectrum. It is very unusual for different compounds even having the same molecular weight (molar mass) (unless they are geometric isomers) to produce the same pattern of ions. This different mass spectral behavior coupled with the high resolving power of capillary GC (typically more than 100,000 plates) means that it is extraordinarily unlikely for two compounds to elute at the same time and also have the same mass spectrum. This combination of separation and fragmentation makes GC-MS a powerful tool for screening and confirmation of identity.

By comparison the most common mode of ionization used for LC-MS is atmospheric pressure ionization (API), either electrospray ionization (see Section 2.1.15) or atmospheric pressure chemical ionization (APCI; see Section 2.1.8). In both modes each analyte ionizes to produce just one ion, typically the $(M + H)^+$ where M is the molecular weight (molar mass). Thus, with API several compounds having the same molecular weight (molar mass) but completely different structures would all produce the same ion. This markedly reduces the specificity of detection. In addition the resolution of LC is typically 10,000 to 20,000 plates which is an order of magnitude lower than GC. These two factors combine to severely limit the performance of LC-MS as a means of detecting drugs in sport. The lack of characteristic ionization is the main problem but this can be overcome by adding an additional layer of complexity to the mass spectrometer. One way of achieving this is to use a so-called triple-quadrupole instrument (see Section 3.1.1.1) which allows the ions separated in the first quadrupole to pass into a second quadrupole where they undergo collisions with an inert gas such as argon. The ions undergo collision induced dissociation (CID; see Section 3.2.3) and produce a series of product ions. The product ions are separated in the third quadrupole. The mass spectrum of the product ions is characteristic of the precursor ion. Although compounds with different chemical structures having the same molecular weight (molar mass) will give the same precursor ion, the fragmentation of each of these structurally dissimilar ions will be different. Thus, by using an additional layer of complexity in the mass spectrometer one can obtain characteristic mass spectra (MS-MS spectra) from the compounds of interest. In this way LC-MS-MS is able to produce the level of specificity needed for drug screening and confirmation. MS-MS spectra can also be produced using ion trap mass spectrometers which achieve a similar objective using MS-MS in time rather than in space (see Section 3.1.2.1). Now instruments are available that combine MS-MS in time with MS-MS in space to take advantage of the benefits of both systems (see examples in Section 3.1.3).

In the general scientific community the technique of LC-MS-MS is most widely used for the detection and identification of proteins and WADA laboratories are now moving into this area. The reason for this has been the rapid developments in biotechnology which have meant that naturally occurring bioactive compounds such as erythropoietin (EPO) and human growth hormone (hGH), which were previously only available in minute quantities, are now readily available. The large-scale use of EPO in sport was confirmed in the 1998 Tour de France, the so-called "Tour of Shame." Although the current detection methods used for the detection of EPO and hGH in sports doping rely on immunological techniques the rapidly improving sensitivities of LC-MS-MS will soon be applied to the detection and confirmation of doping of these and other peptide hormones. Already proteomics methodology using enzymatic digestion and LC-MS-MS has been applied by WADA laboratories to the detection of doping with hemoglobin-based oxygen carriers (HBOCs) [4]. These compounds, which are chemically modified versions of bovine or human hemoglobin, have been developed as so-called "blood substitutes." They are intended for use in emergency situations where pure matched whole blood is not available but have the clear potential to enhance the aerobic performance of an athlete. Our laboratory has been screening for HBOCs for some years but as yet no violations have been detected by our or any other WADA accredited laboratory.

A very specialized form of mass spectrometry called carbon isotope ratio mass spectrometry is now being used by many WADA laboratories in their quest to detect doping with endogenous anabolic steroids such as testosterone. Such doping is difficult to detect as testosterone occurs naturally in both males and females. Up until recently the only way of detecting such doping was by measuring the ratio of the concentrations of testosterone to epitestosterone. The ratio of testosterone to epitestosterone (T/E ratio) is relatively constant for an individual and rarely exceeds 6 naturally. The T/E ratio measurements are made using conventional bench top GC-MS systems. Administration of exogenous testosterone increases the T/E ratio not only because of the additional testosterone present but also because of suppression of epitestosterone. Doping could only be proven by carrying out follow up studies over some months. Fortunately for sports drug testers there is a subtle difference between natural and synthetic testosterone and that is in the ratio of ^{12}C to ^{13}C present. The isotopic abundance of

^{13}C is approximately 1% of that of ^{12}C but there are small differences depending on the mechanisms whereby the compound is produced. Synthetic testosterone is slightly depleted in ^{13}C compared to natural testosterone. These small differences can be detected using a gas chromatograph coupled to a mass spectrometer which is dedicated to such measurements. The eluate from the GC is combusted in a furnace and the resulting CO_2 passed to a small magnetic sector mass spectrometer which simultaneously measures the ions of mass 44, 45, and 46 using three Faraday cups (see Section 2.3.2). The precision of the technique is such that differences of less than 1 part in 10,000 can be detected. The technique was first used at the Sydney 2000 Paralympics to permit immediate penalties to be imposed for testosterone abuse.

The pressure on sports drug testing laboratories to test for more drugs without markedly increasing the cost of drug testing is a major problem. Multi-analyte GC-MS and LC-MS screens using the latest available technologies are being used to achieve the required detection limits. The expansion of the availability of protein-based drugs for the improvement of human health will lead to further applications of MS techniques developed for proteomics into the area of sports drug testing. Unfortunately there are always athletes who will try almost any means, including risking their health by the inappropriate use of legitimate pharmaceuticals, to enhance their athletic performance and win.

REFERENCES

1. WADA 2008 Prohibited List. http://www.wada-ama.org/rtecontent/document/2008_List_En.pdf.
2. G. J. Trout and R. Kazlauskas. *Chem. Soc. Rev.*, **33**(2004), 1–13.
3. M. Thevis and W. Schanzer. *Mass Spectrom. Rev.*, **26**(2007): 79–107.
4. C. Goebel, C. Alma, C. Howe, R. Kazlauskas, and G. Trout. *J. Chromatogr. Sci.*, **43**(2005). 39–46.

9

OCEANOGRAPHY

R. Timothy Short, Robert H. Byrne, David Hollander, Johan Schijf, Strawn K. Toler, and Edward S. VanVleet

The primary goal of oceanography, the scientific study of the marine environment, is a fundamental understanding of past, present, and future oceanic processes that have major impacts on the entire global ecosystem. Oceanography is interdisciplinary and typically is divided into four subdisciplines: *biological oceanography*, *chemical oceanography*, *geological oceanography*, and *physical oceanography*. Biological oceanography involves investigations of ocean plants and animals, and their ecological relationships, whereas chemical oceanography is concerned primarily with the interaction and cycling of elements and compounds within the oceans, and the exchange of chemicals across ocean boundaries. Geological oceanography involves studies of geologic structure, tectonics, and history of the ocean basins and margins, whereas physical oceanography is the study of physical properties and processes in the oceans, in particular, investigations and modeling of ocean circulation and the influence of the oceans on global heat exchange. Naturally, there is often extensive overlap among the subdisciplines.

Mass spectrometry (MS) has traditionally played a major role in both the chemical and geological subdisciplines of oceanographic research, but more recently has been introduced into the field of biological oceanography. It has also been used to a limited extent in physical oceanography, primarily for carbon dating and ocean circulation experiments that use ^{14}C tracers. The advent of in situ mass spectrometers

Mass Spectrometry. Edited by Ekman, Silberring, Westman-Brinkmalm, and Kraj

may also lead to physical oceanographic applications in the form of in situ sensors to detect chemical tracers that are used to monitor water mass flow and formation.

The challenges to achieving an integrated and systematic understanding of the oceans are substantial. The vast volume of the oceans, along with their highly heterogeneous, dynamic, and complex processes, makes adequate monitoring and sampling a daunting task. Considerable effort is devoted to (a) understanding physical and biogeochemical processes within the water column and at the sediment–water and air–water interfaces, (b) quantifying exchanges between the coastal and open ocean, and (c) determining the influence of terrestrial and freshwater sources on oceanic processes. Coupling of ocean behavior to global climate change is also a challenging and increasingly important topic of study. Development of methods to effectively characterize biological distributions, and the health and response of biological populations to environmental change (both natural and anthropogenic), is a very active area of research. For example, the declining health of the globe's coral reefs is of intense international scientific concern and may have a serious economic impact on tourism industries in certain areas. A recent area of discovery that has also captured the general public's attention is the biology and chemistry of deep-water hydrothermal vents, some of which are referred to as "black smokers." Black smokers contain high concentrations of reactive chemicals such as hydrogen sulfide, iron, and manganese. The high chemical concentrations in hydrothermal waters, often at pressures exceeding 300 atmospheres and temperatures up to 350°C or more, create dark precipitates when they are discharged into low temperature (~4°C) seawater. Characterization of these extreme environments and the biota that thrive there is a rapidly growing area of research.

Mass spectrometers have been used at some level in all of these types of investigations because of their unsurpassed sensitivity and specificity, their multicomponent analytical capability and, in some cases, their ability to provide precise and accurate isotope ratios. Traditional methods of analysis typically involve the collection of water and sediment samples, or biological specimens, during field expeditions and cruises on research vessels (R/Vs), and subsequent delivery of samples to a shore-based laboratory for mass spectrometric analyses. The recent development of field-portable mass spectrometers, however, has greatly facilitated prompt shipboard analyses. Further adaptation of portable mass spectrometer technology has also led to construction of submersible instruments that can be deployed at depth for in situ measurements.

A variety of oceanographic applications that employ mass spectrometric techniques are described below. Applications involving laboratory systems are discussed, followed by recent measurement procedures involving portable and submersible mass spectrometers. This discussion, although not definitive or all-encompassing, will hopefully illustrate the wide-ranging impact that mass spectrometry has had, and continues to have, in the field of oceanography. Finally, new national and global ocean observing initiatives will be described briefly, along with a discussion of their possible impact on the future development and use of in situ mass spectrometry in oceanography.

Organic molecules produced by marine organisms can be used to infer a wide variety of oceanic processes and pathways. High resolution gas chromatography-mass spectrometry (GC-MS) (see Chapter 4) provides a valuable tool for undertaking these studies. Molecules best suited for this type of work are those that are moderately volatile and can be passed through a GC column into a mass spectrometer. Although most molecules (including alkanoic acids, alkanols, sterols, amino acids, and simple sugars) generally require derivatization to increase their volatility, GC-MS analyses are applicable to a wide range of compounds found in the oceans. Ion fragmentation produced by electron impact can often provide definitive structural information to help identify and confirm the presence and concentration of these compounds.

Many organic molecules are produced only by a single marine species or group of related species and can be used diagnostically to infer the input of organic matter solely from those organisms. These biological marker, or "biomarker," compounds can sometimes survive unaltered in the environment for millions of years or more, and can provide unequivocal information on the contribution of these species to dissolved and particulate chemicals in the water column, as well as sedimentary environments. There are many examples of how we can use naturally produced organic molecules to gather information on various oceanic processes. As common examples, organic molecules can be used (1) to trace feeding relationships among members of a simple food chain, (2) to distinguish between terrestrial, marine, microbial, and anthropogenic sources of organic matter, and (3) to infer oceanic sea surface temperatures and help reconstruct paleoenvironment and paleoclimate conditions.

A variety of *trace elements* are essential micronutrients for organismal growth. Understanding the role of such elements in biological processes, such as primary production (e.g., phytoplankton growth), is a very active area of oceanographic research. Inductively coupled plasma mass spectrometry (ICP-MS; see Chapter 2, Section 2.1.5) is used extensively for trace element analysis of marine samples because of its ability to provide rapid, high precision multielement analyses of solids, solutions, and slurries. Trace elements can often be quantified to parts-per-trillion (ppt) levels using ICP-MS, and nearly all elements in the Periodic Table are measurable in at least a qualitative or semiquantitative manner. Quantitative analysis of trace elements in solutions as concentrated and complex as seawater is, however, challenging. Direct MS analysis of seawater gives rise to both spectroscopic and nonspectroscopic interferences. In the first category, isobaric interferences can result from polyatomic ions formed within the plasma from major seawater constituents. Second, the seawater matrix as a whole can produce signal instability (poor ion transmission in regions of excessive charge density) and long-term signal degradation (build-up of deposits on cones and ion lenses). Although a simultaneous remedy for the problems in both categories can be as simple as a 10- to 20-fold dilution of the sample, at that dilution the concentrations of most trace elements in seawater are generally far below the detection limit of quadrupole ICP-MS (see Section 2.2.3).

Isobaric interferences (especially those arising from the plasma itself, e.g., ArO^+ on Fe) can be eliminated using cool-plasma conditions, sometimes in combination with a shield torch. This option is not suitable for seawater samples because a cool plasma, in the presence of a heavy matrix, cannot fully ionize elements with high first ionization potentials, notably Zn, Cd, and Hg. Protocols have thus been established for analysis of 10-fold diluted seawater on instruments with sufficiently high resolution to separate most of the affected isotopes from their isobaric interferences [1]. To circumvent the issue entirely, others have used online chemical extraction to separate analytes of interest from the seawater matrix, thereby suppressing both categories of interferences. Such methods require that the extraction technique yields near-complete analyte recovery. This requirement can be obviated using isotope dilution analysis for certain elements. Specific problems in isotope dilution analysis, namely mass bias effects and the proliferation of elemental (as opposed to polyatomic) isobaric interferences, can be alleviated by using a multicollector ICP-MS instrument. As a means of promoting higher multielement sample throughput than can be achieved using a multicollector ICP-MS, isotope dilution methods with online extraction are being developed for high resolution ICP-MS instruments. A novel solution to many ICP-MS analysis problems has emerged with the invention of collision-cell ICP-MS [2]. These instruments are equipped with a cell within the ion path that can be filled with an inert gas at low pressure. Isobaric interferences are actively eliminated inside the cell through disruption and kinetic energy dispersion (using He), or charge exchange (neutralization) and protonation (using H_2) of the offending polyatomic ions. Detection limits on these instruments are reportedly low enough to permit direct trace element analysis of undiluted seawater.

Since the theoretical work of Harold Urey in the 1930–40s [3] (awarded the 1934 Nobel Prize in Chemistry) and the development of the isotope ratio mass spectrometer by Nier in 1940 [4], *stable isotope* measurements have played a significant role in earth and ocean science. The relatively large mass differences between stable isotopes of hydrogen (D/H), boron ($^{11}B/^{10}B$), carbon ($^{13}C/^{12}C$), nitrogen ($^{15}N/^{14}N$), oxygen ($^{18}O/^{16}O$), and sulfur ($^{34}S/^{32}S$) lead to small but measurable variations in their relative natural abundances. The distributions and isotopic fractionations of hydrogen, boron, carbon, nitrogen, oxygen, and sulfur are controlled by biological, chemical, physical, and geological processes in both oceanic and terrestrial environments. As such, through a fundamental understanding of stable isotope distributions and fractionations, the isotopic compositions of substances can be used to evaluate formative environmental processes. Stable isotope analyses of sedimentary materials (organic and inorganic) have provided researchers the ability to reconstruct environmental processes and climatic conditions throughout the geologic record. For example, Urey's seminal work on temperature-dependent isotope fractionation in the water-carbonate system [5] laid the foundation from which Emiliani [6] was able to interpret the role of temperature on the sedimentary oxygen isotopic record preserved in tests of foraminifera, and the history of glacial–interglacial changes in the ocean.

Light stable isotope ratio mass spectrometers require gas samples for analysis and a dual-inlet system that allows sequential comparisons of sample and standard gases. Early analytical limitations of dual-inlet systems necessitated manual introduction of samples and standards and thus required off-line preparation prior to analysis. This process was time consuming, required large sample sizes, and severely limited the rates of sample throughput. In spite of these limitations, isotope ratio MS has led to a fundamental understanding of the processes that underlie the light stable isotope distributions and fractionations in solids, liquids, and gases throughout a broad range of terrestrial and oceanic environments (see, e.g., Faure [7] and Hoeff [8]). Development of ancillary equipment such as a carbonate acidification device for online preparation of CO_2 gas has significantly reduced the amount of material needed for analysis. Increased accuracy and sample capacity in measurements of small samples (single foraminifera tests) have improved analytical resolution and thereby the unique capabilities that isotopic analyses provide in assessments of contemporary oceans and paleoenvironmental reconstructions.

Compound-specific isotopic analyses of carbon, nitrogen, and hydrogen have revolutionized oceanographic research in carbon cycling, microbial ecology, metabolic pathways and biosynthesis, ecosystem analysis, and paleoenvironmental and paleobiological studies. Coupling of a gas chromatograph (GC) or a high performance liquid chromatograph (HPLC) to an isotope ratio mass spectrometer via a combustion or pyrolysis interface has enabled isotopic measurements of individual organic compounds. Occasional co-elution of distinct compounds, which can produce mixed isotopic signals from multiple sources, is the principal limitation on measurement accuracy. Increasing information regarding the sources and isotopic compositions of individual organic compounds is producing an improved understanding of the role of microorganisms in coupled biological-chemical processes in the oceans and its extreme environments (e.g., deep-water hydrothermal vents).

Radiocarbon analysis, or carbon dating, is an extremely useful tool for determining the age of collected samples. The technique is based on the determination of the relative abundance of radioactive ^{14}C (5730 year half-life). ^{14}C measurements have been greatly facilitated by the advent of accelerator mass spectrometry (AMS) (see Section 2.2.7), in which ^{14}C atomic abundances are directly counted, rather than determined via ^{14}C decay rates. AMS sample preparation typically involves conversion of samples to elemental carbon, compression of the carbon into aluminum cathodes, and insertion into the accelerator ion source. Negative ions created in the cesium sputter source are accelerated into the AMS system, generating positive ^{12}C, ^{13}C, and ^{14}C ions that are magnetically separated and detected simultaneously. Relative signal intensities are converted to a radiocarbon age. High precision and accuracy can be obtained using sample sizes as small as several hundred micrograms of carbon.

In order to provide AMS analyses to the broad ocean sciences research community, the National Ocean Sciences Accelerator Mass Spectrometry Facility (NOSAMS) was established at Woods Hole Oceanographic Institution (Massachusetts) in 1989. Studies performed there include identification of sources of carbon-bearing materials in the water column and sediment, dating of sedimentary samples, investigations of paleocirculation patterns (e.g., from observations of differences in ^{14}C relative abundances in planktonic and benthic foraminifera, and coral cores and cross sections), as well as studies of modern oceanic carbon cycling and circulation. In fact, much that is known about advective and diffusive processes in the ocean comes from measurements of chemical tracers, such as ^{14}C, rather than from direct measurements of water mass flow.

In addition to the dissolved elements and compounds in the oceanic water column, a wide variety of water column chemicals are found in *marine organisms* and organic detritus. For example, a milliliter of surface seawater can contain on the order of 10 million viruses, 1 million bacteria, 100,000 phytoplankton, and 10,000 zooplankton [9]. With the advent of soft ionization processes for mass spectrometry systems, scientists have been able to study these marine organisms at molecular level. The use of electrospray ionization (ESI; see Section 2.1.15), atmospheric pressure chemical ionization

(APCI; see Section 2.1.8), and matrix-assisted laser desorption/ionization (MALDI; see Section 2.1.22), which in some cases is preceded by liquid chromatography, has enabled the analysis of polar, semipolar, and nonvolatile compounds (including proteins) from marine organisms. Coupling of soft ionization techniques with very high resolution mass spectrometric techniques, such as time-of-flight (TOF; see Section 2.2.1) and ion cyclotron resonance (ICR; see Section 2.2.6) has led to unprecedented molecular-level characterization of compounds. These novel techniques have been used to complete the first phytoplankton genome, to trace organic matter and nutrient pathways within both individual organisms and microbial webs (e.g., bacteria, phytoplankton, zooplankton), and to detect toxins produced by harmful algal blooms, (e.g., "red tide"), which can pass through food webs and affect and kill zooplankton, fish, birds, and even marine mammals, such as manatees.

Global climate change is influenced to a large degree by *gases* whose biogeochemical cycles include marine sources and sinks. As such, gas flux measurements within the ocean and across ocean boundaries are essential to the formulation of accurate predictive climatic models. In addition, temporally and spatially resolved measurements of dissolved gas concentrations can be used to monitor many physical, chemical, and biological processes (e.g., primary production and respiration). Mass spectrometry has been used in recent years for highly sensitive and precise determinations of dissolved gas concentrations, especially because of its ability to provide data for multiple gases from a single water sample. Very precise and accurate measurements have been developed for analysis of discrete samples (e.g., isotope dilution MS analysis of the headspace of equilibrated water samples), but such methods often require highly labor-intensive sample preparation.

Membrane introduction mass spectrometry (MIMS) is an alternative technique for dissolved gas analysis that can provide high-frequency, real-time measurements of dissolved gases, albeit typically with lower precision and accuracy. MIMS systems eliminate the need for headspace equilibration. Dissolved gases, and light-weight nonpolar volatile organics enter the MS ion source by pervaporation through a semipermeable membrane (e.g., polydimethyl siloxane—PDMS). Recent progress in the development of small, portable, and rugged mass spectrometers, and membrane interfaces that reduce MS vacuum system gas loads compared to direct sampling interfaces, has led to shipboard MIMS systems [10], and submersible MIMS systems [11, 12], capable of real-time, spatially resolved, high-density dissolved gas measurements. These portable MS systems eliminate sample collection, storage, and transport steps, in which there is a high risk of contamination and sample degradation. While shipboard systems can provide low detection limits and a high degree of spatial and temporal resolution for a variety of biogeochemically important gases, submersible systems offer the additional advantage of remote and autonomous operation (although there is often a slight decrease in sensitivity for these smaller, more portable systems). Submersible linear quadrupole (see Section 2.2.3) and ion trap (see Section 2.2.4) MIMS systems have been deployed on moorings, tethered depth-profiling rosettes, remotely operated and autonomous aquatic vehicles [13], and under the control of SCUBA divers. MS systems deployed on aquatic vehicles equipped with a global positioning system (GPS), can be used to simultaneously create two- or three-dimensional chemical maps of a wide variety of analytes.

The oceanographic community is initiating capabilities for permanent, continuous, and pervasive monitoring of oceanic processes, similar in concept to what has been established by the meteorological and atmospheric scientific communities. Programs such as the Global Ocean Observing System (GOOS), the European Sea Floor Observatory Network (ESONET), and the U.S.-based Integrated Ocean Observing System (IOOS) are currently under development and will likely revolutionize the field of oceanography. These observing systems will consist of coastal and regional cabled and global components. These proposed networks promise to provide ocean monitoring on an unprecedented spatial and temporal scale to address important scientific issues, including (a) climate variability, ocean food web dynamics, and biogeochemical cycles; (b) coastal ocean dynamics and ecosystems; (c) turbulent mixing and biophysical interactions; and (d) global and plate-scale geodynamics. The regional and coastal cabled observatories will provide spatially distributed networks of high-power and high-bandwidth-communication nodes, and should offer an exciting opportunity for long-term distributed monitoring using in situ MS. Networks of long-duration fixed or profiling mass spectrometers would provide previously unobtainable data on temporal and spatial scales relevant to a variety of important physical and biogeochemical processes. Although there are significant obstacles to the establishment of a distributed long-term in situ MS presence in the oceans (e.g., autonomous maintenance, biofouling, and calibration issues), the potential impact of a versatile, high-sensitivity MS sensor network is exciting and intriguing.

REFERENCES

1. M. P. Field, J. T. Cullen, and R. M. Sherrell. Direct Determination of 10 Trace Metals in 50 μL Samples of Coastal Seawater Using Desolvating Micronebulization Sector Field ICP-MS. *J. Anal. Atom. Spectrom.*, **14**(1999): 1425–1431.
2. H. Louie, M. Wu, P. Di, P. Snitch, and G. Chapple. Determination of Trace Elements in Seawater Using Reaction Cell Inductively Coupled Plasma Mass Spectrometry. *J. Anal. Atom. Spectrom.*, **17**(2002): 587–591.
3. H. C. Urey. F. G. Brickwede, and G. M. Murphy. An Isotope of Hydrogen of Mass 2 and its Concentration (Abstract). *Phys. Rev.*, **39**(1932): 864.
4. A. O. Nier. A Mass Spectrometer for Routine Isotope Abundance Measurements. *Rev. Sci. Instr.*, **11**(1940): 212–216.
5. H. C. Urey. The Thermodynamic Properties of Isotopic Substances. *J. Chem. Soc.*, (1947): 562.
6. C. Emiliani. Pleistocene Temperatures. *J. Geol.*, **63**(1955): 538–578.
7. G. Faure. (1986). *Principles of Isotope Geology*. New York: Wiley.
8. J. Hoeff. (1980). *Stable Isotope Geochemistry*. Berlin: Springer-Verlag.
9. K. S. Johnson, K. H. Coale, and H. W. Jannasch. Analytical Chemistry in Oceanography. *Anal. Chem.*, **64**(1992): 1065–1075.
10. P. D. Tortell. Dissolved Gas Measurements in Oceanic Waters Made by Membrane Inlet Mass Spectrometry. *Limnol. Oceanogr. Methods*, **3**(2005): 24–37.

11. G. P. G. Kibelka, R. T. Short, S. K. Toler, J. E. Edkins, and R. H. Byrne. Field-Deployed Underwater Mass Spectrometers for Investigations of Transient Chemical Systems. *Talanta*, **64**(2004): 961–969.

12. R. T. Short, D. P. Fries, M. L. Kerr, C. E. Lembke, S. K. Toler, P. G. Wenner, and R. H. Byrne. Underwater Mass Spectrometers for in situ Chemical Analysis of the Hydrosphere. *J. Am. Soc. Mass Spectrom.*, **12**(2001): 676–682.

13. R. Camilli, and H. F. Hemond. *Trends Anal. Chem.*, **23**, no. 4(2004): 307.

10

"OMICS" APPLICATIONS

Simone König

10.1. INTRODUCTION

Genomics, transcriptomics, proteomics, or metabolomics are terms for research areas that developed in the course of high-impact analytical improvements. Ever since it became possible with the advent of electrospray ionization (ESI; see Section 2.1.15) and matrix-assisted laser desorption/ionization (MALDI-MS; see Section 2.1.22) to investigate large numbers of biomolecules within reasonable time frames and with acceptable financial and personal input, scientists have wanted to study the total complement of any given kind of molecule with the goal of elucidating the function of the whole system (Fig. 10.1). Hans Winkler, professor of botany at the University of Hamburg, Germany, is credited with using the term "genome" for the first time in 1920, describing the whole hereditary information of an organism. However, it became popular only when techniques such as gene chips arrived. The capabilities of biological MS in high throughput analyses are largely responsible for the wave of further -omics, in particular proteomics and metabolomics.

Each of the -omics (Fig. 10.2) areas focuses on one group of biomolecule which is in itself structurally consistent. Each group requires quite different methods of preparation, handling, and analyzing. For very complex groups of biomolecules such as proteins, which can carry various modifications, subomics have formed. They deal with

Mass Spectrometry. Edited by Ekman, Silberring, Westman-Brinkmalm, and Kraj

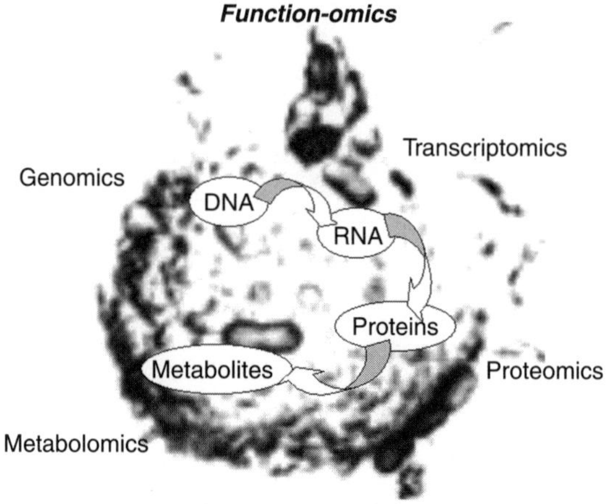

Figure 10.1. "Omics" technologies investigate the different biomolecules in a cell, tissue, or other defined biological system in order to elucidate function.

> An **omics** is an English neologism referring to a field of study in biology and biochemistry, ending in the suffix -omics, such as genomics or proteomics.
>
> The related neologism **omes** are the objects of study, such as genomes or proteomes, respectively.
>
> Omes stems from the Greek for "all," "every," or "complete."

Figure 10.2. Definition of -omics and -omes. From Wikipedia.

phosphorylated (phosphoproteomics) or glycosylated (glycomics) proteins. Table 10.1 describes a number of the commonly encountered -omics as well as terms that are not widely used. This development has led to the foundation of a number of new scientific journals, such as *Proteomics*, *Genome Research*, or *OMICS: A Journal of Integrative Biology*, in recent years. It is obvious that an "omification" takes place also in areas not immediately requiring the analytical techniques described here.

Modern mass spectrometers can be utilized in many different ways in all "analytomics" (-omics requiring analytical technology) although some instrumental configurations have advantages for a particular purpose, such as triple-quadrupoles (see Chapter 3, Section 3.1.1.1) in quantitation or Fourier transform ion cyclotron resonance (FT-ICRs; see Section 2.2.6) for high mass accuracy. A common theme in all -omics is that it becomes increasingly difficult to make meaningful comparisons of large numbers of spectra, chromatograms, or other data manually and, therefore, bioinformatical input such as multivariant statistical methods is necessary to assist in visualizing how samples relate to each other and extract statistically secured information.

TABLE 10.1. "Omics" and Their Research Objects: Established Terms and Examples for the -omics Flood

-omic	Study Object (Totality of Class of Molecules in a Given Biological System)
Genomics	**DNA**, genes
Transcriptomics	**mRNA** complement
Ribonomics	mRNA that binds with proteins
RNomics	RNA
Structural genomics	gene localization, mapping, and sequencing
Functional genomics	gene function and regulation
Pharmacogenomics	correlation of genetic basis, disease and drug treatment
Toxicogenomics	interaction of toxic environment and genome
Chemogenomics	effects of chemicals on genes or the encoded products
Nutrigenomics	influence of nutrition on genome
Epigenomics	influence of environmental factors on development of genes (e.g., DNA methylation)
ORFeomics	DNA sequences that begin with the initiation codon ATG, end with a nonsense codon, and contain no stop codon (code for part or all of a protein)
Cardiogenomics	cardiac genome
Clinomics	clinical genomics
Proteomics	**proteins, peptides**
Peptidomics	peptides
Structural proteomics	protein structure and interaction
Functional proteomics	protein function with respect to structure
Spliceomics	alternative splicing protein isoforms
Phosphoproteomics	**phosphorylated proteins and peptides**
Glycomics/ Glycosilomics	**glycosylated biomolecules**, mainly proteins and peptides, **carbohydrates**
CHOmics	carbohydrates
Kinomics	protein kinases
Neuroproteomics	proteins in neuronal systems
Crystallomics	protein crystals
Proteogenomics	summarizes the efforts in genomics and proteomics for the comprehensive description of functional processes
Metabolomics	**metabolites**
Metabonomics	quantitative measurement of metabolic responses to changes (drugs, disease, environment)
Lipidomics	**lipids**
Metallomics	metals and metalloid species
Cellomics, cytomics	study of cell phenotype and function
Chromonomics	chromosomes
Expressomics	expressed entities
Functomics	functional entities
Interactomics	molecular interactions (systems biology)
Complexomics	molecular complexes

(Continued)

TABLE 10.1. *Continued*

-omic	Study Object (Totality of Class of Molecules in a Given Biological System)
Reactomics	biological processes
Regulomics	regulation of biological processes
Phenomics	phenotype
Behavioromics	behavior
Physiomics	physiological dynamics and functions of whole organisms
Chronomics	biological time
Biomics	ecology
Ecomics	ecosystems
Taxomics	taxonomy
Predictomics	complete sets of predictions
Textome	body of scientific literature accessible to text mining
Bibliomics	bibliography

10.2. GENOMICS AND TRANSCRIPTOMICS

Study object: DNA, RNA, oligonucleotides gene microarrays
Techniques: PCR (polymerase chain reaction) mass arrays (Sequenom.com)

Biomolecular MS and in particular MALDI-TOF-MS (see Sections 2.1.22 and 2.2.1) permit the routine analysis of oligonucleotides up to 70-mers, intact nucleic acids, and the direct detection of DNA products with no primer labels with an increase in analysis speed and mass accuracy especially in contrast to traditional DNA separation techniques such as slab gels or capillary electrophoresis. Applications focus on the characterization of single nucleotide polymorphisms (SNPs) and short tandem repeats (STRs). Precise and accurate gene expression measurements show relative and absolute numbers of target molecules determined independently of the number of PCR cycles. DNA methylation can be studied quantitatively.

A commercial mass array (Sequenom.com; Mass Array application notes) is based on the annealing of a primer adjacent to the polymorphic site of interest. The process starts with template amplification following genomic extraction. Dephosphorylation using arctic shrimp alkaline phosphatase removes any residual amplification nucleotides and prevents their future incorporation and interference with the primer extension assay. Then, primer, DNA polymerase, nucleotides, and terminators (dNTPs, ddNTPs) are added to initiate the primer extension reaction, which generates allele-specific primer extension products that are generally one to four bases longer than the original primer (Fig. 10.3a, b). Nucleotide mixtures are selected to maximize mass differences for all potential products. Appropriate deoxynucleotides are incorporated through the polymorphic site until a single dideoxynucleotide is added and the reaction terminates (Fig. 10.3c). Since the termination point and number of nucleotides is sequence specific, the mass of the extension products generated can be used to identify the possible variants

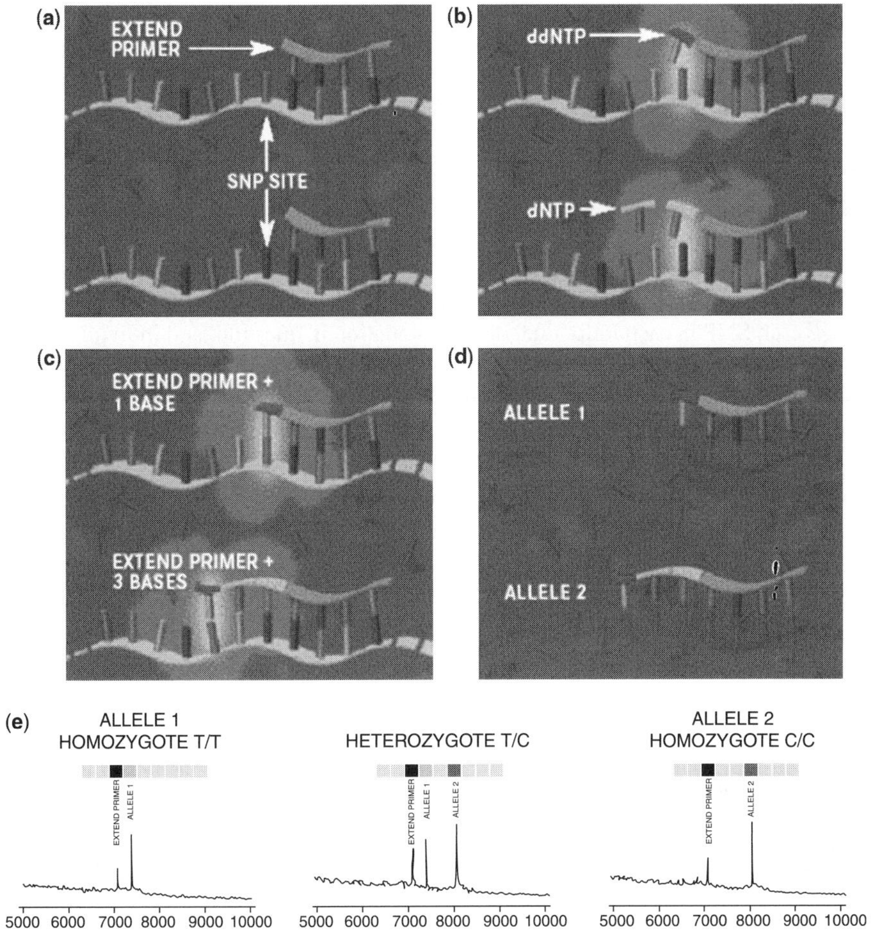

Figure 10.3. Mass array. (a) Primer binding; (b) primer extension; enzyme, ddATP and dCTP/dGTP/dTTP addition; (c) primer terminates; (d) primer extension products ready for MALDI-MS; (e) MS spectrum of primer extension products. Each addition of a nucleotide to the primer extension product increases the mass by 289 to 329 Da, depending on the nucleotide added. The mass difference is easily resolved by MALDI-TOF, which has the ability to detect differences as small as 3 Da. Printed by kind permission of Sequenom. (See color insert.)

without errors (Fig. 10.3d). The resultant mass of the primer extension product is then analyzed and used to determine the sequence of the nucleotides at the polymorphic site (Fig. 10.3e).

Comparative sequence analysis based on MALDI-TOF-MS analysis of nucleic acids cleaved at specific bases and reference sequences used to construct in silico cleavage patterns enable cross-correlation of theoretical and experimental mass signal patterns. Observed signal pattern differences are indicators of sequence variations and form the basis for SNP analysis with numerous applications, such as the detection of variations associated with inherited susceptibility to disease. With reference sequences available for the human genome and constantly expanding genomic sequence information of various other species, the method is a tool for identifying sequence variations such as single-base substitutions, insertions, and deletions. Performing multiple reactions in a single well (multiplexing) allows increasing the throughput and reducing the cost per genotype.

Rapid, accurate SNP validation can be carried out using a sample-pooling technique—allele frequency—that rapidly screens and confirms the presence of an SNP and its allelic frequency in patient populations. Conventional technologies typically analyze each SNP in each individual of the population in question, and individual results are then consolidated to yield the overall SNP allele-frequency distribution. MS-based technology is able to determine SNP allele frequencies with high precision in pooled samples, thereby replacing hundreds of individual measurements with one consolidated analysis. To that end, DNA is isolated, purified, and quantitated. A pool of DNA is formed from a high number of different individuals, amplified, and mass measured.

10.3. PROTEOMICS

Study object:	Proteins, peptides, amino acids
Techniques:	Two-dimensional gel electrophoresis, Multidimensional chromatography, MS protein identification, MS profiling

The statement that genomic analysis is sufficient for disease-related research has long been proven wrong. In contrast, the translation products, the proteins, are the executers of cell functions and are of major interest. The example of the caterpillar and the butterfly that have the same genotype, but a quite different phenotype, is often used to exemplify the need for proteomics. However, the study of proteins is complicated by a number of facts:

- One gene codes for more than one protein (e.g., in bacteria, for about three in yeast, six in human)
- Alternative splicing leads to different isoforms of the same protein (Fig. 10.4)
- Hundreds of posttranslational modifications create heterogeneity
- Different sites in one protein may be modified without functional relevance
- Variations within the modifying group (e.g., oligosaccharide) create microheterogeneity
- There may be directed and unspecific protein cleavage
- Sample handling may introduce modifications
- The dynamic range of protein expression is $>10^{10}$
- Proteins cannot be amplified for analysis and must be obtained from the biological system in sufficient amounts

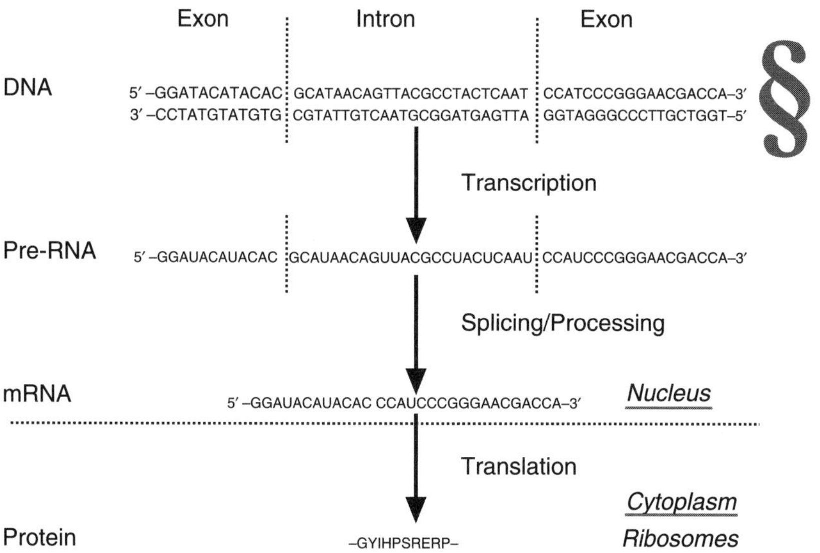

Figure 10.4. Translation of the genetic blueprint into proteins. Alternative splicing leads to isoforms. After translation proteins are often modified to become functional.

The dynamic range of protein expression represents a main obstacle since abundant proteins are seldom of interest and others such as transcription factors are only present in a few copies. There is no detector that is able to visualize all proteins at the same time so that prefractionation and the investigation of subproteomes is required. In fact, pre-MS sample preparation techniques exploiting electrophoretic, chromatographic, or chemical properties of the analyte are often the bottleneck of proteomics.

In general, protein mixtures are purified from a given tissue or cell culture and separated into fractions of one or very few proteins. 2D-PAGE is often used for this, providing thousands of protein spots on a polyacrylamide gel matrix. Spots can then be excised and enzymatically digested in the gel. The resulting peptides are extracted and subjected to MS mapping and sequencing, leading to protein identification. (Fig. 10.5).

For reproducible expression analysis and protein quantification MS methods based on isotopic labeling are available. They were designed in conjunction with two or more dimensional chromatographic peptide separation coupled online to MS and require advanced bioinformatics input to analyze the complex data sets in a reasonable time frame. This is also true for the alternative fluorescence-based technology of differential gel electrophoresis (DIGE; Fig. 10.6) with tailor-made software which allows statistical validation of multiple data sets.

A newer trend in MS born out of the need to adapt to requirements from clinicians and diagnostics is analyte profiling. There, tests have to be simple and reproducible, preferably on crude samples such as serum or urine, and diagnostic results have to be absolutely correct. Another point is that the view that one disease can be associated with or

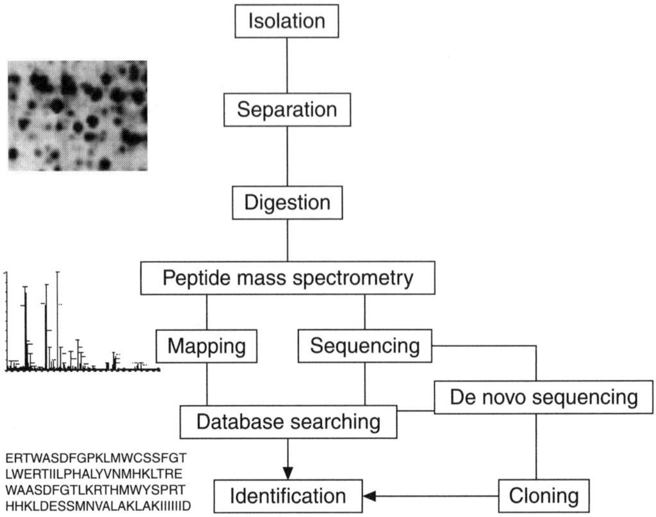

Figure 10.5. Proteomic workflow via protein separation, digestion, MS analysis, and database search.

recognized by one or very few biomarkers has been complemented by a systems biology approach. It is stated that any disease or other biological process will manifest in more than one change. For instance, an old person can be differentiated from a young person in many properties and the fact that someone has grey hair or no hair does not necessarily

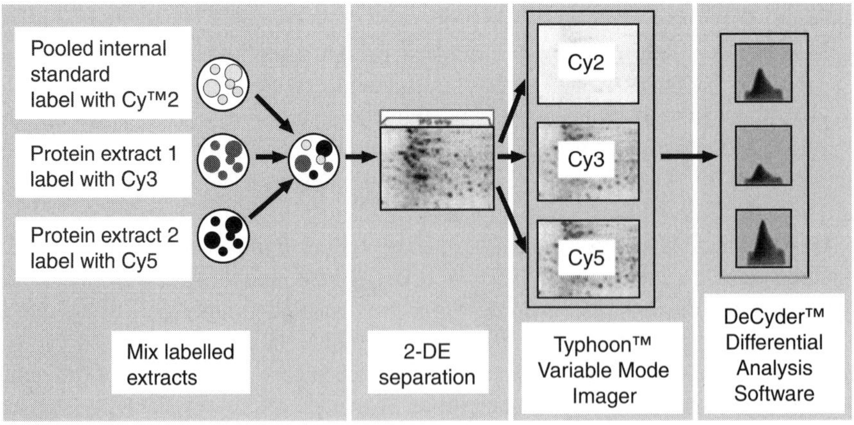

Figure 10.6. Principle of DIGE analysis: separation of control and treated sample on one gel and statistical validation using more than three repeated experiments. Printed by kind permission of GE Healthcare (formerly Amersham Biosciences). (See color insert.)

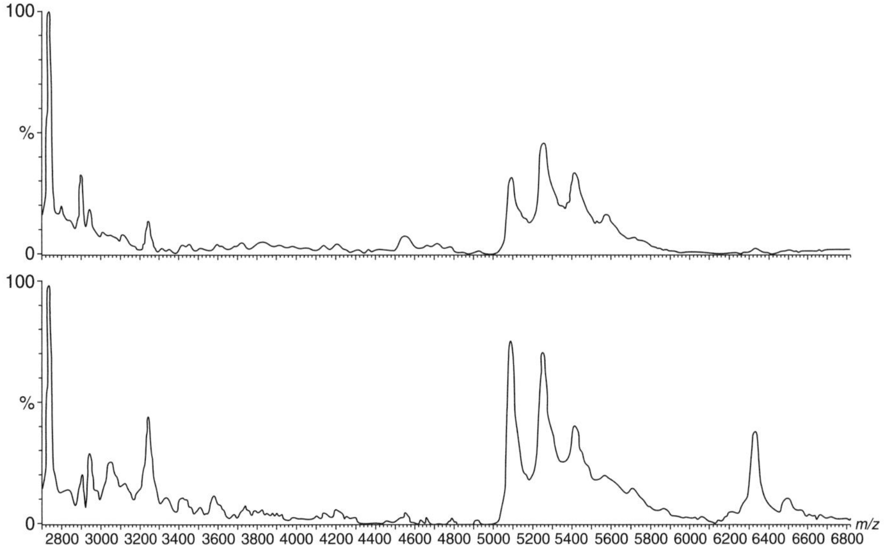

Figure 10.7. Comparison of two serum MS profiles where the appearance of the peak at m/z 6350 Da could be viewed as indicator. However, the samples were identical and were prepared in the same manner in parallel in order to show reproducibility of the handling procedure demonstrating the difficulties in sample workup.

mean that he or she is old. Therefore, instead of molecule-specific diagnostics, profiles and images of biological systems are compared to find disease markers. MS can be used as a detector for profiling of proteins or peptides and reproduced differences in profiles could be associated with a property of the sample. However, MS techniques suffer from the inherent disadvantage that they generate, in general, qualitative, not quantitative results due to ion suppression and ionization efficiency effects. In addition, slight variations in sample handling and preparation might have an influence on the spectral output (Fig. 10.7). Therefore, MS profiling has not yet reached the confidence level to be suitable for use in diagnostics.

10.4. METABOLOMICS

Study object:	Small molecular weight (molar mass) organic compounds, inorganic ionic species
Techniques:	^{1}H and ^{13}C NMR, also coupled to LC
	GC and HPLC, also coupled to MS

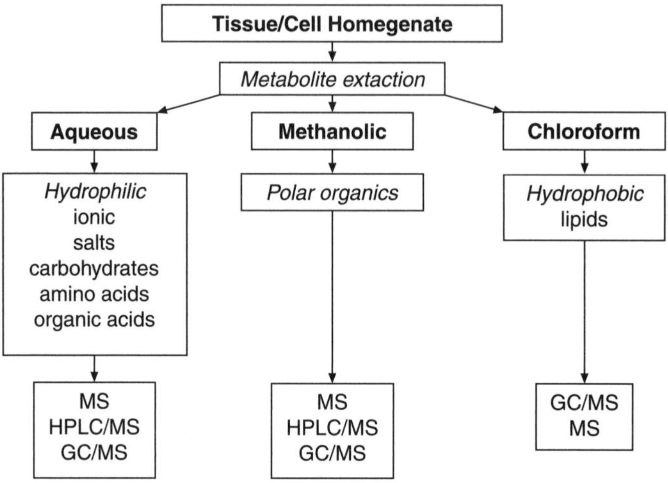

Figure 10.8. Classes of metabolites and analytical access.

With the advances in genomics and proteomics researchers realized that they needed to study metabolites in the same systematic fashion to get a more complete insight into the complex biological processes. This is essentially done with the established techniques, but with much bioinformatical input in order to handle the large data sets (principle component analysis, hierarchical clustering, self-organizational maps, neural network). Also the need for diagnostic tests is a major driving force although metabolites tend to be poor markers. Concentrations may fluctuate simply due to diet.

As in the other -omics, analyses may be directed at a specific metabolite, at all metabolites in a given system in a shot-gun approach, or at accessible groups of molecules in profiling experiments. In that also the technology varies. In addition, the chemistry of different metabolites is very heterogeneous since it involves hydrophobic lipids, hydrophilic carbohydrates, ionic inorganic species, and other secondary natural products and already the choice of solvent in metabolite extraction dictates which types of molecules will be present (Fig. 10.8). Therefore, total metabolome profiling is not possible, because no analytical method will be able to accommodate all the different molecule classes at once.

11

SPACE SCIENCES

Robert Sheldon

11.1. INTRODUCTION

The adjective "space" in the chapter title loosely means "extraterrestrial" and could include planetology, the study of other solid bodies in the solar system, such as Mars, Comet Halley, or asteroid Ceres. While MS is vital to all planetary exploration, these devices function much the same way as laboratory MS, except that they are remotely operated, use less power, and are considerably more expensive. But "space" can also have the more restricted meaning of "outside the ionosphere of any planet, but inside the solar system," which will be the area discussed in this chapter. The properties and challenges of this region are very different from the lab, although the science turns out to be often the same.

In this region of space, matter is very tenuous, so for example, at 1 AU (AU = astronomical unit = the average distance from the Sun's center to the Earth's center = 149,597,871 km) distance from the Earth the density is approximately three atoms per cubic centimeter, which would pass for ultra-high vacuum in the lab! Furthermore, all those atoms are elemental ions, both because of their likely origin in the solar atmosphere, and because the Sun is a prolific source of ultraviolet, hydrogen Lyman-alpha radiation that rips apart and ionizes molecules. For example, interstellar hydrogen atoms entering the solar system with a 13 eV ionization potential and traveling some 30 km/s can only make it in to about Jupiter's orbit before being ionized by the

Mass Spectrometry. Edited by Ekman, Silberring, Westman-Brinkmalm, and Kraj

Sun. Helium, with the highest ionization energy of any element, 25 eV, barely makes it to 1 AU before being ionized by the Sun. Therefore, in space, any observation of neutral atoms, or even molecular ions, must be of a temporary population arriving from a nearby planetary body or comet.

Accordingly, space mass spectrometers are different from laboratory MS in that they have no need for an ionizer or a vacuum pump. As we will discuss below, these two advantages are far outweighed by the disadvantages of space, which are responsible for the paradoxical observation that space mass spectrometers costing millions of dollars have only recently achieved a mass resolution of $m/\Delta m = 100$! So the main challenge facing space MS is getting a technology that works and is reliable enough to last 5 to 10 years in a brutal environment without human intervention [1]. But perhaps we should first ask what information a space MS gives us that is important and unique.

11.2. ORIGINS

When the United States launched its first satellite, Explorer I, in 1957, Van Allen's biggest surprise was finding that his cosmic ray experiment saturated in the previously unknown radiation belts surrounding the Earth [2]. Space was full of energetic ions that were dangerous for astronauts and electronics alike. However, what makes them dangerous is primarily their energy, and only secondarily their composition, so mass analysis was not emphasized. Even today, the field of megaelectronvolt ion composition is dominated by cosmic ray "high energy" physicists rather than mass spectrometer "plasma" physicists, with little overlap of techniques, science, or goals of the two communities. So despite the obvious connection to the plasma origin of cosmic rays posed by Fermi a half century ago, we will confine the discussion in this chapter to techniques that measure ions with energies less than 100 keV or so.

In the heady decade of the 1960s, satellites that orbited above the atmosphere but below the radiation belts were found to be useful for communications, weather observations, and of course, intelligence gathering. The United States had a fleet devoted to monitoring nuclear explosions and compliance with nuclear test ban treaties [3]. They quickly found that in this plasma environment, electronvolt ions and photoelectrons caused annoying sparks on nonconductive glass optics, which could easily be mistaken for nuclear explosions. Thereafter, plasma diagnostic packages became standard for nearly all satellites, commercial and defense-related, to monitor the environment. In low Earth orbit (LEO), oxygen often dominates over hydrogen as the main component of the plasma, depending on location, auroral activity, and time of year. So these first plasma environmental diagnostics evolved to become crude mass spectrometers, with a mass resolution of about four to separate the three main constituents of Earth plasmas: hydrogen, helium, and oxygen ions. This type of diagnostic was used to explain the sudden degradation of NASA's Chandra telescope X-ray detectors due to kiloelectronvolt ions, or the Jeans escape paradox of the disappearance of atmospheric helium. (Heavy atmospheric atoms can be lost more easily as ions than as neutrals. Even molecular nitrogen ions have been observed escaping Earth's gravity! [4])

In the decade of the 1960s, satellites explored more of the Earth environment, mapping beyond the radiation belts to the magnetically confined ions surrounding the Earth, the magnetosphere [5]. They discovered the $\sim$1 keV/nucleon ions streaming from the Sun, the solar wind [1, 6], and its plasma interaction with the Earth's magnetic field, the magnetosphere.

The magnetosphere was found to have a sharp boundary at the front, and a long magnetic geotail trailing invisibly behind the Earth like a windsock, many times longer than the distance to the Moon. And most of these discoveries were not predicted, partly because the conditions of space could not be duplicated in the laboratory, and partly because the physics of plasma "magnetohydrodynamics" was more complex even than Navier-Stokes hydrodynamics. Minor details, such as the mass composition of the plasma and its origins, had to wait for the theory to catch up to the gross morphological observations.

Now the composition of the solar wind, like the Sun itself, was expected to be 90% hydrogen and 10% helium, with less than 1% oxygen and minor constituents, all determined telescopically [7]. Optical spectroscopy, however, cannot determine isotopic composition, so that models of nucleosynthesis in the Big Bang, stars, or supernovas could only be tested on Earth rocks, on meteorites, and after 1968, on Moon rocks [8]. This type of extrapolation from meteoritic to galactic composition had built in pitfalls, such as ^{40}Ar being the most common isotope of argon on the Earth (due to radioactive decay), but rare in space. This was shown in the 1970s, when the later *Apollo* missions flew an aluminum foil experiment designed like a window shade that was unrolled on the Moon to collect solar wind noble gases for later isotopic analysis on the Earth [9]. For the majority of the isotopes, however, the largest mass in the solar system, the one that dominates the cosmochemistry chronology of all the planets, the Sun, had never been measured.

The first few decades of space exploration went by without any MS better than $m/\Delta m \sim 4$ because understanding the solar wind flow, its density, its pressure, or its temperature did not require mass resolution. Not until 1984 did a space MS fly with a resolution of about 10 [10], and it was 1994 before that increased to 100 [11]. Suddenly for the first time, the solar wind isotopes of carbon, oxygen, magnesium, silicon, and iron were known, and the models could be tested. Even then, space MS had difficulty measuring rare isotopes, so that it wasn't until 2005 that solar wind samples were returned to Earth inside ultra-pure silicon wafers (the ill-fated *Genesis* mission [12]) to determine the important triple ratios of $^{16}O:^{17}O:^{18}O$.

Such a measurement can tell us about the chemical evolution of oxygen, such as whether the isotopes differentiated via a thermal cycle in which lighter ^{16}O fractionates from the heavier ^{18}O, much as Vostok ice-core oxygen ratios reveal the Earth's prehistoric climate. From this fixed point of the Sun's oxygen ratios, we can then trace the history of water in other planetary bodies since their birth in the solar nebulae through the subsequent cometary bombardment [13]. In NASA's search for water on the Moon, important for the establishment of a future Moon base, such isotopic ratios will determine whether the water is a vast mother lode or just a recent cometary impact residue.

So we see that space MS functions much as laboratory MS, unlocking secrets about origins, the origin of plasmas, the origin of planets, water, and maybe even the Holy Grail, life itself. Why is this capability unique to mass spectroscopy? Because elemental isotopic ratios are barely affected by the forces that homogenize plasmas or destroy the evidence from the early solar system: the meteoritic bombardment, the atmospheric chemical changes, or the motion of tectonic plates. The same properties that make laboratory MS so useful in forensics also make space MS valuable in NASA's Origins theme. There are, however, yet more uses for mass spectroscopy related to its ability to track dynamic changes.

11.3. DYNAMICS

The European Space Agency's *Ulysses* space probe showed that there are two kinds of solar wind: fast wind coming from the poles of the Sun, and slow wind coming from regions closer to the equator. These two types of wind were analyzed with a space MS with resolution R $\sim$10 and found to have differing magnesium-to-oxygen ratios, indicative of the differing temperatures of their differing origins in the solar atmosphere [14]. As these two populations leave the Sun, the faster wind piles up against the slower, causing a huge solar system traffic jam referred to as the corotating interaction region (CIR) with copious shock-accelerated ions. Not only do CIRs produce energetic ions hazardous at Earth, but they modify the entire solar system, causing Forbush decreases in the cosmic ray flux, which in turn, modulates the ^{14}C production at Earth [15] (and some have even suggested that it modulates the tropospheric climate). After several decades of study, models are just now being tested that can predict these CIRs starting with optical observations of the Sun. Testing of these models requires far flung spacecraft to pinpoint the CIR boundaries at widely spaced positions in the solar system. Perhaps not surprisingly, plasma measurements of density or magnetic field cannot reveal unambiguously which part of the traffic jam is being observed, whereas composition measurements can. In this case, minor ions in the solar wind act as tracers, like smoke particles in a wind tunnel that reveal the dynamics of the underlying fluid. It was for exactly this purpose that the Active Magnetospheric Particle Tracer Experiment (AMPTE) launched barium and lithium canisters in 1984 to track ion flows with space MS [16].

And because these minor ions have different gyro-radii in the magnetic field carried by the solar wind, they sample different spatial scales and can reveal even greater secrets about the structures in the solar wind: the discontinuous shock interface, the interaction with the Earth's shock, or reconnecting plasma flux tubes. Already we have been able to sort out the structure of a solar eruption that swept past the Earth on January 10, 1997 based on these trace elements [17]. Even greater secrets remain to be revealed when space MS can operate at higher cadence. Currently *Ulysses* requires 13 minutes to complete a mass-energy spectrum, during which time a 400 km/s solar wind has traveled three-quarters of the distance from the Earth to the Moon. Future space MS with cadences of a few seconds will provide the data needed to calibrate the rapidly growing field of MHD space plasma simulations.

11.4. THE SPACE MS PARADOX

As with many things built for space, the first reaction on hearing the cost is surprise that such expensive MS have such limited abilities. One answer is that space is most unforgiving. Many ambitious designs were launched in the 1960s that never returned useful data. In this section we look at some of the unique difficulties encountered by space MS, as well as some of the techniques that have worked.

Solar wind ion densities of 3/cc traveling 400 km/s give respectable fluxes of 120 million/s-cm^2 of solar wind, but this is against a background photon flux about $1.2\,\mathrm{kW/m^2} = 0.12\,\mathrm{W/cm^2} \sim 10^{18}\,\mathrm{eV/s\text{-}cm^2} \rightarrow$ 10 billion more photons than particles. Most detectors sensitive to the electronic excitation caused by a 1 keV/nucleon ion traveling just faster than the electron Bohr velocity, would be equally sensitive to a 1 eV electron traveling at the same speed, whose energy corresponds to a solar UV photon. Therefore, the first attempts to measure solar wind saw only photons. Even a carbon black surface with an albedo $<0.1\%$ would require a minimum of three bounces before the UV photon flux were suppressed enough to see the ions. From bitter experience, space MS have electrostatic ion deflection systems that pass the ions through a curved, blackened entrance system designed to reduce the UV flux by about a billion. Such a deflection system naturally selects only ions of the correct energy to pass through the entrance voltages, which makes the instrument both a mass analyzer and an energy analyzer, further reducing the measurable flux.

A second problem not often appreciated in lab MS is the effect of energy spread, which in both TOF-MS (see Section 2.2.1) and magnetic sector MS (see Section 2.2.2), directly degrade the mass resolution. In lab MS, this is solved by the ionizer, which produces ions from cold neutrals with a thermal spread of less than an 1 eV, subsequently post-accelerated up to 1000 eV, resulting in a $\Delta E/E < 0.1\%$. But in space, the solar wind ions come pre-accelerated to 1 keV and at arbitrary angles to the instrument, resulting in energy spreads of $\Delta E/E > 100\%$. Since we extract them from the photon flux with an energy filter, we can pick an energy bandwidth, but the reduction in spread comes with a reduction in sensitivity. As we argue later, this reduced sensitivity limits the typical space MS bandwidth to $\Delta E/E > 5\%$.

A third problem with space MS is the necessity to pack it into a small volume. A rule of thumb for satellite instrumentation is that the final package will have the density of water, 1 gm/cc. Since typical scientific satellite payloads weigh in the neighborhood of 100 kg, and room has to be provided for 5 to 15 instruments, typical weight allotments come out in the 5 to 10 kg range, or 5 to 10 liters of volume. Ten liters is a cube with 21 cm on a side, which has to accommodate not just instrument volume, but power supplies, computer boards, and packaging. Space MS that rely on spatial separation of differing masses, such as magnetic sector MS, would have a maximum lever arm of about 10 cm. This volume limitation proves to be a fundamental limitation for high resolution magnetic MS (Fig. 11.1). For example, in 1994 a state-of-the-art toroidal magnetic sector space MS that tried to pack both UV deflection and a magnet into the same volume with position sensitive multichannel analyzer readout ultimately achieved only mass resolution R $\sim$5 [12].

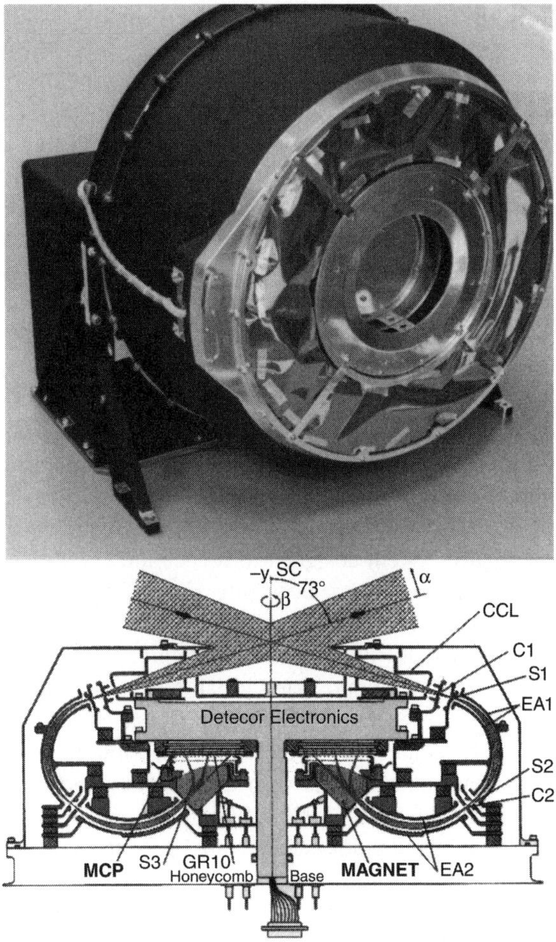

Figure 11.1. POLAR/TIMAS photo and schematic. (Courtesy W. Petersen, NASA.) (See color insert.)

In laboratory MS, one has a trade-off between sensitivity and mass resolution, so for example, a magnetic sector MS can have increased mass resolution if the entrance slit is made narrower. However, the already low fluxes of space do not allow that solution to the resolution dilemma. Beginning with a $1.2 \times 10^8/\text{s-cm}^2$ solar wind flux, less than 1% is heavier than helium, reducing the interesting count rate to $10^6/\text{s-cm}^2$. One must also not permit the hydrogen to create a scattering background to the heavies, which necessitates more collimation and baffles. To filter out the photons in the small distances allowed on a satellite, a cylindrical or spherical deflection system 10 cm across will have a gap or slit only a few millimeters wide. Unlike laboratory MS with capillary feeds, space MS are wide open to the environment, but the collimation and UV suppression make it difficult for a space MS to have more than a few square millimeters of "geometric factor," further reducing the flux to $10^5/\text{s}$. The solar wind has a much wider energy range than the 5% energy passband resulting in the need to step the deflection plates in voltage so as to sweep through the peak. Duty cycles of 1% to 5% are common depending on details of the energy resolution, reducing the count rate to 10^3/s. Finally, few space MS have the good fortune to be pointed continuously at the Sun, but are often on spin-stabilized spacecraft, further reducing the duty cycle another factor of 100. With 10 counts per second dedicated to carbon, oxygen, sulfur, magnesium, silicon, and iron, and accounting for the inevitable background of scattered protons, photons, and penetrating cosmic radiation, it is not too surprising that mass spectra require minutes to hours of integration time.

We have only covered the signal-to-noise problem; several others must be solved simultaneously. Since space is a vacuum, one cannot cool the electronics or power supplies with a fan, but must ensure that thermal contact direct the heat to the spacecraft radiators. Solid state detectors (SSD) (see Section 2.3.5), uncommon in laboratory MS, are often used in space to get an additional energy signal from the ion impact, and these detectors must not go above 30°C. Likewise, fast electronics are often power hungry, and all that power must be dissipated as heat. More than one space MS has failed for thermal reasons.

Not only must space MS be compact, low power, and autonomously operated, but they must survive launch by rocket. The trend over the past few decades has been toward solid-fueled rockets or boosters that have a much rougher ride than liquid-fueled rockets. Over-zealous specifications often require that space MS survive 15 g of random shake acceleration, which is about like lifting the instrument 10 cm and dropping it on the floor repeatedly. All those shims in a magnetic sector MS must be capable of being realigned in space, perhaps with stepper motors, which is what ESA had to fly in its 2011 comet mission [19]. Likewise, carbon foil technology took an additional 10 years to fly after it had been developed in the laboratory, primarily to ensure that it survived launch.

Finally, there are the social limitations of being crammed onto a single satellite platform with ten other science teams. The magnetometer team trying to measure 100 pT fields does not appreciate the stray fields from the mT magnets needed for a magnetic sector MS. The electric field team trying to measure millivolts per meter fields objects to the kilovolts per millimeter sweeping voltages on the deflection plates. And no one wants to be around should that 30 kV post-acceleration voltage arc over. Even with the best social etiquette, and the best measurements, there remains the unavoidable Darwinian telemetry competition for the data sent back to Earth. Just about every space scientist has a "fish story" about the data that got away.

For all these reasons and more, space MS has been an expensive challenge, but one with many accomplishments.

11.5. A BRIEF HISTORY OF SPACE MS

11.5.1. Beginnings

The lowly beginnings of space MS began in the early 1960s with Russian "ion traps" or Faraday cups (see Section 2.3.2) flown on the outside of a satellite [1]. In the United States, these were later outfitted with a grid that could be biased to different voltages, thereby acting as a crude energy filter, or a "retarding potential analyzer [20]." The advantages of Faraday cups were their integral flux measurements, their insensitivity to photons, and their general rugged construction, though they had no mass capability, and produced poor differential energy spectra.

The next step was flying a SSD or channeltron (see Section 2.3.3.2) with deflection plates to eliminate the UV photon flux [21]. These devices gave much better energy spectra, with 1% to 5% $\Delta E/E$ resolution achieved with close spacing between the deflection plates. In "cool" solar wind, the fact that all species travel at nearly the same velocity meant that helium had four times as much energy as hydrogen, permitting these devices to observe helium and sometimes oxygen as a bump on the hydrogen tail [22]. However, shock heated plasma cannot be analyzed this way, nor could rare species that did not rise above the noise in the hydrogen tail, which is a dynamic range limitation of the technique.

Several magnetic sector MS (see Section 2.2.2) were flown, but the small size of the magnets, as well as the desire to keep the spacecraft "magnetically clean," limited the maximum energy analyzed to $E < 15$ keV, and the resolution $R < 7$ [18]. Even with such low resolution, it was still sufficient to separate oxygen from carbon, the third and fourth most abundant species in the solar wind, because they were no longer contaminated by the long thermal tail on the hydrogen peak. Nevertheless, the real revolution in mass resolution awaited TOF-MS (see Section 2.2.1). Because this instrument revolutionized space MS in such an elegant fashion, it deserves more careful study.

11.5.2. Linear TOF-MS

In 1984, the AMPTE mission launched the first carbon-foil TOF-MS into space, which would have been the second, had the *Challenger* shuttle disaster not delayed the *Ulysses* launch until 1991 (Fig. 11.2) [23]. The photons were filtered out by a traditional blackened deflection system, which directed the ions toward the 2 $\mu g/cm^2$ thick foil mounted on an 85% transparent grid almost a square centimeter in area. The grid provided the support needed to survive the launch. The foil thickness permitted >2 keV/nuc ions to pass through and hit a SSD some 10 cm away. To ensure that the ions made it through the foil and also through the "dead layer" on the SSD (caused by the upper electrode), the foil and the entire TOF section were floated at ~20 kV to post-accelerate the ions. Electrons sputtered off the carbon foil became the start, whereas electrons sputtered off the SSD became the stop pulse for the TOF.

These sputtered electrons were detected by multichannel plate detectors (see Section 2.3.3.2) having a pulse width of about 1 ns, which when combined with a 3 ns variation due to C-foil location, gave a timing error in the ~70 ns flight time of <5%. Since time-of-flight is proportional to the square root of the mass, this doubles the mass uncertainty to 10%, for a resolution of R ~10. Lowering the post-acceleration would increase the TOF, which helps the mass resolution, but would also increase the "straggle" from energy loss in the foil, so that there is an optimum voltage roughly in the tens of kilovolts region. Making the foil thinner would reduce the straggle, but at the cost of producing fewer start electrons, so that the 0.5 $\mu g/cm^2$ foils flown on later space MS returned no signal for hydrogen [24]. Once again, 1 to 2 $\mu g/cm^2$ C-foils appear about ideal.

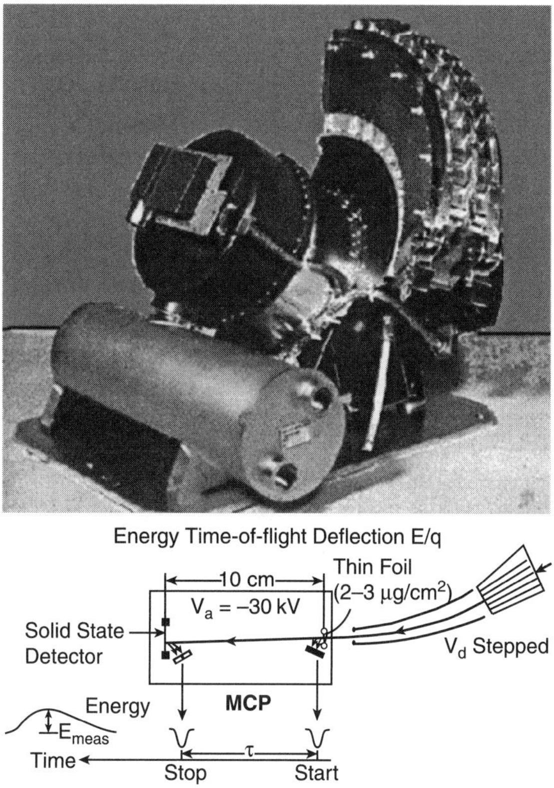

Figure 11.2. ULYSSES/SWICS photo/schematic. (Courtesy G. Gloeckler, NASA.) (See color insert.)

Since the voltage on the deflection plates determined the energy per charge of ions, E/Q, a TOF measurement giving the velocity could be combined to give the mass per charge, m/z. Unlike laboratory MS, ions in space can have many different charge states, so that there remains an essential ambiguity in the signal. For example, the second, third, and fourth most abundant species in the solar wind will overlap, with He^{2+} having the same m/z ratio as C^{6+} and O^{8+}, to within a few tenths of a percent. This was solved by measuring the total energy with the SSD with a resolution of 10%. Despite the increase in noise caused by having to subtract off the large post-acceleration voltage, the fact that charge is a small integer permitted an accurate determination of both charge and mass. Plots of M versus m/z clearly resolved solar wind species up to magnesium and silicon, effectively doubling the mass resolution to twenty or so (Fig. 11.3).

Although the SSD could not be used for TOF determination because the pulse height analysis used to extract the energy from the SSD had a shaping constant of around a microsecond, the presence of a SSD signal along with a start and stop signal permitted triple coincidence logic to virtually eliminate background electronic noise. This was crucial in enabling the instrument to collect spectra in the heart of the Van Allen radiation belts, as well as during solar energetic particle events that follow solar

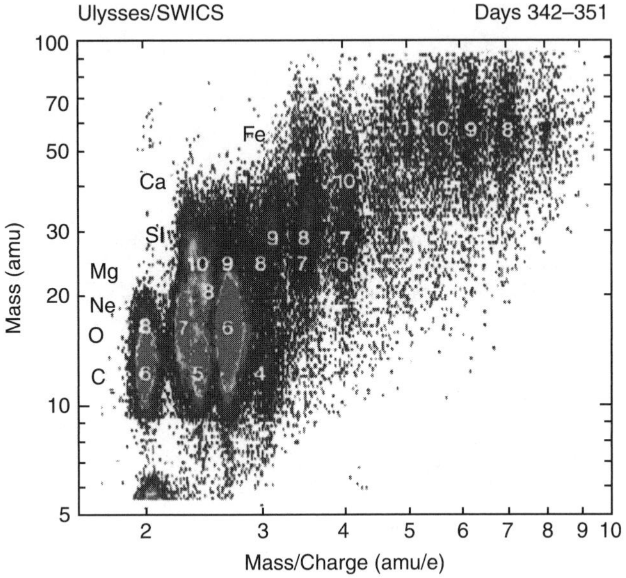

Figure 11.3. SWICS M vs M/Q analysis to enhance R. (Courtesy G. Gloeckler, NASA.) (See color insert.)

flares. In addition, these three rates permitted internal calibration of the instrument, so that MCP efficiencies, age degradation, or carbon foil damage could be tracked as well.

Also note that carbon foils are excellently suited for the low density space environment, using an asynchronous timing technique capable of measuring every rare particle that enters. Such a technique will saturate only when the count rate exceeds the reciprocal "window" time, the maximum time allowed for a stop signal before resetting the trigger. In the solar wind, the heaviest ions were expected to be iron, or about eight times slower than hydrogen. This converts to a $\sim$350 ns window, for a saturation rate of 3 MHz. The SSD shaping amplifier had a "pulse pile up" saturation at a slightly lower rate, around 300 kHz. With a few square millimeters of geometric factor, and an energy sweep of 32 steps, AMPTE only began to saturate in the highly compressed region behind the Earth's bowshock, the magnetosheath, where the density peaked around 50/cc [25].

11.5.3. Isochronous TOF-MS

These instruments were so well optimized that it became clear no further improvements were possible using more advanced materials or electronics. Much attention was directed to the C-foil, and the inherent energy straggle that directly reduced the mass resolution. Despite extensive testing of metals, plastics, and composites, resulting in several theory papers, nothing was found that had better energy straggle properties [26]. But if the TOF could be made independent of energy, then perhaps the straggle could be finessed. In the laboratory, TOF-MS "reflectrons" had been demonstrated that achieved first-order correction in the energy (see Section 2.2.1). But Laplace's equation for a vacuum electric potential Φ, $\nabla^2\Phi = 0$, says that if one dimension is reflecting with a positive second derivative, one or both of the other dimensions must be diverging, with a negative second derivative. In practice, this means that the ions in a reflectron must be highly collimated to avoid being diverted into the walls by the reflecting field. However, collimation is inversely proportional to energy spread, so the C-foil destroys both! Would it be possible, we reasoned, to place the C-foil as close as possible to the reflectron to minimize the divergence of the beam?

In this case, the mirror should be a parabolic electric field, so that like Galileo's chandelier the period of oscillation would be independent of the energy amplitude, $\Phi = z^2$. As it turned out, two simple solutions to Laplace's equation exist, $\Phi = z^2 - y^2$ in planar geometry, and $\Phi = 2z^2 - (x^2 + y^2)$ in cylindrical geometry. Both of these solutions diverge badly, but a computer glitch saved the day. The first, planar solution is seen to be the equation for a hyperbola, so one quarter of a hyperbola consisting of a "V" held at ground, and a "U" at positive 20 kV was simulated on an ancient HP2000 computer (Fig. 11.4). The computer had only 64 k of memory, and was unable to store the entire model in memory at once. As a consequence, particles often did not collide, but passed through potential structures. When we started particles from the point of the "V" upward toward the "U," they reflected and scattered, passing through the "V" potential until ultimately striking the bottom of the simulation box. Surprisingly, it had very little effect on the TOF; the isochronicity of the particles was preserved. When we analyzed the effect, we found that the greatest contribution to the TOF came at the closest approach to the "U," not down near the point of the "V." Galileo's chandelier, after all, wasn't swinging in a parabola but in a circle, yet it made little difference to the TOF. With this insight, we stopped worrying about the scattering of the particles, put in a transparent grid on the "V" and collected them wherever they went, thus making the first practical application to an isochronous TOF and achieving a R $\sim$100.

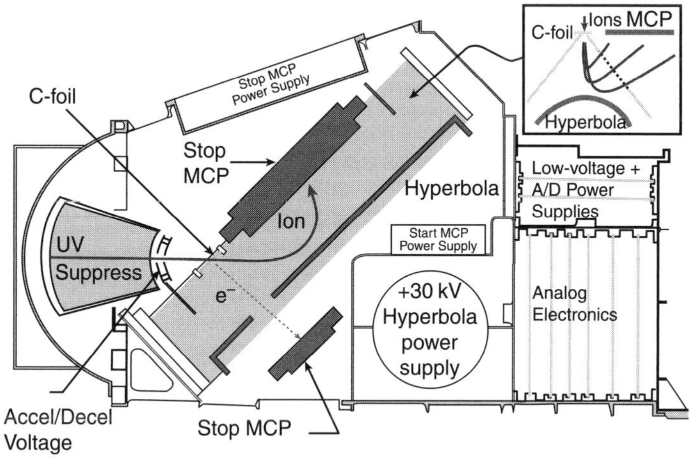

Figure 11.4. WIND/MASS schematic. (Courtesy R. Sheldon, NASA.) (See color insert.)

11.6. GENESIS AND THE FUTURE

This instrument flew on three missions: SOHO, WIND, and ACE, where it has returned information about solar wind isotopes confirming the meteoritical abundances for the proto-solar nebulae [27], as well as identifying boundaries in the solar wind. With more than enough resolution to detect ^{17}O, why then was the GENESIS mission funded to return samples of the solar wind [22]? Well, the problem turned out to be the C-foil once again.

When the ions penetrate the foil, they achieve an equilibrium charge state that depends on their speed. The 2 to 10 keV/nuc achieved in solar wind plasmas, even after post-acceleration, resulted in an exit population that was more than 90% neutral. These neutralized ions did not reflect, but impacted the hyperbola and produced electrons, ions, or just scattered around inside the instrument. This created a high background that we could suppress only slightly, so that there always existed a long TOF "tail" to every peak. A rare isotope to the right of the main peak was then invisible, and the effective dynamic range for an isotope such as ^{17}O was about 1%, but its abundance was only 0.1% of ^{16}O. Consequently GENESIS was conceived as a sample-and-return mission to solve the rare abundances in the laboratory [29].

The future, perhaps, is exemplified by the ROSETTA mission, which uses an ion trap and electrostatic gating rather than carbon foils to achieve a R ~3000. While the foils are excellent in producing nanosecond timing on fast particles with very simple electronics, advances in power supplies and switching means that future instruments will be capable of both the nanosecond gates and the kilovolt energies without the C-foils. Then all the advantages of C-foils will be obtained without the energy straggle, scattering, or neutralization disadvantages. We are developing such instruments in our lab, and foresee space MS that can exceed R >5000.

REFERENCES

1. M. Neugebauer and R. von Steiger. The Solar Wind, in *The Century of Space Science*, J. A. M. Bleeker, M. Huber, J. Geiss (eds.), Springer, 2001.
2. Newell, E. Homer. *Beyond the Atmosphere: Early Years of Space Science*, 1980.
3. E. W. Hones, Jr, T. Pytte, and H. I. West, Jr. Associations of Geomagnetic Activity with Plasma Sheet Thinning and Expansion—A Statistical Study, *J. Geophy. Res.*, **89**(1984): 5471–5478.
4. B. Klecker, E. Moebius, D. Hovestadt, M. Scholer, and G. Gloeckler. Discovery of Energetic Molecular Ions (NO+ and O2+) in the Storm Time Ring Current, *Geophys. Res. Lett.*, **13**(1986): 632–635.
5. N. Wilmot, Hess. *The Radiation Belt and Magnetosphere*, 548 pp., Blaisdell Publ. Co., Waltham, MA, 1968.
6. A. Bonetti, H. S. Bridge, A. J. Lazarus, E. F. Lyon, B. Rossi, and F. Scherb. Explorer X Plasma Measurements, *Proceedings of the Third International Space Science Symposium*, Space Research III, W. Priester (ed.), North-Holland Publ. Co., Amsterdam, The Netherlands, 1963, 540–552.

7. J. E. Ross and L. H. Aller. The Chemical Composition of the Sun, *Science*, **191**(1976): 1223–1229.

8. E. Anders and N. Grevesse. Abundances of the Elements—Meteoritic and Solar, *Geochimica et Cosmochimica Acta*, **53**(1989): 197–214.

9. F. Bühler, P. Eberhardt, J. Geiss, J. Meister, and P. Signer. *Science*, **166**, no. 3912 (1969): 1502. [DOI: 10.1126/science.166.3912.1502]

10. G. Gloeckler, et al., The Charge-Energy-Mass Spectrometer for 0.3–300 keV/e Ions on the AMPTE CCE, *IEEE Transactions on Geoscience and Remote Sensing*, **GE-23**(1985): 234–240.

11. D. C. Hamilton, G. Gloeckler, F. M. Ipavich, R. A. Lundgren, R. B. Sheldon, D. Hovestadt. New High Resolution Electrostatic Ion Mass Analyzer Using Time of Flight, *Rev. Sci. Instr.*, **61**, no. 10(1990): 3104–3106.

12. D. S. Burnett, et al., The Genesis Discovery Mission: Return of Solar Matter to Earth, *Space Sci. Rev.*, **105**, no. 3–4(2003): 509–534.

13. R. C. Wiens, G. R. Huss, and D. S. Burnett. The Solar Oxygen Isotopic Composition: Predictions and Implications for Solar Nebula Processes. *Meteoritics and Planetary Science*, **34**(1995): 99–108.

14. J. Geiss, G. Gloeckler, and R. von Steiger. Origin of the Solar Wind from Composition Data, *Space Science Reviews*, **72**, no. 1–2(1995): 49–60.

15. R. E. Lingenfelter. Production of Carbon 14 by Cosmic-Ray Neutrons, *Rev. of Geophysics and Space Physics*, **1**(1963): 35–55.

16. S. M. Krimigis, G. Haerendel, R. W. McEntire, G. Paschmann, and D. A. Bryant. The Active Magnetospheric Particle Tracer Explorers (AMPTE) Program, In *ESA Active Expts. in Space*, (1983): 317–325.

17. F. R. Wimmer-Schweingruber, O. Kern, and C. D. Hamilton. On the Solar Wind Composition During the November 1997 Solar Particle Events: WIND/MASS Observations, *Geophys. Res. Lett.*, **26**, no. 23 (1999): 3541–3544.

18. D. T. Young, J. A. Marshall, J. L. Burch, T. L. Booker, A. G. Ghielmetti, and E. G. Shelley. A Double-Focusing Toroidal Mass Spectrograph for Energetic Plasmas. II: Experimental Results, *Nucl. Instr. and Meth.*, **A258**(1987): 304.

19. H. Balsiger, et al., Rosetta Orbiter Spectrometer for Ion and Neutral Analysis-Rosina, *Advances in Space Research*, **21**, no. 11(1998): 1527–1535.

20. H. S. Bridge, A. J. Lazarus, E. F. Lyon, B. Rossi, and F. Scherb. Plasma Probe Instrumentation on Explorer X, *J. Phys. Soc. Japan*, **17**, Supplement A-III(1962): 1113–1121.

21. M. Neugebauer. Pioneers of Space Physics: A Career in the Solar Wind, *J. Geophys. Res.*, **102** (A12)(1998): 26,887–26,894.

22. M. A. Coplan, et al., Ion Composition Experiment, *IEEE Trans. Geosci. Electron.*, **GE-16**, no. 3(1978): 185–191.

23. G. Gloeckler, J. Geiss, H. Balsiger, P. Bedini, J. C. Cain, J. Fisher, L. A. Fisk, A. B. Galvin, F. Gliem, and D. C. Hamilton. The Solar Wind Ion Composition Spectrometer, *Astronomy and Astrophysics Supplement Series*, **92**, no. 2(1992): 267–289.

24. D. T. Young, B. L. Barraclough, D. J. McComas, M. F. Thomsen, K. McCabe, and R. Vigil. CRRES Low-Energy Magnetospheric Ion Composition Sensor, *J. Spacecraft and Rockets*, **29**, no. 4(1992): 596–598.

25. R. von Steiger, S. P. Christon, G. Gloeckler, and F. M. Ipavich. Variable Carbon and Oxygen Abundances in the Solar Wind as Observed in Earth's Magnetosheath by AMPTE/CCE, *Astrophys. J., Part 1*, **389**(1992): 791–799.
26. R. Kallenbach, M. Gonin, P. Bochsler, and A. Bürgi. Charge Exchange of B, C, O, Al, Si, S, F and Cl Passing Through Thin Carbon Foils at Low Energies: Formation of Negative Ions, *Nuclear Instruments and Methods in Physics Research Section B*, **103**(1995): 111–116.
27. R. Karrer, P. Bochsler, C. Giammanco, F. M. Ipavich, J. A. Paquette, and P. Wurz. Nickel Isotopic Composition and Nickel/Iron Ratio in the Solar Wind: Results from SOHO/CELIAS/MTOF, *Space Science Reviews*, **130**, no. 1–4(2007): 317–321.
28. D. Rapp, et al., The Suess-Urey Mission (Return of Solar Matter to Earth), *Acta Astronautica*, **39**, no. 1–4(1996): 229–238.
29. R. C. Wiens, D. S. Burnett, C. M. Hohenberg, A. Meshik, V. Heber, A. Grimberg, R. Wieler, and D. B. Reisenfeld. Solar and Solar-Wind Composition Results from the Genesis Mission, *Space Science Reviews*, **130**, no. 1–4(2007): 161–171.

12

BIOTERRORISM

Vito G. DelVecchio and Cesar V. Mujer

12.1. WHAT IS BIOTERRORISM?

Bioterrorism is the intentional use of pathogenic microorganisms or toxins to cause disease or death in plants, animals, and humans. Deliberate bioterrorism attacks are often perpetuated by individuals, groups, or hostile governments for financial, political, or ideological purposes.

12.2. SOME HISTORICAL ACCOUNTS OF BIOTERRORISM

Incidents of bioterrorism date back as early as 600 BC when infectious materials were recognized for their potential impact to conquer adversaries and incapacitate or kill the enemies. One account tells the story of how the Assyrians poisoned wells with rye ergot, caused by the fungus *Claviceps purpurea*. The fungus produces alkaloids and lysergic acid (LSD), a powerful hallucinogen that interferes with neurotransmitter function. Ergot poisoning causes delusions, paranoia, myoclonic twitches, seizures, and cardiovascular problems that can lead to death [1]. In the fourteenth century, the Tartars (Mongols) used catapults to hurl cadavers of plague victims into the city of Kaffa, presently known as Feodosiya in the Ukraine, initiating an epidemic. After the

Mass Spectrometry. Edited by Ekman, Silberring, Westman-Brinkmalm, and Kraj

city surrendered, refugees carried plague germs with them to Genoa, Italy, which some historians speculated was the cause of "Black Death" in Europe [2].

In the eighteenth century, during the French-Indian War, Sir Jeffrey Amherst suggested the deliberate use of smallpox to diminish the native Indian population hostile to the British. One of his subordinate officers provided the Native Americans with blankets from smallpox patients, resulting in a large outbreak of smallpox among the Indian tribes in the Ohio River Valley [3]. There are many more examples of biological and chemical warfare use during the past several hundred years and in the last century. For additional information on the history of bioterrorism, the reader is referred to the following reviews: Atlas [4], Christopher et al. [5], Klietmann and Ruoff [6], and Tucker [7].

More recently, anthrax has been used as a biological weapon in the United States and a total of 22 cases were identified. Six fatalities occurred due to inhalation of the causal agent, *Bacillus anthracis*. Use of microorganisms for agroterrorism as well as infection of companion animals, and the potential development of genetically engineered agents have made the twenty-first century more vulnerable than past centuries.

12.3. GENEVA PROTOCOL OF 1925 AND BIOLOGICAL WEAPONS CONVENTION OF 1972

Previous treaties directed towards limiting the proliferation and use of biological and chemical weapons have been ratified. On June 17, 1925, the Geneva Protocol was signed calling for the prohibition of the use in war of asphyxiating, poisonous, or other gases, and of bacteriological methods of warfare. However, the protocol did not address verification or compliance, and despite the agreement several countries developed biological weapons after the protocol's ratification [8]. In 1972, the "Convention on the Prohibition of the Development, Production, and Stockpiling of Bacteriological (Biological) and Toxin Weapons and on Their Destruction," known as BWC (Biological Weapons Convention), was developed. This treaty prohibits possession of biological agents except for prophylactic, protective, or other peaceful purposes (Stockholm International Peace Research Institute).

12.4. CATEGORIES OF BIOTHREAT AGENTS

Recent bioterrorist events have shown the need to immediately detect and identify biological threat agents. Rapid, accurate identification of such agents is important to confirm that a bioterrorism event has occurred and to determine appropriate measures that should be implemented to protect public health. The Centers for Disease Control and Prevention (http://www.cdc.gov/) as well as the National Institute of Allergy and Infectious Diseases (http://www2.niaid.nih.gov/) have classified threat agents as Category A to Category C Priority Pathogens. Category A agents have the greatest potential for adverse public health impact with mass casualties, requiring broad-based public health preparedness, including improved surveillance, laboratory diagnosis, and stockpiling of specific medications. Furthermore, Category A agents have a moderate to high potential for large-scale dissemination or a heightened general public awareness that could cause fear and civil disruption [9] (Rotz et al., 2002). Category A agents and their associated diseases (in parenthesis) include: *Variola major* (smallpox), *Bacillus anthracis*, *Yersinia pestis* (plague), *Clostridium botulinum* (botulism), *Francisella tularensis* (tularemia), and Filoviruses/Arenaviruses such as Ebola and Marburg viruses (viral hemorrhagic fever).

In contrast, Category B agents have some potential for large-scale dissemination with resultant illness, but generally cause less illness and death and therefore would be expected to have lower medical and public health impact. Agents in this category showed limited requirements for stockpiled therapeutics compared to those in Category A. Category B agents and their associated diseases (in parenthesis) include *Coxiella burnetii* (Q fever), *Brucella* spp. (brucellosis), *Burkholderia mallei* (glanders), *Burkholderia psuedomallei* (melioidosis), Alphaviruses (encephalitis), *Rickettsia prowazekii* (typhus fever), toxins such as ricin and staphylococcal enterotoxin B (toxic syndromes), *Chlamydia psittaci* (psittacosis), food safety threats such *Salmonella* spp. and *Escherichia coli* O157 : H7, and water safety threats such as *Vibrio cholerae* and *Cryptosporidium parvum*. Biological agents that are believed not to present a high bioterrorism risk to public health but could emerge as future threats are classified as Category C agents. Examples are Nipah virus and hantavirus.

12.5. CHALLENGES

The key to effectively counter bioterrorism agents lies in the development of new rapid diagnostic tests, new vaccines and immunotherapies for prevention, and new drugs and biologics for treatment [10]. Proteomics (see also Chapter 10) plays a key role in the development of these countermeasures [11]. For instance, the monitoring of proteomic differences in bacterial and viral pathogenesis will facilitate the direct comparison of strain variability, severity of infection, environmental influences, and the effects of genetic manipulations. From a biodefense perspective, proteomic profiling approaches will enable the identification of unique protein signatures as well as differences in virulence among strains of biothreat agents which are useful for rapid detection as well as development of potential therapeutics.

Proteomics is the study of the protein complement of the genome and refers to the total number of proteins expressed by the genome at a particular time or under a specific set of conditions. Some commonly used experimental tools include two-dimensional gel electrophoresis (see Section 4.2.2.1), MALDI-TOF-MS (see Sections 2.1.22 and 2.2.1) and LC-MS/MS (see Chapters 3 and 4). Other tools, such as SELDI (surface-enhanced laser desorption/ionization)-TOF-MS (see Section 2.1.22) and FTICR (Fourier transform ion cyclotron resonance)-MS (see Section 2.2.6) are also being used depending on the objectives of the study being conducted. Prior to the advent of proteomics, the global analysis of the proteome was considered a gargantuan task that typically took years to finish. However, the facile separation of proteins using two-dimensional gel electrophoresis and the rapid identification of gel-separated proteins using

MALDI-TOF-MS have considerably reduced the time to completion of this task, to about a month or less depending on the bacterial subproteome (i.e., membrane, cytosol, nuclear fraction, exosporium, secretome) being analyzed. In addition, the genomes of some of the priority pathogens have been completely sequenced and annotated or they are priorities for sequencing efforts facilitating the functional analysis of their proteomes. LC-MS/MS is also used in conjunction with 2DE/MALDI-TOF-MS to identify high or low molecular weight (molar mass) and hydrophobic proteins not amenable using MALDI-TOF-MS.

12.6. MS IDENTIFICATION OF BIOMARKER PROTEINS

MALDI-TOF-MS greatly facilitates the characterization of important biomarker proteins for strain identification, diagnosis, and vaccine development and therapeutic agent development. For instance, the wild-type virulent strain of *B. melitensis* (16M) was distinguished from the attenuated vaccinal strain (Rev 1) by comparing their differentially expressed proteins [12]. Computer assisted analysis and MALDI-TOF-MS identified a total of 40 overexpressed and 20 underexpressed proteins in Rev 1 that distinguished this strain from 16M. Likewise, proteomic profiling of other *Brucella* species, including *B. canis*, *B. suis*, *B. ovis*, and *B. neotomae*, has distinguished one *Brucella* species from the other [13].

Identification of the various members of *B. cereus* group, including *B. anthracis*, has been accomplished by proteomic analysis using MALDI-TOF-MS and LC-MS/MS of their spores. Thus, it is possible to distinguish *B. anthracis* from *B. cereus* and *B. thuringiensis* by the analysis of proteins present on the outermost surface of the endospore. LC-MS/MS can also be applied to the identification of proteins that are differentially expressed under various culture conditions and during the course of infection. Comparative expression profiling of biomarker proteins can be investigated by LC-MS/MS using isotope-coded affinity tags (ICAT). This method permits a relative quantitation of the differences in abundance of differentially expressed proteins between strains. After labeling the samples with light or heavy ICAT, they are combined, digested with trypsin, affinity purified, and subjected to LC-MS/MS. The full scan spectrum obtained after LC-MS/MS will show the ratio of the light and heavy ICAT-tagged peptides, thereby allowing a direct comparative quantitation of the same protein from two samples. The ratio of the same protein labeled with light and heavy ICAT will indicate the difference in their expression levels. Such comparative expression profiling is important for identifying targets for therapeutics and vaccine development. Identification of highly differentially expressed proteins between strains can also be useful for developing simulants for biological warfare operational simulations. A summary of the different mass spectrometry approaches that can be used for biomarker identification and subsequent development of polymerase chain reaction (PCR)-based kits for strain identification is presented in Fig. 12.1.

There are numerous other reports that illustrate the potential use of MALDI-TOF-MS for the rapid identification of various strains of *B. anthracis*. For instance, by using a new hybrid ion-trap MALDI-TOF instrument, in situ proteolytic digestion of

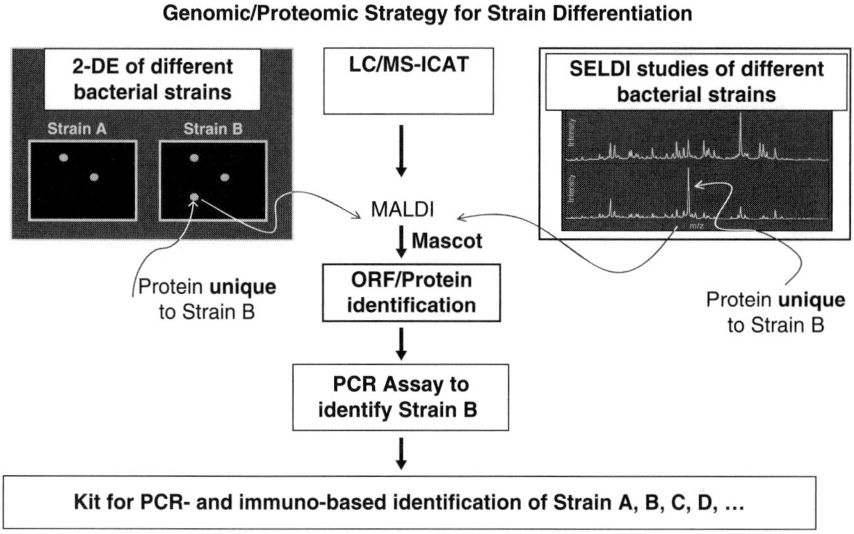

Figure 12.1. Different MS strategies for identification of biomarker proteins. (See color insert.)

small, acid-soluble proteins led to the identification of five different *B. anthracis* strains from their mixed spores in less than 20 min [14, 15]. As more genomic information becomes available for other biothreat agents, the MS approach could be very useful for multiple biodefense applications and for emergency first responders.

In addition to MALDI-TOF and LC-MS/MS, SELDI-TOF-MS can also be used to determine expression profiling of various biological samples, such as serum or plasma for early detection of infection. Serum proteomic profiling assay, for instance, has been used to distinguish patients with acute SARS (severe acute respiratory syndrome) from patients with fever and influenza with 100% accuracy [16]. A major limitation of SELDI-TOF-MS, however, is that it cannot be used for direct amino acid sequence identification of the biomarker proteins, necessitating further sample fractionation and protein purification.

12.7. DEVELOPMENT OF NEW THERAPEUTICS AND VACCINES USING IMMUNOPROTEOMICS

The advent of immunoproteomics made possible the identification of highly immunogenic proteins that can be used for vaccine development. Proteins that have the greatest potential for eliciting a protective immune response are collectively referred to as the pathogen's immunome. Immunoproteomics has been utilized to characterize the immunome of *B. anthracis* for the development of a safer and equally efficacious vaccine. The immunoreactive proteins are first identified by using 2DE Western blot analysis in conjunction with mass spectrometry. In *B. anthracis*, for example, antisera from humans post-infected with anthrax were used to probe Western blots of its various

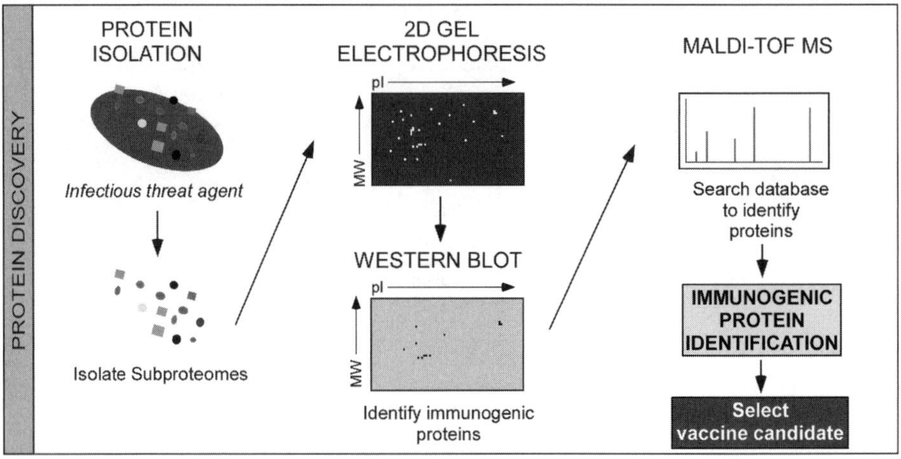

Figure 12.2. Identification of immunogenic proteins by 2DE/MALDI TOF-MS and Western blots probed with antisera from immunized animals or humans. (Reprinted from Khan, A. S. et al., 2006. Proteomics and Bioinformatics Strategies to Design Countermeasures Against Infectious Threat Agents. *J. Chem. Inf. Model.* 46: 111–115. With permission.) (See color insert.)

subproteomes [17]. Other immunoproteomic studies have also been reported for *F. tularensis* [18] and *Shigella flexneri* [19]. Once immunoreactive proteins are identified, these proteins are evaluated for their potential as vaccine candidates using a multifaceted algorithm [11]. An overall strategy in proteomics-based vaccine discovery is presented in Fig. 12.2.

Currently the same approach is being used to identify protective antigens from other biothreat agents, including *B. melitensis*, *Salmonella* spp., *Shigella* spp., Norovirus, Ebola, and others. The major goal is to translate proteomic discoveries into useful counter-bioterrorism products such as vaccines and therapeutics. The potential benefits of using the same proteomics approaches are enormous particularly when applied to other pathogenic bacterial and viral pathogens.

REFERENCES

1. M. B. Phillips. 2005. Bioterrorism: A Brief History. *Focus on Bioterrorism*. Northeast Florida Medicine (www.DCMSonline.org).
2. M. Wheelis. Biological Warfare at the 1346 Siege of Caffa. *Emerg. Infect. Dis.* **8**(2002): 971–975.
3. A. G. Robertson and L. J. Robertson. From Asps to Allegations: Biological Warfare in History. *Mil. Med.*, **160**(1995): 369–373.
4. R. M. Atlas. Bioterrorism: From Threat to Reality. *Annu. Rev. Microbiol.*, **56**(2002): 167–185.
5. G. Christopher, T. Cieslak, J. Pavlin, and E. Eitzen. Biological Warfare: A Historical Perspective. *JAMA*, **278**(1997): 412–417.

6. W. F. Klietmann and K. L. Ruoff. Bioterrorism: Implications for the Clinical Microbiologist. *Clin. Microbiol. Rev.*, **14**(2001): 364–381.
7. J. B. Tucker. Historical Trends Related to Bioterrorism: An Empirical Analysis. *Emerg. infect. Dis.*, **5**(1999): 498–504.
8. R. P. Kadlec, A. P. Zelicoff, and A. M. Vrtis. Biological Weapons Control: Prospects and Implications for the Future. *JAMA*, **278**(1997): 351–356.
9. L. D. Rotz, A. S. Khan, S. R. Lillibridge, S. M. Ostraff, and J. M. Hughes. Public Health Assessment of Potential Biological Terrorism Agents. *Emerg. Infect. Dis.*, **8**(2002): 225–230.
10. R. R. Drake, Y. Deng, E. Ellen Schwegler, and S. Gravenstein. Proteomics for Biodefense Applications: Progress and Opportunities. *Expert Rev. Proteomics*, **2**(2005): 203–213.
11. A. S. Khan, C. V. Mujer, T. G. Alefantis, J. P. Connolly, U. B. Mayr, P. Walcher, W. Lubitz, and V. G. DelVecchio. Proteomics and Bioinformatics Strategies to Design Countermeasures Against Infectious Threat Agents. *J. Chem. Inf. Model.*, **46**(2006): 111–115.
12. M. Eschenbrenner, M. A. Wagner, T. A. Horn, J. A. Kraycer, C. V. Mujer, S. Hagius, P. Elzer, and V. G. DelVecchio. Comparative Proteome Analysis of *Brucella Melitensis* Vaccine Strain Rev 1 and a Virulent Strain, 16M. *J. Bacteriol.*, **184**(2005): 4962–4969.
13. V. G. DelVecchio and C. V. Mujer. 2004. Comparative Proteomics of *Brucella* Species. In *Brucella Molecular and Cellular Biology*, eds. I. Lopez-Goni, and I. Moriyon, Chapter 6, pp. 103–115, Horizon Bioscience, Norfolk, UK.
14. B. Warscheid and C. Fenslau. A Targeted Proteomics Approach to the Rapid Identification of Bacterial Cell Mixtures by Matrix-Assisted Laser Desorption/Ionization Mass Spectrometry. *Proteomics*, **4**(2004): 2877–2892.
15. B. Warscheid, K. Jackson, C. Sutton, and C. Fenselau. MALDI Analysis of *Bacilli* in Spore Mixtures by Applying a Quadrupole Ion Trap Time-of-Flight Tandem Mass Spectrometer. *Anal. Chem.*, **75**(2003): 5608–5617.
16. X. Kang, Y. Xu, X. Wu, Y. Liang, C. Wang, J. Guo, Y. Wang, M. Chen, D. Wu, Y. Wang, S. Bi, Y. Qiu, P. Lu, J. Cheng, B. Xiao, L. Hu, X. Gao, J. Liu, Y. Wang, Y. Song, L. Zhang, F. Suo, T. Chen, Z. Huang, Y. Zhao, H. Lu, C. Pan, and H. Tang. Proteomic Fingerprints for Potential Application to Early Diagnosis of Severe Acute Respiratory Syndrome. *Clin. Chem.*, **51**(2005): 56–64.
17. V. G. DelVecchio, J. P. Connolly, T. G. Alefantis, A. Walz, M. A. Quan, G. Patra, J. M. Ashton, J. T. Whittington, R. D. Chafin, X. Liang, P. Grewal, A. S. Khan, and C. V. Mujer. 2006. Proteomic Profiling and Identification of Immunodominant Spore Antigens of *Bacillus Anthracis*, *Bacillus Cereus* and *Bacillus Thuringiensis*. *Appl. Environ. Microbiol.*, **72**(2006): 6355–6363.
18. J. Havlasova, L. Hernychova, P. Halada, V. Pellantova, J. Krejsek, J. Stulik, A. Macela, P. R. Jungblut, P. Larsson, and M. Forsman. Mapping of Immunoreactive Antigens of *Francisella Tularensis* Live Vaccine Strain. *Proteomics*, **2**(2002): 857–867.
19. X. Peng, X. Ye, and S. Wang. Identification of Novel Immunogenic Proteins of *Shigella Flexneri* 2a by Proteomic Methodologies. *Vaccine*, **22**(2004): 2750–2756.

13

IMAGING OF SMALL MOLECULES

Małgorzata Iwona Szynkowska

Imaging of small molecules has found a wide application in all fields where the chemical composition and visualization of surfaces and interfaces play an important role. Applications range from fundamental scientific research to failure analysis and production control, in many different fields, such as materials science, microelectronics, paint adhesion, coating of metals, geochemistry, and surface analysis. Lately it has also appeared to be very promising in life sciences, especially in medicine and drug development. The information obtained from imaging studies is useful to scientists, engineers, or manufacturers because it helps to understand and predict the behavior of the materials they are working with in order to verify theories or make better products and procedures.

In the literature these studies are classified as imaging mass spectrometry (IMS) and defined as the investigation of the chemical profile of a sample surface with a submicron lateral resolution and chemical specificity. The main aim is to use the power of mass spectrometry techniques to create chemical images showing the distribution of compounds ranging in size from atomic ions and small molecules to large proteins.

The ideal imaging method should enable simultaneous detection and identification of known and unknown compounds present in the studied sample. During analysis, analyte atoms and molecules are removed from the sample surface, in most cases only

Mass Spectrometry. Edited by Ekman, Silberring, Westman-Brinkmalm, and Kraj

from a few monolayers. The generation of chemical images with mass spectrometry requires:

- Sample preparation
- Analyte transfer to the gas phase and its ionization
- Separation in a vacuum
- Analyte ion detection (Fig. 13.1)

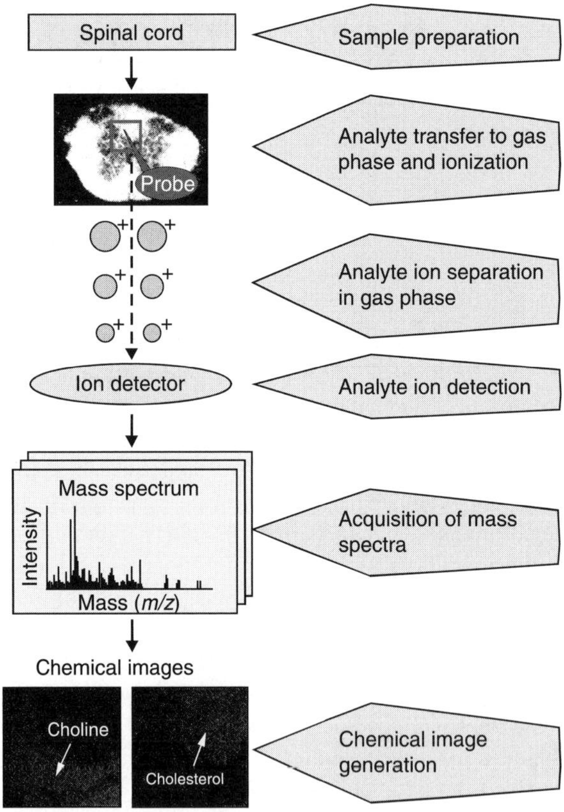

Figure 13.1. Imaging mass spectrometry. The acquisition of chemical images with mass spectrometry involves multiple, often elaborate, steps, beginning with sample preparation and ending with generation of the chemical image. The figure illustrates the process of imaging of a spinal cord section using a TOF-SIMS spectrometer equipped with a gold ion source, where the images of distribution of choline and cholesterol are obtained. The sample preparation consists of spinal cord dissection, freezing, cryostat sectioning, single section deposition on a wafer, and drying. (Reprinted from Rubakhin, S. S. et al., 2005. *DDT*, **10**, no 12: 823–837. With permission from Elsevier.) (See color insert.)

An image is collected by scanning the probe within the image area and collecting a mass spectrum at each pixel. For acquisition of the highest quality images the highest possible ionization efficiency of desorbed atoms and molecules is needed. The choice of mass spectrometer used for imaging is based on the sample properties and specific experimental conditions.

Mainly three ionization methods are applied in mass spectrometry:

- Ion-beam-induced desorption for secondary ion mass spectrometry (SIMS)
- Laser desorption, including laser desorption/ionization (LDI) and matrix-assisted laser desorption/ionization (MALDI)
- Electrospray ionization (ESI).

SIMS (see Section 2.1.18) and MALDI (see Section 2.1.22) are already well-established techniques in imaging mass spectrometry, whereas imaging with ESI (see Section 2.1.15) is quite a new, promising approach [1]. The selection of an imaging mass spectrometry method depends on the required m/z range, lateral resolution, mass resolution, sensitivity, and selectivity. SIMS is widely applied for characterization of atomic ions and small molecules below 500 Da, although in some cases, when state-of-the-art ion sources are used, 2 to 3 kDa species can be detected and imaged. By comparison, MALDI provides information on higher molecular masses, so it has found a broad application in analyses of proteins and nucleotides of several hundred thousand daltons.

In this chapter a number of the recent applications of imaging of small molecules with the use of SIMS imaging will be presented.

13.1. SIMS IMAGING

Molecular surface characterization of a variety of materials has been made possible in recent years by the emergence of static SIMS and related methodologies [2]. SIMS imaging in combination with time-of-flight secondary ion mass spectrometry (TOF-SIMS) (see Section 2.2.1) and a liquid metal ion source (LMIS) can simultaneously detect all elemental and small molecular masses with very high sensitivity and lateral resolution <100 nm. Lateral resolution refers to the smallest distance between two points on the sample surface that can be visibly distinguished. Using dynamic SIMS it is possible to acquire images as a function of time; however, elemental information may be obtained as a function of depth into the sample. The measurements in dynamic mode result in erosion and chemical damage of the surface.

TOF-SIMS has been employed for the characterization of a wide range of materials, including metallic, salt, organometallic, organic, and polymeric substances, as well as for electronics, catalysts, and forensic samples. The ability to image molecular ions with submicrometer spatial resolution makes TOF-SIMS well suited to analysis of pharmaceuticals and biological cells, as well as for use in biotechnology and molecular electronics.

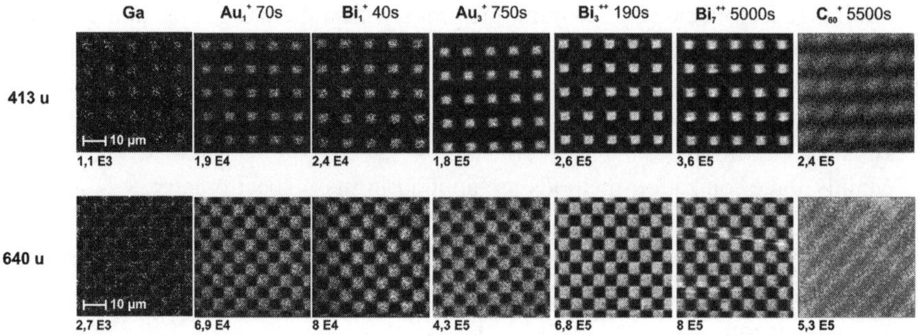

Figure 13.2. TOF-SIMS images of blue ($m = 413$ u) and green ($m = 641$ u) pigments of color filter array. Above each image the primary ion gun and the measurement time is displayed. Corresponding signal intensity of emitted secondary ions from an analyzed surface is given under suitable image. (Reprinted from Kollmer, F. 2004. *Appl. Surf. Sci.*, 231–232: 153–158. With permission from Elsevier.) (See color insert.)

In comparison to other surface-sensitive imaging techniques, TOF-SIMS has a number of remarkable advantages, including:

- High surface sensitivity
- Trace component detection
- Hydrogen detection
- Isotope detection
- The ability to obtain detailed molecular information [2]

The development of a new generation of ion sources like SF_5 and C_{60} (buckminsterfullerene), as well as gold and bismuth cluster ion sources (Au_n, Bi_n), shows the improvement in secondary ion efficency when compared to monoatomic liquid ion guns (Ga, Cs, In, Au, Bi). The greater signal intensity results in better image resolution and contrast. The increased yield produced by cluster ion sources opens new possibilities for molecular SIMS imaging, especially for higher molecular weight (molar mass) fragments from organic, medical, and biological samples.

Figure 13.2 presents a comparison of images of color filter array (CFA) obtained using gallium, C_{60}^+, gold, and bismuth cluster guns as primary ions. In the case of the cluster ions emitted by LMIS the significant increase in the quality of images connected with the growing number of secondary ions emitted from an analyzed surface is observed [3].

13.2. BIOLOGICAL APPLICATIONS (CELLS, TISSUES, AND PHARMACEUTICALS)

The surface sensitivity and lateral resolution of TOF-SIMS suggest that this method has great potential for imaging substances in cells and tissues, as well as in cellular and subcellular monitoring of pharmaceuticals.

It has been applied to elemental imaging of the distribution of cancer therapy drugs in single cells and molecular imaging of tissue, liposomes, paramecia, rat pheochromocytoma (PC12) cells and subcellular membranes, among others. Figure 13.3 shows images of cholesterol and phospholine/phosphate from a freeze-dried mouse brain section at different increased magnifications, obtained at submicrometer resolution using TOF-SIMS with Au_3^+ ions as primary ions. These studies are particularly important from the point of view of neurodegenerative diseases as there is a well-documented connection between cholesterol and Alzheimer's disease [4], and sulfatides have recently been indicated as potential early (negative) markers for this disease [5]. The results illustrate unique possibilities of TOF-SIMS to provide spatial information about specific lipids and other organic substances in biological samples, and changes of their distribution at different stages of disease. TOF-SIMS development in imaging of biochemical samples may be improved by finding new sample preparation methods or by the application of existing ones, such as chemical imprinting, freeze-fracturing, or freeze-drying [6].

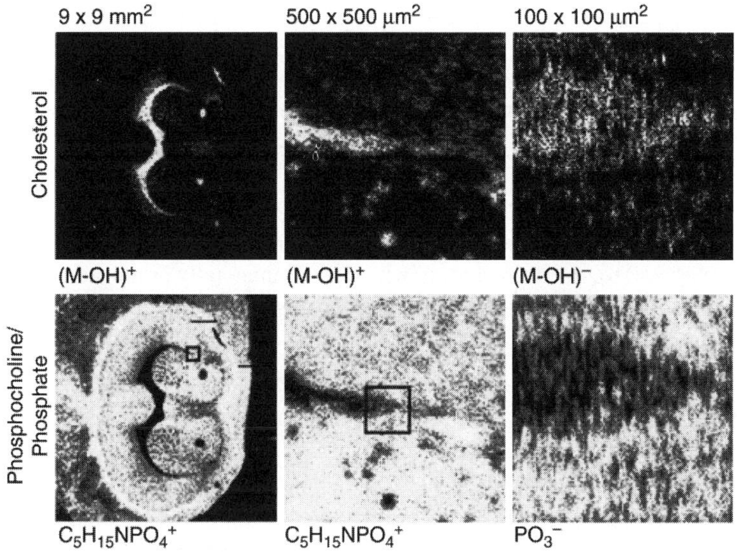

Figure 13.3. TOF-SIMS images showing the spatial signal intensity distribution from cholesterol and phospholine/phosphate of a mouse brain section at successively increased magnifications. The 9×9 mm^2 and 500×500 μm^2 images were obtained from measurements of positive secondary ions, showing the spatial intensity distribution of the (M-OH)$^+$ peak for cholesterol (369 u) and of the phosphocholine peak ($C_5H_{15}NPO_4^+$) with maximum image resolution 3 to 5 μm. The 100×100 μm^2 images were obtained from measurements of negative ions, showing the spatial intensity distribution of the (M-H)$^-$ peak for cholesterol and the phosphate peak (PO_3^-) with maximum image resolution 0.2 to 0.3 μm. (Reprinted from P. Sjovall et al., 2004. *Anal. Chem.*, 76: 4271–4278. With permission from the American Chemical Society.) (See color insert.)

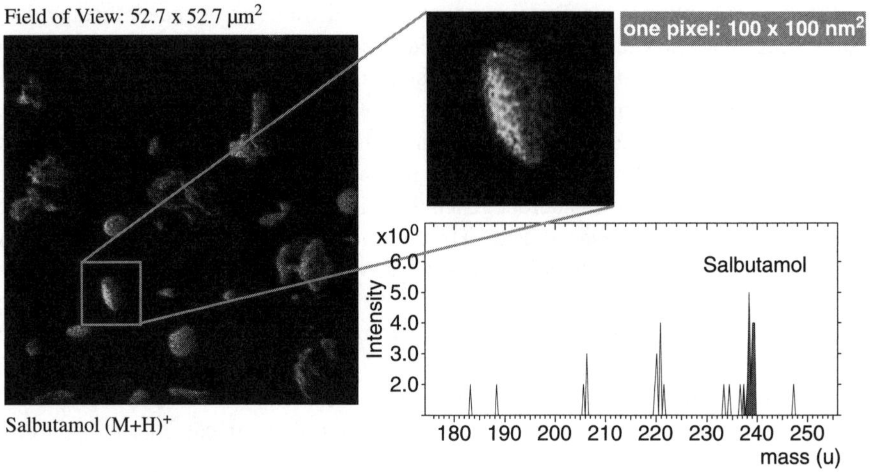

Figure 13.4. TOF-SIMS image and mass spectrum of molecular ion of salbutamol. A large area containing many beads (illustrated left) and the image of one bead enlarged (right). The pixel size in the image is 100×100 nm, and the mass spectrum from the pixel selected shows significant intensity for the salbutamol molecular ion peak. Calculation shows that the amount of salbutamol in this pixel area was in the range of 2×10^{-20} mol. (Reprinted from Kollmer, F. 2004. *Appl. Surf. Sci.*, 231–232: 153–158. With permission from Elsevier.) (See color insert.)

Figure 13.4 presents the mass spectrum and distribution of the molecular ion of salbutamol (mass 240 u), the asthma drug, which is commercially produced as a coating on micron-sized sugar beads. For analysis a Bi_3^{++} cluster gun was used. Before measurements the sample was sprayed onto a silicon substrate in order to disperse the beads [7].

13.3. CATALYSIS

TOF-SIMS has lately become appreciated as a tool that provides the truly surface-specific information that is so important for understanding how a catalyst works. Catalysts may be metals, oxides, sulfides, zeolites, carbides, organometallic compounds, or enzymes. For example, in the case of metal/support catalysts chemical reactions proceed on the surface. Therefore, knowledge about their surface properties is necessary to optimize the activity and stability of such systems. The homogeneity and uniform distribution of the active species are very important in catalytic performance. The quality of the results obtained depends not only on the standard of instruments, sample topography is very essential too, especially for powder catalysts. Grains may take various shapes, so their surface is not plain. It causes a considerable deterioration of mass and lateral resolution during measurements. An initial tableting of powder samples appears a promising way of improving resolution [8].

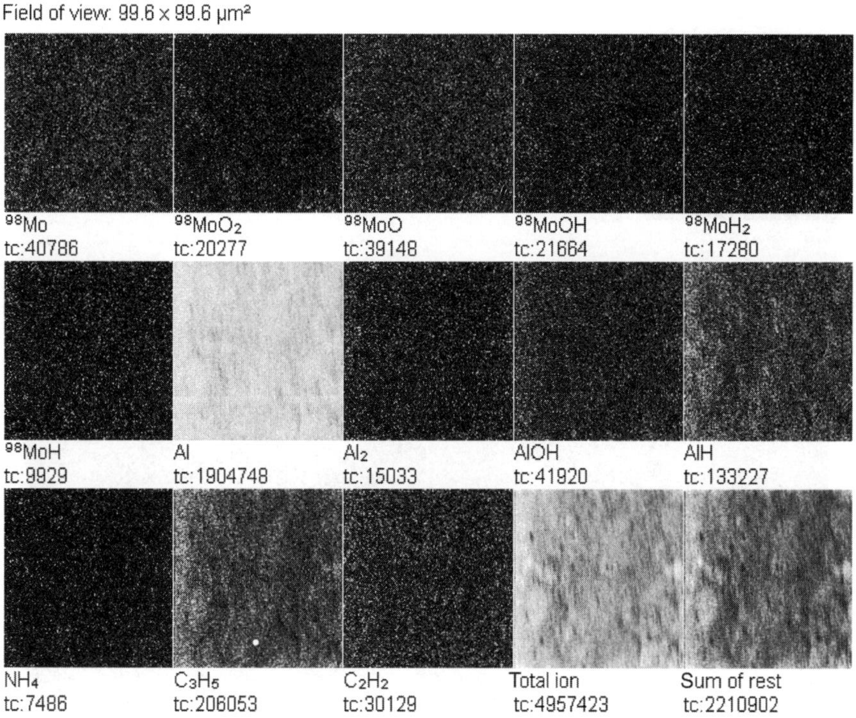

Figure 13.5. TOF-SIMS surface image of catalyst 10% Mo/Al_2O_3 calcinated at 500°C in air atmosphere (size $99.6 \times 99.6\ \mu m^2$). (See color insert.)

TOF-SIMS images (Figs. 13.5 and 13.6) illustrate the ability to detect changes in the dispersion (uniform or presence of metal clusters) of the active phase in supported-oxide catalysts. Figure 13.5 shows nearly uniform distribution of molybdenum. The surface contamination with NH_4^+ ions coming from a precursor, which were not removed during the catalyst preparation process, is also observed. Cobalt clusters in the range of several micrometers are clearly visible in Fig. 13.6.

13.4. FORENSICS

The last few years have shown increasing applications of TOF-SIMS in forensic science. Preliminary studies in the visualization and analysis of fingerprints indicate that the TOF-SIMS method opens new perspectives for the examination of fingermarks, especially in the imaging of fingermarks in various ions.

Figure 13.7 shows the image of fingerprint lines obtained from a glass surface. There is no clear signal from Si secondary ions coming from a finger, because Si is the component of the glass. Despite that, fingerprint lines are observed by reason of Si-oils (mass 73 and 147) present because of natural secretion. In the image of a

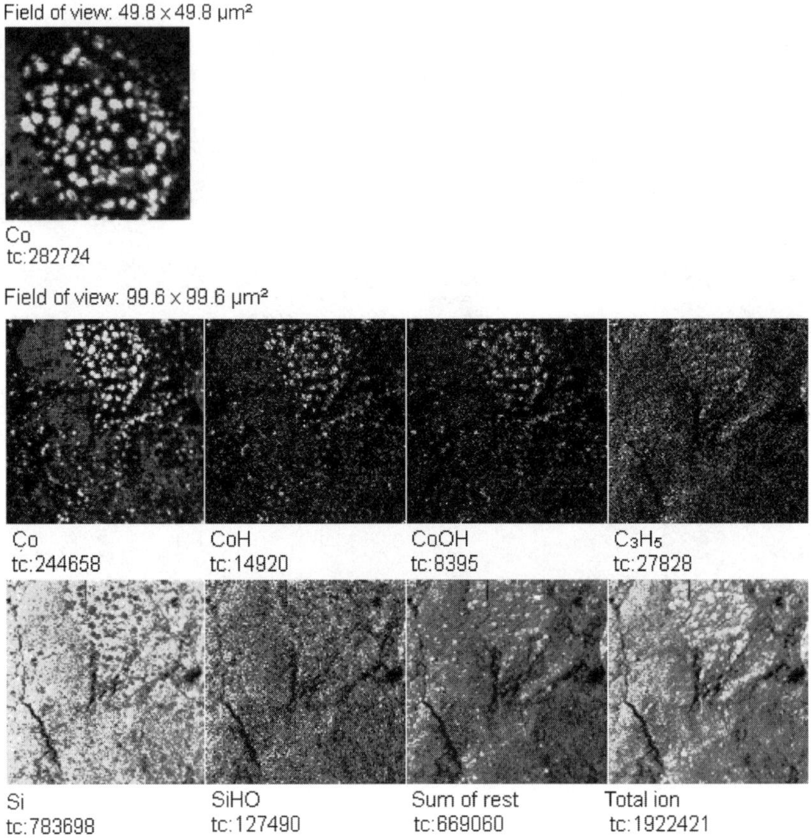

Figure 13.6. TOF-SIMS surface image of catalyst 15%Co/SiO_2 calcinated at 500°C in air atmosphere (size 99.6 × 99.6 μm^2) and Co image zoom (49.8 × 49.8 μm^2). (See color insert.)

fingerprint polluted by As_2O_3 taken from a brass base (Fig. 13.8) ^{75}As ions are visible in spite of low intensity caused by a small amount of arsenic on the finger. It indicates that even after crime residues have been precisely wiped from hands, some pollutants still remain on the fingers and they are noticeable in TOF-SIMS imaging. It can make it possible to relate the TOF-SIMS fingerprints to other evidence found at the crime scene, for example, chemicals (including drugs), beverages, or gunshot residues discovered on the fingerprints. Also, the aging fingerprint process can be examined [9].

13.5. SEMICONDUCTORS

Surface imaging is widely carried out to obtain information about the existence and distribution of trace impurities, alterations that occur when surfaces are contaminated

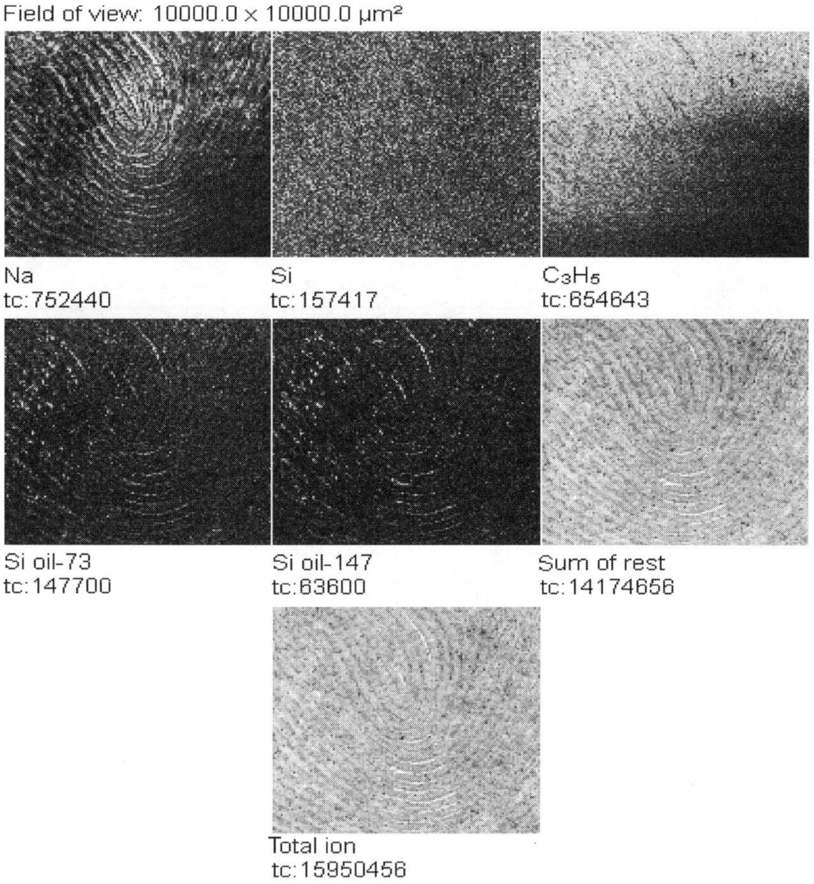

Figure 13.7. The image of a fingerprint taken from a glass sheet surface (10,000 × 10,000 μm). (Reprinted from Szynkowska, M. I. et al., 2007. Preliminary Studies Using Scanning Mass Spectrometry (TOF-SIMS) in the Visualisation and Analysis of Fingerprints, *Imaging Sci. J.*, 55: 180–187. With permission from Maney Publishing.) (See color insert.)

or subjected to different types of modification. The detection and quantification of trace metals are an important analytical task in the semiconductor industry. Figure 13.9 presents the image of elements and molecules found on a silicon semiconductor specimen area.

13.6. THE FUTURE

The future of SIMS as an imaging method depends on instrumental and chemical developments. The progress will be focused on special instruments dedicated to very

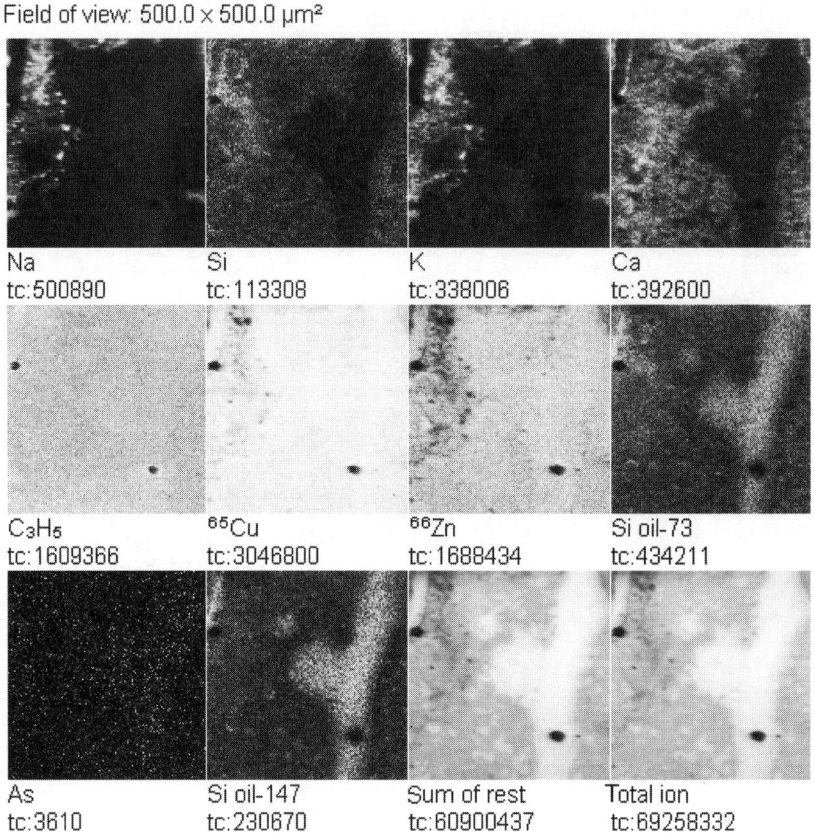

Figure 13.8. The image of a fingerprint after contact with As_2O_3 taken from a brass sheet surface (500×500 μm). (Reprinted from Szynkowska, M. I. et al., 2007. Preliminary Studies Using Scanning Mass Spectrometry (TOF-SIMS) in the Visualisation and Analysis of Fingerprints, *Imaging Sci. J.*, 55: 180–187. With permission from Maney Publishing.) (See color insert.)

special fields of application, for example, for biological samples equipped with suitable preparation chambers for freeze drying and sectioning. There are also several areas that should be taken under further development: progress in sample preparation, molecular ion yield enhancements (cluster ion beams or matrix-enhanced SIMS), improvement of lateral resolution, quantitative analysis, application of computer simulations, and studies in new fields [2, 10–13]. Nanotechnology is one of the areas where wide static SIMS applications can be investigated; however, it requires considerable progress in the development of ultrafine focused ion sources or in the manipulation of nanosamples [14]. Summing up, SIMS-based imaging techniques are very valuable tools for imaging of a variety of small molecules, but their full potential still remains to be explored.

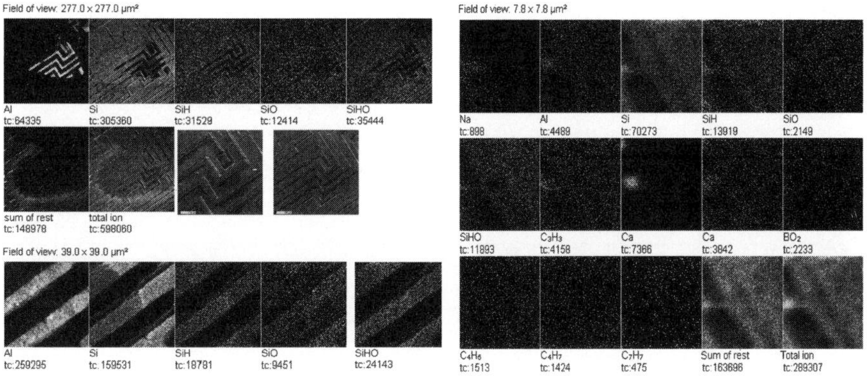

Figure 13.9. TOF-SIMS images of a silicon wafer taken from the fields 277 × 277 μm², 155 × 155 μm², 39 × 39 μm², and 7.8 × 7.8 μm². On the highest magnification image the existence and distribution of trace impurities of ions such as Na, Ca, BO_2, and C_xH_y are observed. (See color insert.)

REFERENCES

1. S. S. Rubakhin, J. C. Jurchen, E. B. Monroe, and J. V. Sweedler, *DDT*, **10**, no. 12(2005): 823–837.
2. J. C. Vickerman and D. Briggs (eds.), *TOF-SIMS: Surface Analysis by Mass Spectrometry*. Huddersfield, UK, IM Publications and Surface Spectra Limited, 2001.
3. F. Kollmer, *Appl. Surf. Sci.*, **231–232**(2004): 153–158.
4. L. Puglielli, R. E. Tanzi, and D. M. Kovacs, *Nat. Neurosci.*, **6**(2003): 345–351.
5. X. Han, A. M. Fagan, H. Cheng, J. C. Morris, C. Xiong, and D. M. Holtzman, *Ann. Neurol.*, **54**(2003): 115–119.
6. P. Sjovall, J. Lausmaa, and B. Johansson, *Anal. Chem.*, **76**(2004): 4271–4278.
7. Materials ION-TOF GmbH, www.ion-tof.com.
8. J. Grams, M. I. Szynkowska, and C. P. Norris (eds.), *Focus on Surface Science Research*. Huntington, NY: Nova Science Publishers, 2005.
9. M. I. Szynkowska, K. Czerski, J. Grams, T. Paryjczak, and A. Parczewski, *Imaging Sci. J.*, **55**(2007): 180–187.
10. M. L. Pacholski and N. Winograd, *Chem. Rev.*, **99**(1999): 2977–3005.
11. N. Winograd, *Appl. Surf. Sci.*, **203–204**(2004): 13–19.
12. M. Belu, D. J. Graham, and D. G. Castner, *Biomaterials*, **24**(2003): 3635–3653.
13. J. Xu, S. Ostrowski, C. Szakal, A. G. Ewing, and N. Winograd, *Appl. Surf. Sci.*, **231–232**(2004): 159–163.
14. A. Benninghoven, P. Bertrand, H. N. Migeon, and H. W. Werner (eds.), *Secondary Ion Mass Spectrometry, SIMS XII*. Amsterdam: Elsevier, 2000.

14

UTILIZATION OF MASS SPECTROMETRY IN CLINICAL CHEMISTRY

Donald H. Chace

14.1. INTRODUCTION

The definition of "clinical" is *as involving or concerned with the direct observation and treatment of living patients* or *of, relating to, based on, or characterized by observable and diagnosable symptoms of disease* (*Webster's New World College Dictionary*, Third edition). The key concepts presented in this definition are that there is (1) an observation and (2) it relates to the diagnosis of a disease. From the laboratory perspective this "medical observation" is the laboratory test. These test results are used by health care professionals in the diagnosis of disease or monitoring of its treatment. It is useful to distinguish the term "clinical" from the term "research." In research, there is no "official" diagnosis derived from the study, it is not evidence based and there is no burden of proof of validity. Clinical laboratory analyses require an extensive validation, ongoing quality assurance, and regulatory monitoring. These extra precautions in laboratory medicine are designed to prevent laboratory error and potential harm to a patient.

Clinical tests are targeted to measure the concentration of different sets of compounds, such as small molecules (e.g., amino acids, fatty acids, organic acids, steroids) and peptides and proteins (e.g., thyroid stimulating hormone, hemoglobin A1C) and oligonucleotides (e.g., DNA, RNA, SNPs). The presence, absence, or altered concentrations of a diagnostic compound or compounds may indicate the presence of a disease, type and severity of a disease, risk factors for disease, what is the basis for

Mass Spectrometry. Edited by Ekman, Silberring, Westman-Brinkmalm, and Kraj

disease (inherited or acquired), treatment monitoring, adverse interactions, infections, etc. Some clinical tests cannot measure a specific biomarker directly; rather they are targeted towards measuring the activity of enzymes or other processes. Abnormal levels of biomarkers may or may not be produced by inherited disease process. For example, an elevation of an amino acid, phenylalanine, in blood may indicate an inherited disorder, phenylketonuria, or effect of supplemental nutrition containing phenylalanine. Clinical test results are an important part of the diagnostic process made by a physician. The physician uses this information together with other observations to determine the next course of action. In some cases, a lab test may be the only means of detecting diseases that have no physical signs in early stage—the stages of highest potential for a positive outcome.

14.2. WHERE ARE MASS SPECTROMETERS UTILIZED IN CLINICAL APPLICATIONS?

Traditionally, mass spectrometers for clinical applications were found in specialty or reference laboratories with experts in particular areas of medicine. For example, MS systems were found at university or medical centers with specialized biochemical genetics, metabolism, or endocrinology labs. More recently, public health labs and larger commercial labs have begun utilizing mass spectrometry. Many of these new applications are replacements of older technology for classical tests such as steroid analyses. In some cases it adds new diagnostic tests such as acylcarnitine analyses and newborn screening. It will be less common in the near future not to see a mass spectrometer as one of the tools in clinical laboratory medicine. Certainly in areas such as newborn screening, a mass spectrometer will likely be part of the clinical analysis in many countries throughout the world and perhaps in the majority.

14.3. MOST COMMON ANALYTES DETECTED BY MASS SPECTROMETERS

Historically, the target analytes in clinical mass spectrometric applications were small, volatile compounds that could be analyzed by GC-MS (see Chapter 4). With time, new chemical preparation techniques and derivatization schemes broadened the scope of these metabolites to include fatty acids, amino acids, intermediates of glucose oxidation, phospholipids, steroids, neurogenic amines, nucleic acids, etc. The molecular weights (molar masses) after derivatization were less than 1000 Da, a mass range easily within the limits of most conventional mass spectrometers.

One problem with GC-MS, in addition to being labor intensive and having particularly long analysis times, was that higher molecular weight (molar mass) components or compounds with preformed cations (such as cholines or carnitine) are easily hydrolyzed and cannot be analyzed effectively using GC-MS. With the advent of new ionization techniques for LC effluents (see Section 4.1.2), such as electrospray ionization (see Section 2.1.15), more volatile and larger molecular mass compounds could be analyzed, such as complex lipids, preformed cations, peptides and proteins, and oligonucleotides. Many of the compounds have mass values beyond the mass range of the instruments. However, they also have multiple charges. Because a mass spectrometer measures m/z and not just m, the measured mass is divided by the number of charges. A protein of a mass of 10,000 Da with 10 charges will have ions detectable at m/z 100. Essential software is necessary to deconvolute protein mass spectra such that the actual mass is determined rather than the observed m/z results.

If the scope of mass spectrometry is "limitless," why are the applications of clinical MS almost completely small molecules? The answer is that most clinical tests analyze small molecules, biomarkers that are either metabolites or steroids and, hence, mass spectrometers would target those first. Perhaps a more complete answer would also include that methods must be very robust, easily reproduced in different labs, reliable, and subjected to an extensive array of validation tests. Although peptide and protein analysis is increasing rapidly in clinical labs, the MS approaches to these assays is lagging behind somewhat. MS techniques targeting these peptides and proteins exist, but they are primarily in the research stage, with few systems and methods subjected to the clinical rigors of validation. Once the necessary validations occur and methods simplified, it will only be a short time before MS is used routinely in clinical proteomics.

14.4. MULTIANALYTE DETECTION OF CLINICAL BIOMARKERS, THE REAL SUCCESS STORY

Consider one small molecule, phenylalanine. It is an essential amino acid in our diet and is important in protein synthesis (a component of protein), as well as a precursor to tyrosine and neurotransmitters. Phenylalanine is one of several amino acids that are measured in a variety of clinical methods, which include immunoassay, fluorometry, high performance liquid chromatography (HPLC; see Section 4.1.2) and most recently MS/MS (see Chapter 3). Historically, screening labs utilized immunoassays or fluorimetric analysis. Diagnostic metabolic labs used the amino acid analyzer, which was a form of HPLC. Most recently, the tandem mass spectrometer has been used extensively in screening labs to analyze amino acids or in diagnostic labs as a "universal detector" for GC and LC techniques. Why did MS/MS replace older technological systems? The answer to this question lies in the power of mass spectrometer.

Clinical analysis of significant utility for volatile organic acids started with the introduction of gas chromatography. There are hundreds of organic acids in urine and its analysis requires chromatography. Biomarkers of significance are detected by their retention time. However, many compounds co-elute and sometimes are unresolved. The introduction of mass spectrometry as the detector enabled the acquisition of mass spectra to help identify peaks (and actually characterize unresolved peaks by specific mass spectral features). The mass spectrum also served as a "confirmatory test" for resolved chromatographic peaks as two methods of identification that are fundamentally different could be generated (chromatographic retention time and a mass spectrum). With the addition of enhanced chromatographic resolving systems, improved derivatization schemes, and a variety of new columns, the GC-MS became a very powerful and important tool in examination of metabolic elimination products that went beyond organic acids (to include steroids, amino acids, and sugars). Today, GC-MS is still one of the most important tools in the diagnosis of disorders of metabolism for volatile organic compounds in clinical chemistry.

Until the 1990s, organic acid analysis by GC-MS was the most significant and widely performed analysis. However, during the 1980s new methods were developed to improve how mass spectrometry analyzed products from liquid chromatography. This technology, known as electrospray ionization, enabled the analysis of less volatile, more ionic species that characterize many metabolites and biopolymers, such as peptides, proteins, complex carbohydrates, and polynucleotides. A mass spectrometer could detect the effluent of liquid chromatography with the same benefits as GC-MS. However, this use did not seem to become the staple of amino acid analysis, perhaps in part because of the cost involved. Electrospray MS systems were four to five times more expensive than the simple MS systems used in GC-MS. Further, there was little technical expertise in the technology and it was relatively new.

Another technology that was improved, known as tandem MS (see Chapter 3), was made easier to use and priced at more affordable levels with more compact systems (compared to older systems, which were very expensive and large). Tandem MS could approximate some features of a chromatographic separation followed by mass detection. Rather than separate compounds by a chromatographic step, a tandem MS can separate compounds by mass, fragment them, then detect products. Depending on the type of analysis, one could detect a variety of different compounds without necessarily utilizing chromatography. With the ionization of liquid specimens being easier and more reliable, more polar compounds could be analyzed directly from blood in a relatively short time. It was fortuitous that compounds such as alpha amino acids and L-carnitine and acylcarnitines fragment in very specific ways for the entire family of compounds. This was capitalized upon in metabolic screening applications where not just one but many different compounds could be measured in a single analysis in a couple of minutes.

If we return to the phenylalanine example we can assume that MS/MS could measure other amino acids in addition to phenylalanine. If tandem MS could be adapted to analyze phenylanine in blood spots, this technology may be useful to newborn screening labs. For a lab only measuring phenlyalaline the cost of the MS/MS system would be prohibitive. What about labs that measure other amino acids, such as methionine, leucine, and tyrosine for other metabolic diseases? Closer examination of one disease PKU shows that the defect is in the conversion of Phe to Tyr. It would make sense, therefore, to measure both Phe and Tyr and calculate a ratio of the two metabolites. MS/MS could measure many of these compounds to both expand the disorders screened as well as improve the diagnostic efficacy of the assay. For labs measuring multiple amino acids in screening, the costs of the system become more justifiable. Addition of perhaps a new assay for different metabolites but nevertheless important disease may have been the justification to implement the technology.

This in fact is what happened. Tandem MS has clearly been shown to be the only technology to screen for disorders of fatty acid oxidation and could also detect many disorders of organic acid metabolism. Tandem MS has the ability to detect both compound classes (amino acids and acylcarnitines) and after demonstrating that both classes could be prepared in the same manner, the MS/MS analysis of blood spots for newborn screening applications was born.

14.5. QUANTITATIVE PROFILING

Not only could MS/MS be utilized to measure multiple analytes, a so-called profile, it could also quantify individual components. Like other clinical tests, a standard is used as a means of quantitative comparison where known amounts of a standard compound, usually related in structure, are added to the specimen and processed with it. Further, various quantitative evaluations such as standard curves can be generated by adding known amounts of the analyte in question relative to a fixed amount of a reference standard. With mass spectrometry, the concept of the standard is ideal in a chemical sense. Many quantitative methods using mass spectrometry utilize a technique known as isotope dilution mass spectrometry (IDMS). The technique utilizes stable isotope analogs of the analyte of interest. The stable isotope is usually identical to the compound in question related to structure and elemental composition except that it is enriched with a stable isotope such as deuterium, carbon 13, nitrogen 15 or oxygen 18. The degree of enrichment will determine the mass of the stable isotope analog. For example, in the detection of phenylalanine extracted from a blood spot, the standard utilized is a ring-labeled carbon 13 molecule where the 6 carbons (carbon 12) in the aromatic ring are replaced with carbon 13. The mass value shifts (increases) by 6 Da. Chemically, however, the labeled and unlabeled phenylalanine are similar. Extraction, derivatization, and detection (if the appropriate masses are utilized) are identical.

The analytical limits of detection mass spectrometry versus immunoassays can be quite different for many analytes. Immunoassays are inherently sensitive when the antibody is optimally designed. It is especially useful for detection of low concentrations of analytes in a complex matrix. However, in immunoassays where the selectivity is poor, the quantification, albeit measured by levels of detection, is compromised and inaccurate. As mass spectrometer and sample preparation techniques continue to improve, it is quite likely this one limitation will disappear. A general rule of thumb in clinical testing is that mass spectrometers tend to be more selective than other assays but not necessarily as sensitive. Note, however, that this tends to be a dynamic relationship in that a sample present in a complex sample may not be detected with high sensitivity in some clinical methods because of cross reactivity and signal quenching. The ability of a mass spectrometer to selectively identify a compound by mass or take advantage of features of mass to be more selective can make a test have a much higher "practical" or clinical sensitivity.

14.6. A CLINICAL EXAMPLE OF THE USE OF MASS SPECTROMETRY

Today, mass spectrometers will soon be used to screen millions of infants worldwide in both modern and developing nations. It demonstrates the power and progress of MS technology in clinical applications. The method is described briefly here, with appropriate illustrations to support the concepts presented previously and demonstrate through one example how a mass spectrometer is used in medicine.

Essential amino acids are alpha amino acids. This means that the amino group is attached to the alpha carbon of an amino acid. When subjected to a source of high energy and collisions produced in a collision cell in a tandem mass spectrometer, alpha amino acids (that are derivatized as butyl esters) will form a stable molecule that has a mass of 102 Da. That fragment molecule is illustrated in Fig. 14.1. Note that the fragment does not contain the variable "R" group that distinguishes one amino acid from another. By using a tandem mass spectrometer, we can separate all

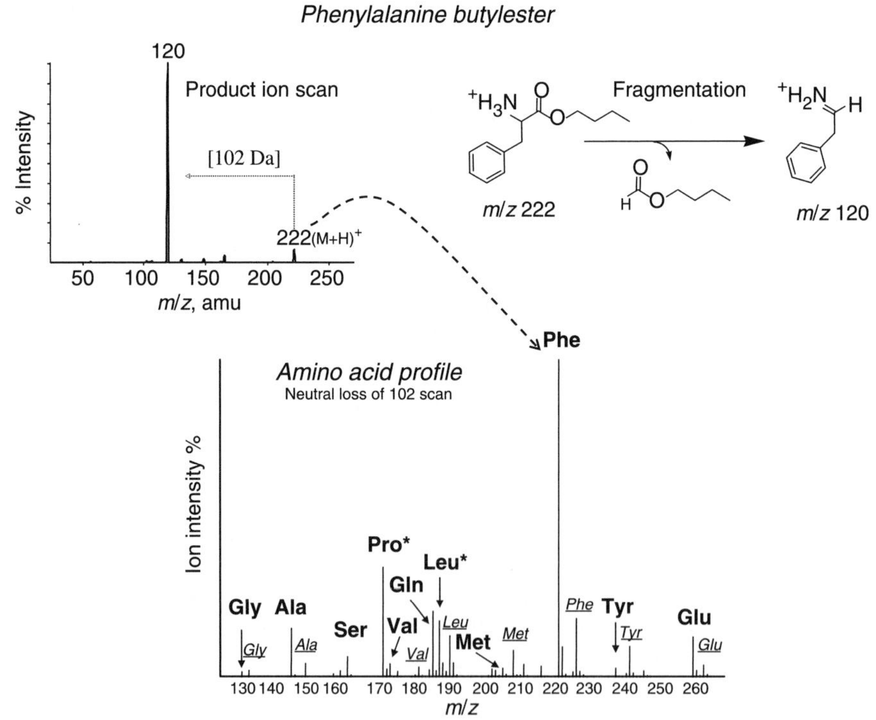

Figure 14.1. Tandem mass spectra of amino acids. Top left panel shows the product ion spectra of the buylester of phenylalanine. Fragmentation that explains the neutral loss of 102 Da is shown in the top left panel of the figure. The bottom panel is a neutral loss scan from m/z 125 to 270 which includes many common amino acids. The profile is obtained by an analysis of a blood spot from a patient with PKU.

intact amino acids by their mass-to-charge ratio and subject each of these amino acids to fragmentation. The product of this process is this stable 102 Da molecule. The remaining fragment (after loss of the 102 Da molecule) is an ion with a mass that is 102 Da less than the precursor (original unfragmented molecule). The tandem MS can be set up to examine only those precursor ions that lose 102 Da. This scan is known as the neutral loss scan of 102 Da. This process is quite selective, as shown in Fig. 14.1. The numerous peaks are amino acids (consider that this is a simple methanol extract of hundreds of compounds in blood).

In addition to neutral loss scans, mass spectrometers can be used to detect other compounds in a different manner. Acylcarnitines are fatty acid esters of carnitine. The masses of acylcarnitines differ by the size of the fatty acid attached to it. The tandem mass spectrometer can detect these selectively as well because they all produce a similar product, in this case an ion rather than a molecule. Because it is an ion, it can be detected by the second mass separation device. The ion has a mass of 85 Da and is common to all acylcarnitines. Performing a precursor ion scan of 85 Da (essentially a scan of only molecules that produce the 85 ion) reveals a selective analysis of acylcarnitines, as shown in Fig. 14.2. Additional scans have been added to more selectively detect basic amino acids, free carnitine, short chain acylcarnitines and a hormone, thyroxin (T4) which has amino acid components.

In each of the profiles presented, several internal standards (20) are present and used for quantification. Increases or decreases in the relative concentration of these metabolites indicate disease and as such a positive screen. The complexity of the assay actually resides in the interpretation, which is illustrated in the flow chart shown in Fig. 14.3.

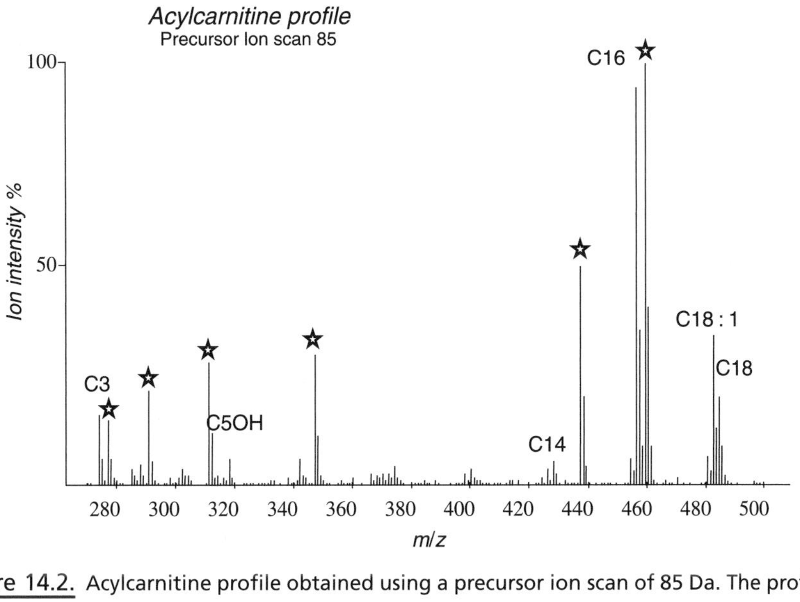

Figure 14.2. Acylcarnitine profile obtained using a precursor ion scan of 85 Da. The profile is from the blood spot of a normal patient.

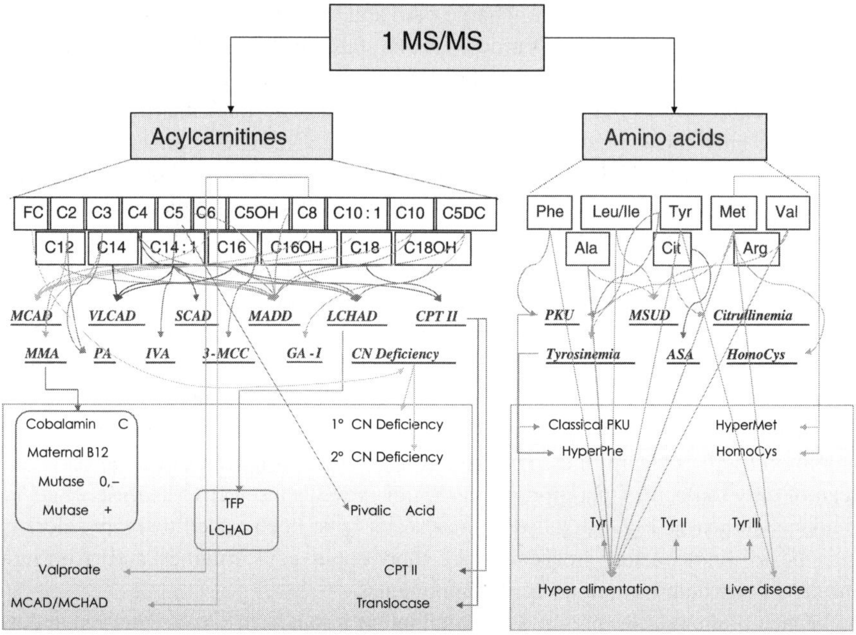

Figure 14.3. Schematic showing the relationship between the Tandem Mass Spectrometer, the metabolites it measures and the diseases or conditions detected. (See color insert.)

With one analysis, a few dozen diseases can be tested and, with less than 2 minutes a test, many hundreds of samples per day per instrument can be tested. This technique has been validated manyfold by its use in numerous screening and metabolic labs and is testimony to its power. Many new assays are being developed that follow a similar approach.

14.7. DEMONSTRATIONS OF CONCEPTS OF QUANTIFICATION IN CLINICAL CHEMISTRY

In addition to explaining verbally the concepts of mass spectrometry, it is also helpful to explain them visually. Two ideas utilized in newborn screening, for example, is the ability of a mass spectrometer to sort molecules by their mass and determine how many of these compounds are present. One illustration uses coins while another uses jelly beans. Instructions on how to prepare and present these experiments are shown below.

14.7.1. Tandem Mass Spectrometry and Sorting (Pocket Change)

Obtain a variety of coins (a handful) in any currency (the example shown here uses the U.S. dollar) (Fig. 14.4). Place the handful of coins on a table in plain view of the

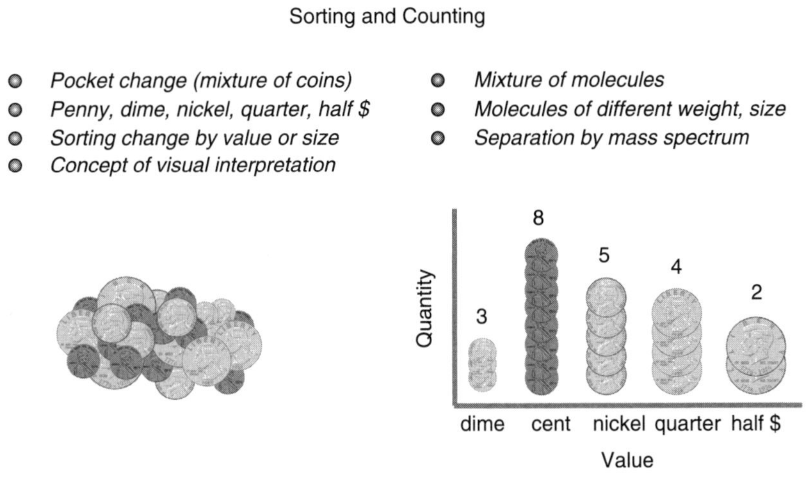
Sorting and Counting
Pocket change (mixture of coins)
Penny, dime, nickel, quarter, half $
Sorting change by value or size
Concept of visual interpretation
Mixture of molecules
Molecules of different weight, size
Separation by mass spectrum
Quantity
3
8
5
4
2
dime
cent
nickel
quarter
half $
Value

Figure 14.4. Illustration of how a mass spectrometer can be compared to counting and sorting coins. (See color insert.)

audience. This represents the mixture of molecules that would be present in a sample. Then tell the audience you are going to sort by the denomination of the coin (1 cent, 5 cents, 10 cents, etc). This is analogous to how a mass spectrometer might separate coins by weight. Interestingly the coins can be sorted by weight as well but may not be in order of denomination. For example, in the U.S. system the quarter is heavier than the nickel, which is heavier than the penny, followed by the dime. But regardless, the coins are sorted and grouped. The next step is quantification. Simply stack the coins on the table and count how many you have of each. Voila. You have just separated and quantified coins as a mass spectrometer might separate and quantify molecules. Note you could add a foreign coin to the mix to demonstrate the ability to detect interfering substances or the ability to do different scan functions to sort out coins from different countries. Explain that some foreign coins and your country's currency may have the same size and weight, but we can still separate visually like a tandem mass spectrometer can separate based on fragments (break apart the coin).

14.7.2. Isotope Dilution and Quantification (the Jelly Bean Experiment)

This demonstration is intended to show how quantification is achieved using isotope dilution principles in mass spectrometry (Fig. 14.5). It also shows the issue of error in clinical measurements, such as precision accuracy and the use of relative relationships of one compound to another.

Materials required include a 100 mL beaker and jelly beans. The jelly beans should be of the smaller variety (about 5 mm in length). Other candies can be used provided they are uniform and are easily identifiable varieties. The experiment can be performed

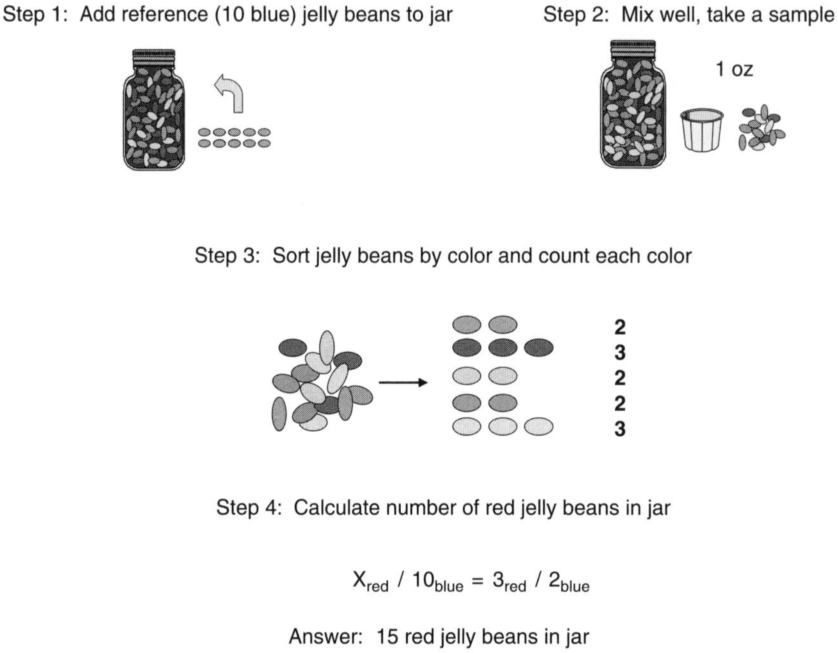
Step 1: Add reference (10 blue) jelly beans to jar
Step 2: Mix well, take a sample
1 oz
Step 3: Sort jelly beans by color and count each color
2
3
2
2
3
Step 4: Calculate number of red jelly beans in jar
X_red / 10_blue = 3_red / 2_blue
Answer: 15 red jelly beans in jar

Figure 14.5. An illustration of how a mass spectrometer measures compounds in blood by using reference standards (Stable Isotopes) using the example of jelly beans. (See color insert.)

with greater numbers of jelly beans and varieties depending upon the complexities and ultimately the accuracy of the measurements.

Obtain five varieties of jelly beans (that differ by color), perhaps red, blue, yellow, green, and white. Put 200 yellow, 150 green, 100 red, and 50 white jelly beans in a container (a large sealable bag also works well). Set aside 100 blue jelly beans in a separate container. Mix the jar containing the four different jelly beans. Bring the materials to the demonstration and ask the audience how many red jelly beans are in the jar. Some may try to guess, while others may try to get an estimate by calculating the number of red ones they see by what they think is the volume. Tell them that you will allow them to add something to the jar and sample it as well but cannot count the entire jar. Tell them this approximates the concept of obtaining a blood specimen and adding a reference standard (blue jelly beans).

After mixing the blue jelly beans in with the other colors, take a 100 mL sample. Hand that sample to two or three people in the audience and have them sort and count each. Have someone record the numbers on a piece of paper. Then have them use the following equation to calculate the number of red jelly beans in the jar.

$$\frac{\text{Red (sampled)}}{\text{Blue (sampled)}} = \frac{\text{Red (in jar)}}{\text{Blue (added to jar)}} \tag{14.1a}$$

We know the numbers of blue and the equation can be rearranged to solve for the number of red in the jar.

$$\text{Red (in jar)} = 100 * \frac{\text{Red (sampled)}}{\text{Blue (sampled)}} \tag{14.1b}$$

These results can be compared to the known amount added (which presumably was not told to the audience before this point). If the error was more than 10% questions can be discussed regarding mixing or insufficient sampling or the power of the assay (numbers of jelly beans). One way to determine whether it was a mixing error is to look at the ratios of two colors, for example, red to green or yellow to white. How did the actual numbers determined match with what was added? The experiment can be repeated with larger sampling or additional jelly beans. Discuss how this relates to isotope dilution and important principles in chemistry, such as sample mixing, sample size, selectivity, and sensitivity.

15

POLYMERS

Maurizio S. Montaudo

15.1. INTRODUCTION

The molecular characterization of a polymeric material is a crucial step in elucidating the relationship between its properties (e.g., mechanical, thermal), its chemical structure, and its morphology. As a matter of fact, the development of a new product stems invariably from a good knowledge of the above relationships. Characterization of polymers is often a difficult task because polymers display a variety of architectures, including linear, cyclic, and branched chains, dendrimers, and star polymers with different numbers of arms.

Furthermore, polymeric samples are usually a mixture of macromolecular chains terminated by different end-groups and thus, an interesting quantity to be determined (in order to characterize the mixture) is the relative abundance of end-groups [1–3]. Other two important quantities are $\overline{M}_n$ and $\overline{M}_w$, the number-average and the weight-average molar masses [1–3].

Since the discovery, in the early 1990s that polymers can be ionized by matrix-assisted laser desorption/ionization (MALDI; see Section 2.1.22) in a fashion similar to proteins, MALDI is rapidly becoming a standard tool in polymer characterization [4–6].

The aim of this chapter is to give to the reader a perspective of the potential of the MALDI technique when applied to polymers and to polymer mixtures. In describing

Mass Spectrometry. Edited by Ekman, Silberring, Westman-Brinkmalm, and Kraj

the mixtures, we shall differentiate between polymer mixtures with the same backbone and mixtures with different backbones. The second kind of mixture is much more demanding.

In order to facilitate the reader, the chapter is split into various sections. Section 15.2 deals with instrumentation, sample preparation, and matrices. remaining sections deal with the analysis of ultrapure polymers, polymer mixtures in which backbones are identical, and polymer mixtures in which backbones are different, respectively. The final section deals with the determination of average molar masses.

15.2. INSTRUMENTATION, SAMPLE PREPARATION, AND MATRICES

The basic instrumentation for MALDI is described in Chapter 2 and thus it will be covered only briefly. The spectrometer is made up of a MALDI ion source, an analyzer, and a detector. The most popular instrument has a time-of-flight (TOF) analyzer (see Section 2.2.1) and a detector in which two multichannel plates (see Section 2.3.3.2) are placed close to each other. The laser pulse is short (a couple of nanoseconds) and the matrix must display a strong absorbance at 337 nm (the laser wavelength). The ionization usually occurs by capture of a proton or alternatively of a metallic ion (Li, Na, K, Cs, Rb, Ag, etc.). In most cases, the "dried droplet" method is utilized for sample preparation. Three solutions are prepared, namely the solutions of matrix, polymeric sample, and salts (cationizing agent) and they are mixed together. The mixture is subsequently spotted onto the MALDI target and the solvent is evaporated. In the liquid state, analyte, matrix, or cationization salt are uniformly distributed. However, crystallization is relatively slow and there is the risk of forming target regions in which uniformity is lost (segregation phenomenon). If segregation occurs, significant variations of peaks, peak intensity, resolution, and mass accuracy are observed by focusing the laser on different regions of the same spot. It is possible to overcome this drawback using the solvent-free method. It consists in immersing the polymer sample in liquid nitrogen, followed by addition of powdered matrix. The resulting mixture is finely ground in a rotating-ball mill. Contrary to the dried droplet, where the solvent evaporation allows for very strong adhesion to the sample holder, in this case, the matrix/analyte powder must be carefully fixed on the MALDI sample holder [5].

The laser induces instantaneous vaporization of a microvolume (called a plume), and a mixture of ionized matrix and analyte molecules is released into the vacuum of the ion source. The relationship between the laser irradiance, I_{laser}, and the number of molecules formed, G_{maldi}, is most peculiar. There exists a threshold irradiance, peculiar to each matrix, below which ionization is not observed. Above this level, the ion production increases in a very strong, nonlinear, manner (often G_{maldi} grows as I_{laser} is raised to the eighth power).

The choice of a matrix tailored for a particular kind of polymer sample is crucial for successful characterization of the sample. Therefore, it is useful to discuss the properties of some common matrices [4–6]. 3-Amino-4-hydroxybenzoic acid and POPOP need high laser power, since they possess a high threshold. Alpha-cyanocynnamic acid is sometimes used for post-source-decay experiments [5], because it yields ions with a (slightly) higher internal energy with respect to the others. Some polymeric samples contain surfactants and other contaminants. It has been shown that, in order to enhance the number of ions produced by MALDI, it is necessary to purify the analyte from these contaminants prior to analysis. Some MALDI matrices, such as all-trans retinoic acid, are particularly sensitive to impurities, whereas for other matrices (like HABA and Dithranol) the loss of efficency is small and hence the latter matrices ought to be preferred with respect to the former, when purification is a problem. Retinoic acid works with polystyrene, but it must be doped, preferably with Ag salts. 5-Clorosalicilic acid gives good MALDI spectra of apolar polymers, whereas trihydroxyacetophenone and nor-harmane are general-purpose matrices [4–6]. Indole-acrylic acid gives good results with esters and carbonates [7].

15.3. ANALYSIS OF ULTRAPURE POLYMER SAMPLES

In ultrapure polymer samples, all chains are terminated in the same way. The MALDI spectrum of an ultrapure polymer resembles a comb and the spacing between the comb's "teeths" equals the mass, M_{repeat}, of the repeat unit. This quantity is often diagnostic and it suggests an almost trivial use of MALDI is the spectral identification of polymers. The reason is that, if one computes the M_{repeat} value for common polymers, most values are different, the number of superpositions being very low [4–6]. The M_{repeat} value is not an integer, due to the fact that various isotopes are present.

An ultrapure polymer is made of chains of the type G1-AAAAAAA-G2, where A is the repeat unit and G1 and G2 are end-groups. One considers the mass number of one of the MS peaks, subtracts the mass of the cation (e.g., H, Li, Na, Ag), and then repeatedly subtracts the mass of the repeat unit, until one obtains the sum of the masses of G1 + G2. For this purpose, a linear best fit can also be used. Tandem mass spectrometry is particularly useful since, from the analysis of ion fragmentation patterns, one can deduce the mass of G1 and, separately, the mass of G2.

15.4. ANALYSIS OF POLYMER SAMPLES IN WHICH ALL CHAINS POSSESS THE SAME BACKBONE

The MALDI spectrum of a polymer sample in which all chains possess the same backbone allows identification of the end-groups present at the chain ends. This type of analysis is referred to as end-group analysis. An example will be helpful. Figure 15.1 reports the MALDI spectrum of a poly(bisphenolA carbonate) (PC for short) sample [7]. It displays a series of peaks from 2 up to 16 kDa, the most intense ones in the region from 5 up to 7 kDa. It also displays peak assignment and an expansion of the spectral region from 3.0 up to 3.7 kDa. Peaks at 3034, 3288, and 3542 are labeled as A and are due to PC chains terminated with phenolcarbonate on both sides. Peaks at 3168, 3422, and 3676 are labeled as B and are due to PC chains terminated with phenolcarbonate on one side and bisphenol-A on the other. Peaks at 3048, 3302, and

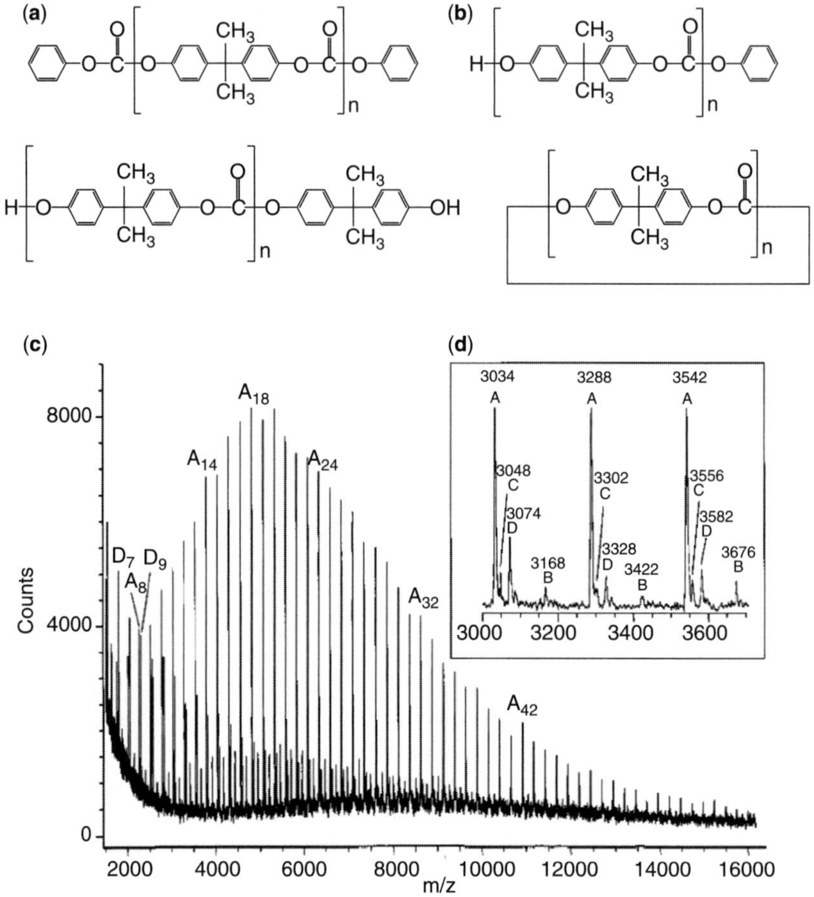

Figure 15.1. MALDI spectrum of a polycarbonate sample along with peak assignment. In the inset, an expansion of the spectral region from 3.0 up to 3.7 kDa is shown. (Reproduced from Puglisi, C. et al., 1999. Analysis of Poly(bisphenol A Carbonate) by Size Exclusion Chromatography/Matrix-Assisted Laser Desorption/Ionization. I. End Group and Molar Mass Determination. *Rapid Communications in Mass Spectrometry,* 13: 2260–2267. With permission of John Wiley & Sons, Inc.)

3556 are labeled as C and are due to PC chains terminated with bisphenol-A on both sides. Peaks at 3074, 3328, and 3582 are labeled as D and are due to cyclic PC chains.

A careful inspection of the spectrum [7] reveals that the intensity of peaks labeled as A grows in the region between 2 and 5 kDa and then it falls slowly, up to 16 kDa. The same pattern is seen for peaks labeled as B and C. On the other hand, the intensity of peaks labeled as D (due to cycles) is very strong in the region between 2 and 3 kDa and falls very quickly as the mass grows, becoming negligibly small at 7 kDa. This behavior is in agreement with the prediction that the abundance of cycles decreases steadily (i.e., it is a monotone function of mass) [1–3].

In practice, from the analysis of the MALDI spectrum, it can be concluded that the sample selected is a mixture of linear and cyclic chains and that the linear chains are terminated in three different ways.

If one assumes that, in this case, ion abundances are (approximately) proportional to chain abundances in the polymeric sample, it can be concluded that chains terminated with phenolcarbonate on both sides are the most abundant ones (since peaks labeled as A are the most intense ones). In this way, a full polymer characterization is achieved.

15.5. ANALYSIS OF POLYMER MIXTURES WITH DIFFERENT BACKBONES

Recording the MALDI spectrum of a mixture of two polymers having different backbones, one finds that MALDI peak intensities reflect in a distorted manner the abundances of the chains and the composition of the blend. In some cases, the distortion is small and thus MALDI is semiquantitative. The main cause is that the ionization efficiency (i.e., the probability of ion production) for the two polymers is not the same. Furthermore, it has been shown that instrumental parameters can affect peak intensities, thus falsifying the composition of the blend. For instance, some authors [5] studied an equimolar mixture of PEG and PMMA, recorded the MALDI spectrum of the mixture and found, on changing instrumental parameters, that the apparent blend composition changed from 100/0 to 50/50 to 0/100.

15.6. DETERMINATION OF AVERAGE MOLAR MASSES

An overview of old and well-established methods is essential when performing a comparison with MALDI, since MALDI possesses some distinct advantages with respect to the former methods. Among conventional methods, size exclusion chromatography (SEC; see Section 4.1.2.5) is often used to measure molar mass (MM), and molar mass distribution (MMD) [1–3]. It is an indirect method, since it needs calibration. More specifically, the quantities that are measured are elution times and these are related to $\overline{M}_n$ and $\overline{M}_w$ by a calibration equation. The most common method for measuring the calibration constants consists in preparing a mixture of five or more polymer samples with the same repeat unit, each having a narrow MM distribution and known MM (so-called SEC primary standards). The mixture is injected in the SEC. The reliability of SEC results strongly depends on the availability of a set of polymers of known MM and narrow MM distribution (primary standards) with the same structure as the polymer of interest.

Viscometry is used to measure average molar masses too [1–3]. It is an indirect method, since the measured quantity is the intrinsic viscosity (IV), which is related to the average molar masses calibration by a peculiar formula, called the Mark–Huwink–Sakurada equation [1–3].

Osmometry and light scattering (LS) can be used to measure $\overline{M}_n$ or $\overline{M}_w$ [1–3]. They are direct methods, since they do not need calibration.

Nuclear magnetic resonance (NMR) is a very popular technique in the field of polymer analysis [1–3]. It can be used to measure $\overline{M}_n$, but this application relies heavily on the presence of NMR signals due to terminal groups (TEG). As the length of the chains grows, the NMR signal due to TEG becomes weak and the accuracy falls. In practice, above 30 kDa the error level on $\overline{M}_n$ is so high that the measurement is useless.

Mass spectrometry can be used to measure the molar mass distribution (MMD) of a polymer sample by simply measuring the intensity, N_i, of each mass spectral peak with mass m_i. This is due to the fact that mass spectrometers are equipped with a detector that gives the same response if an ion with mass 1 kDa or 100 Da (actually any mass) strikes against it. In other words, the detector measures the number fraction and this implies that N_i also represents the number of chains with mass m_i. Thus, the number-average molar mass, $\overline{M}_n$, is given by:

$$\overline{M}_n = \left(\sum_{m_i N_i}\right) \Big/ \left(\sum_{N_i}\right) \tag{15.1}$$

In a similar manner, the weight-average molar mass, $\overline{M}_w$, is given by:

$$\overline{M}_w = \left(\sum_{m_i^2 N_i}\right) \Big/ \left(\sum_{m_i N_i}\right) \tag{15.2}$$

where the summation spans over all masses (from one to infinity).

A word of caution: the ionization process must be "soft," If hard-ionization occurs, chains are no longer intact (fragmentation occurs) and the measurement will be affected by a systematic error towards the bottom (i.e., underestimation of $\overline{M}_n$ and $\overline{M}_w$). Since fragmentation is an annoying concern, some authors [4–6] developed a protocol to avoid it or, at least, to reduce the extent of fragmentation, E_{FR}. They noted that E_{FR} decreases when the laser power is lowered and also when a large excess of matrix is used in sample preparation [4–6]. Thus, the protocol consists in using low laser powers (close to the threshold) and in using a matrix-to-analyte ratio of at least 10,000 : 1.

Many authors [4–6] (we were able to count at least 300 reports) have compared $\overline{M}_n$ and $\overline{M}_w$ values for the polymer sample (obtained using Equations 15.1 and 15.2) with $\overline{M}_n$ and $\overline{M}_w$ obtained by traditional methods for MM determination (i.e., SEC, viscometry, light scattering, osmometry, NMR, etc., as discussed above). In the case of SEC, an additional feature arises, which is very appealing, namely the possibility to compute the entire MMD. However, before comparing it with the MMD obtained by SEC, it must be processed using a suitable transformation algorithm (this processing is seldom done by hand). In fact, MALDI gives the number fraction whereas SEC gives the weight fraction. Furthermore, in the MALDI plot the masses are not logarithmic, whereas in the SEC plot the masses are logarithmic, since they follow the elution volume [4–6].

Most authors found that for narrow-MMD polymer ($\overline{M}_w/\overline{M}_n < 1.10$) the agreement is within 10% to 15% or even better and therefore it is excellent. As a matter of fact, polymers with a narrow MMD can be obtained by anionic or cationic polymerization and their molar mass averages can be readily measured by traditional methods for MM determination.

On the other hand, for broad-MMD polymer, MALDI underestimates both $\overline{M}_n$ and $\overline{M}_w$ [4–6]. This problem is usually called "mass discrimination" [4–6]. Many

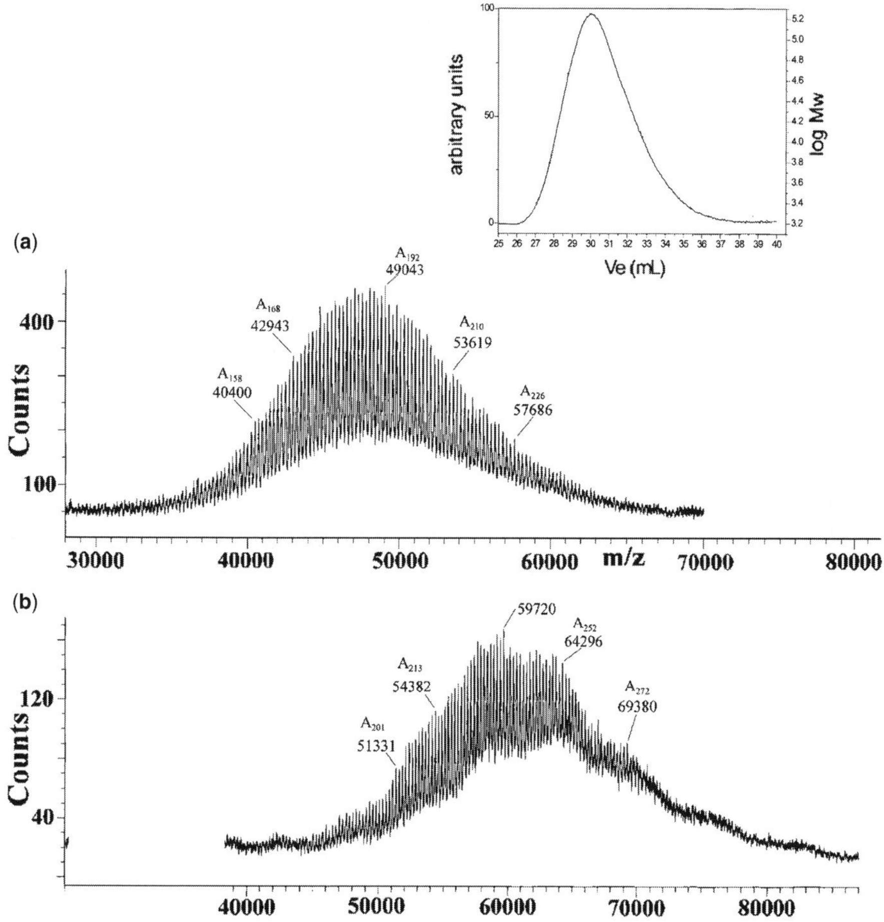

Figure 15.2. The SEC trace for a PC sample along with the MALDI spectra of sample PC2854 (upper spectrum) and of sample PC2780 (lower spectrum). (Reproduced from Puglisi, C. et al., 1999. Analysis of Poly(bisphenol A Carbonate) by Size Exclusion Chromatography/Matrix-Assisted Laser Desorption/Ionization. I. End Group and Molar Mass Determination. *Rapid Communications in Mass Spectrometry,* 13: 2260–2267. With permission of John Wiley & Sons, Inc.)

authors proposed remedies against mass discrimination, but the most effective is the off-line coupling of SEC and MALDI. In the cited experiment, the polymeric solution (typical concentration 10 mg/ml) is injected in the SEC apparatus and selected fractions are analyzed by MALDI. Each MALDI spectrum yields a value, M_F, which is the average molar masses of fraction F. By plotting M_F versus F, it is possible to construct a graph that represents the genuine SEC calibration line for the polymer under scrutiny and thus the goal of MM determination is achieved.

An example can best explain the procedure. A poly(bisphenolA carbonate) sample characterized by a broad-MMD was injected in an SEC apparatus, about 100 fractions were collected, and 24 of them were analyzed by MALDI [7]. Figure 15.2 reports the SEC trace of the PC sample. The trace covers a quite broad range of elution volumes and it is centered at about 30 ml. The polymer starts eluting at about 26 ml and ends at about 38 ml. The MALDI spectra yielded M_F values (see above). Using this information, the SEC trace in Fig. 15.2 is calibrated and the average molar masses turn out to be $\overline{\mathrm{M}}_{\mathrm{w}} = 55{,}800$, $\overline{\mathrm{M}}_{\mathrm{n}} = 23{,}600$.

Some details of the experiment are revealing. The SEC fractions that elute at 28.54 ml and at 27.80 ml will be referred to as PC2854 and PC2780, respectively [7]. Figure 15.2a reports the MALDI spectra of PC2854. There are more than 100 well-resolved peaks, due to PC chains terminated with phenolcarbonate on both sides. The mass range is undoubtedly high. Peaks span from 30 to 70 kDa, the most intense ones being at 50 kDa. Figure 15.2b reports the MALDI spectra of PC2780. The number of MS peaks (about 150) and the mass range are certainly higher than those in the previous one. The most intense peaks fall in the region between 60 and 65 kDa. Note that the spectrum in Fig. 15.2b suffers from poor resolution: peaks do not "pop out," the valley between peaks is shallow and there is substantial overlap among neighboring peaks. On the other hand, it is quite apparent that the first spectrum (Fig. 15.2a) is almost perfectly resolved: the valley between peaks is very deep (in some regions it touches the baseline) and neighboring peaks do not overlap. In Fig. 15.2b, the mass range is higher and the resolution is lower than in Fig. 15.2a. This fact is quite general: above 10 kDa, the resolution depends on the mass and it falls as the mass grows.

REFERENCES

1. H. G. Barth and J. W. Mays. *Modern Methods of Polymer Characterization*, New York: Wiley, 1991.
2. R. H. Boyd and P. J. Phillips. *The Science of Polymer Molecules*, London: Cambridge University Press, 1996.
3. J. L. Koenig. *Spectroscopy of Polymers*, New York: Elsevier, 1999.
4. G. Montaudo and R. P. Lattimer (eds.). *Mass Spectrometry of Polymers*, Boca Raton, FL: CRC Press, 2001.
5. G. Montaudo, F. Samperi, and M. S. Montaudo. Characterization of Synthetic Polymers by MALDI-MS. *Progress in Polymer Science*, **31**, no. 3 (2006): 277–357.

6. M. W. F. Nielen. MALDI Time-of-Flight Mass Spectrometry of Synthetic Polymers. *Mass Spectrometry Reviews*, **18**, no. 5 (1999): 309–344.

7. C. Puglisi, F. Samperi, S. Carroccio, and G. Montaudo. Analysis of Poly(bisphenol A Carbonate) by Size Exclusion Chromatography/Matrix-Assisted Laser Desorption/Ionization. I. End Group and Molar Mass Determination. *Rapid Communications in Mass Spectrometry*, **13**(1999): 2260–2267.

16

FORENSIC SCIENCES

Maria Kala

> When the law requires You to step forward as a witness, be always a person of science. It is not for you to avenge the victim, to save the innocent or destroy the perpetrator. Your sole duty is to bear witness within the limits of Your knowledge and Your professional abilities.
>
> —George Burgess Magrath, Boston, 1906

16.1. INTRODUCTION

Many definitions of "forensic science" exist in the literature. In the most popular sense, the forensic sciences refer to a particular scientific discipline (medicine, toxicology, chemistry) that applies its principles and methods to the needs of law. Therefore, the terms forensic and legal are synonyms. Forensic scientists include pathologists, psychiatrists, toxicologists, criminalists, molecular biologists, and several other specialists. These practitioners, forensic experts, are obliged to explain the smallest details of the methods used, to substantiate the choice of the applied technique, and to give their unbiased conclusions for the benefit of the courts. The final results of the work of forensic scientists influence the fate of a given individual. Very often the final decisions of forensic scientists requires the cooperation of professionally competent scientists of different disciplines. For example, the completion of the cause of death statement by a legal physician for a medical-legal case (e.g., poisoning, suicide, homicide) requires careful consideration of autopsy

Mass Spectrometry. Edited by Ekman, Silberring, Westman-Brinkmalm, and Kraj

findings, toxicological examinations or other type of postmortem studies (bacteriologic, virologic, and immunologic), and circumstances of death.

Forensic toxicology is, quite literally, the use of toxicology in courts of law. This most often refers to analysis of toxicologically relevant compounds (alcohol, drugs, and poisons) in conventional (body fluids and tissues) as well as in alternative (hair, saliva, sweat) matrices and the interpretation of those analytical results to answer questions that occur in judicial proceedings. Forensic toxicology has also played a meaningful role as guardian of the public health. Recognition of the widespread abuse of drugs by toxicologists brought about efforts on both national and international fronts to control the availability of drugs of abuse.

Criminalistics and trace evidence are both terms that apply to all types of physical material that may be circumstantial evidence in the trial of a case. Most often experts who are identified as criminalists, microanalysts, or trace evidence examiners analyze a variety of types of trace evidence. They carry out three types of identification. First is to determine the nature of small items of trace evidence. After this forensic experts compare the trace evidence with known materials for the purpose of determining the origin of the evidence. The third type of criminalistics investigations is performed in order to identify an individual to whom the trace belongs. For this purpose population studies using statistics (especially the probabilistic approach of Bayesian theory) and chemometrics methods are utilized.

In forensic toxicology and criminalistics analytical methods must provide high reliability and accuracy. The combinations of MS with suitable chromatographic procedures are the methods of choice, because they are very sensitive, precise, specific, universal, and relatively fast. The main reasons for their success is that both techniques provide the highest level of confidence in the results by generating two identification parameters (retention time and mass spectrum) and even molecular mass in one process. Capillary electrophoresis (CE) (see Section 4.2.2.3) can also be coupled to MS and used for drug analysis. The preferred mode of ionization for interfacing the CE instrument with MS is electrospray ionization (ESI; see Section 2.1.15). Assays for drugs in biofluids that use CE-MS are, in general, either for drugs taken in relatively high doses or where sample extract clean-up is employed to increase the final analyte concentrations.

Today, GC-MS (see Section 4.1.1) is a golden standard for detection and quantification of drugs and poisons volatile under GC conditions, whereas nonvolatile compounds require LC-MS (see Section 4.1.2). The GC-MS technique is much more popular for identification purposes than LC-MS, because of the easy availability of the reference mass spectra for many xenobiotics and their derivatives, either in printed or computer form. The most popular libraries are the NIST library, which contains the mass spectra of 130,000 compounds, the Wiley Registry of Mass Spectral Data, which contains 390,000 reference spectra, and the Pfleger–Maurer–Weber library, with 6,300 mass spectra and other data, such as chromatographic retention indexes.

LC-MS at present is only complementary to GC-MS. The LC-MS technique is applied for confirmations of positive results obtained by different immunochemical methods and for confirmation of the intake of drug(s). Nevertheless, the number of studies in which procedures for screening a single group of drugs, for example, benzodiazepines, amphetamines, fentanyl analogs (Fig. 16.1), and pesticides, or a subject

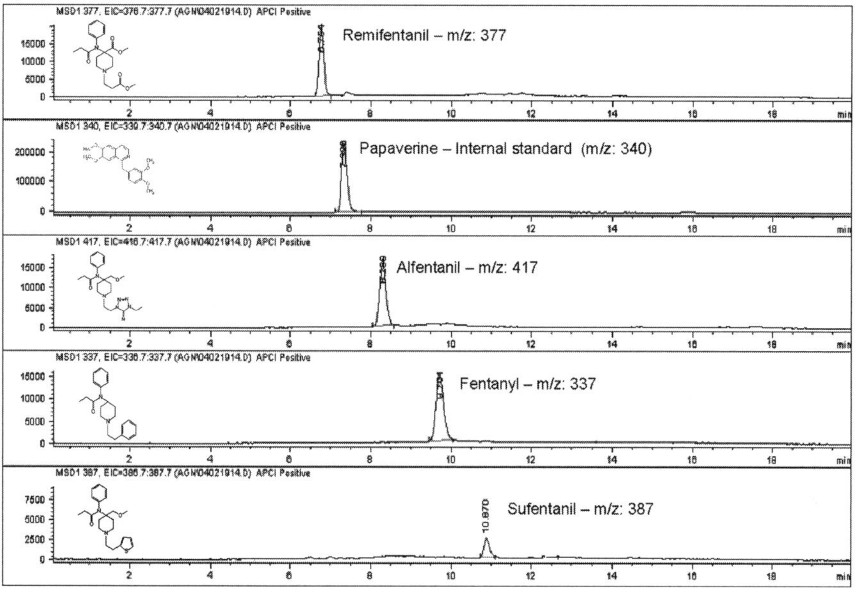

Figure 16.1. Screening method for fentanyl analogs in blood using LC-MS/APCI. Monitored ions $[M+H]^+$ were chosen from full scan spectra. The blood sample was spiked with fentanyl, its three analogs and internal standard to the concentration of 5 ng/mL [6].

category of drugs, for example, drugs potentially hazardous for traffic safety or drugs-facilitated sexual assault are elaborated.

16.2. MATERIALS EXAMINED AND GOALS OF ANALYSIS

Depending on the toxicological or criminalistic problems to be solved and/or the availability of the sample, different matrices must be analyzed by using GC-MS or LC-MS techniques.

Biosamples, taken from living people (mainly blood and urine) and from corpses during autopsy (blood, urine, vitreous humor, bile, and almost any tissue and organ), are analyzed for substances that are toxicologically relevant. Alternative matrices like hair and saliva are also considered for determination of illicit drugs.

A broad spectrum of nonbiological materials is submitted for toxicological and criminalistic analysis. These include, in particular:

- Pharmaceutical preparations, fluids, and powders taken from or found near a patient or body—so-called scene residues—which have forensic relevance.
- Residues of drinks and foodstuffs to which a poison could have been added.
- Samples of river, lake, and well waters suspected of being contaminated by pesticides.

- Clandestine preparations of controlled drugs, including raw materials, by-products, and different wastes.
- So-called "street drugs" for identification of principal components, impurities, adulterants, and cutting agents in order to establish the method of synthesis, to compare the samples seized from users with those from dealers, as well as for comparative analysis or so-called "profiling" to find out the source of the material.
- Explosives and evidence at a fire scene (fire debris) for determination regarding the origin and cause of the fire.

16.3. SAMPLE PREPARATION

Independent of the sample, a suitable sample preparation is necessary before GC-MS or LC-MS analysis. For biosamples this may involve protein precipitation, cleavage of conjugates (e.g., glucuronide metabolites), isolation, clean-up steps, or derivatization of the drugs and their metabolites. Depending on the circumstances and the purposes, two approaches can be distinguished, the directed and the undirected searches. For screening procedures the pH-dependent liquid-liquid extraction (LLE) technique is still the most popular. From each biosample submitted for extraction an acidic-neutral, a basic, and sometimes a strongly basic extract are obtained. Solid phase extraction (SPE) is most often applied to certain compounds. SPE is the extraction technique of choice for glucuronide-conjugated metabolites and quaternary ammonium compounds. Over the years screening procedures using diatomaceous earth (Extrelut column of Merck), polystyrene-divinylbenzene copolymer, and mixed-mode bonded silica as column material in SPE were elaborated.

Drugs and toxicants are metabolized in the human body in such a way that more polar compounds are usually formed. Therefore to decrease the polarity and increase the volatility and thermal stability of the analytes, the derivatization step is an unavoidable requirement for GC analysis. This step enhances the detectability of the analytes and provides very characteristic mass spectra that can be relevant for identification purposes. Most analytes do not require derivatization for LC separation and MS detection.

The solvents used for extraction should be evaporated under a stream of nitrogen to avoid oxidation of metabolites during the analytical process.

Applications of GC-MS or LC-MS techniques to forensic toxicology considerably reduced the volume of specimens necessary for analytical procedure. Recently, the volumes range from 0.2 to 1.0 mL.

16.4. SYSTEMATIC TOXICOLOGICAL ANALYSIS

The screening strategy for systematic toxicological analysis (STA) must be very extensive, because several thousands of drugs or pesticides should be considered. In STA the substance(s) present is (are) not known at the start of the analysis. In such a search all steps of the analytical procedure, especially extraction and detection, must be general procedures that cover a broad spectrum of xenobiotics (Fig. 16.2, Table 16.1). The substances of interest should be isolated at as high a yield as possible and the interfering substances from the biological matrices should be removed. Drugs are low and high dosed which leads to wide ranges of their concentrations in body fluids. The screening procedures are not as sensitive as target drug methods; therefore, drugs present in low concentrations cannot be detected. Qualitative methods should be validated for specificity and limit of detection (LOD) for representative drug examples.

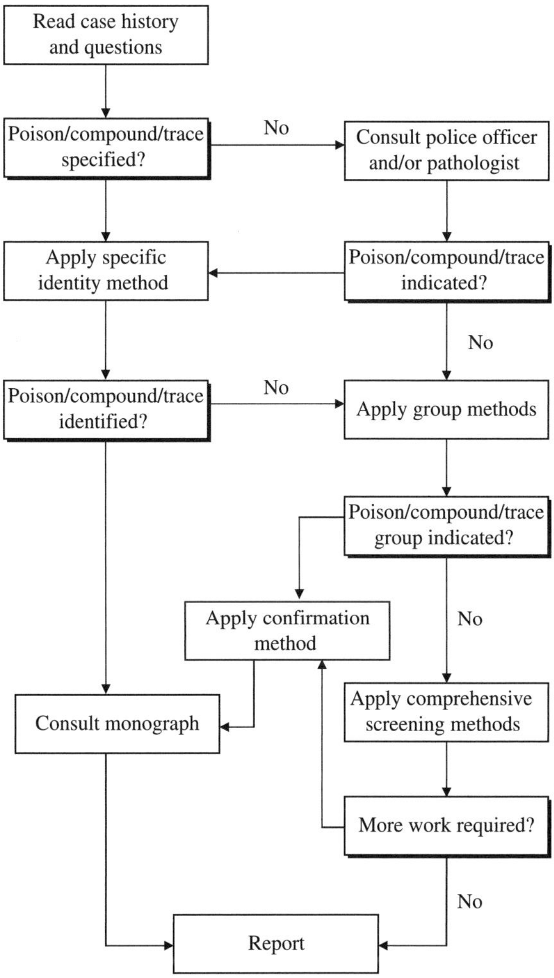

Figure 16.2. Schematic presentation of procedure for identification of chemical compounds and other traces for forensic toxicology or criminalistic purposes [4].

TABLE 16.1. Comparison of Order of Examination for Poison Groups, Target Specimens, and Methods Commonly Used in Systematic Toxicological Analysis

Order	Material	Group of Compound	Screening Method	Sample Preparation	Confirmation Method
1	Lungs, brain, blood	Gases (carbon monoxide, cyanide)	Direct-reading colorimetric indicator tubes	Head-space Microdiffusion	GC
2	Blood, vitreous humor, urine	Ethanol	ADH enzymatic Spectrophotometric	Head-space	GC
3	Blood, brain, lungs	Solvents (methanol, toluene, hydrocarbons)	Color tests GC	Head-space LLE	GC GC-MS
4	Blood	Hemoglobin derivatives (HbCO, MetHb)	Spectrophotometric	Derivatization	GC
5	Blood, urine, liver, other tissues, hair, saliva, injection marks	Drugs	Selective IA TLC, GC-MS	Precipitation, acidic or enzymatic digestion, matrix dissolution LLE or SPE, Derivatization	GC-MS LC-MS
6	Blood, stomach content, urine	Pesticides	GC, GC-MS	LLE, SPE	GC-MS LC-MS
7	Blood, stomach content, tissues	Anions	Color tests, colorimetric methods	Dialysis	ICP-MS
8	Blood, tissues, intestinal content, bone, hair, nails	Metals	Selective tests AAS	Mineralization (wet, dry)	AAS ICP-MS
9	Blood, stomach content, tissues	Miscellaneous compounds	Group tests, specific tests, particular methods	Ion-exchange columns, formation of ion-pairs, SPE, LLE, continuous extraction	GC-MS LC-MS

Note: Consideration—age, occupation, and the site of the incident. Visual examination of evidence for color, odor, sediments, suspensions, and frothing.

16.4.1. GC-MS Procedures

For comprehensive GC-MS screening procedures the full scan EI (see Section 2.1.6) mode is used, because EI produces universally reproducible mass spectra. On the basis of the reference spectra from the Pfleger–Maurer–Weber library the authors elaborated three typical STA procedures. The first was elaborated for detection of most of the basic and neutral drugs in urine after acid hydrolysis, LLE, and acetylation. It included 16 pharmacological categories of drugs. The second procedure was prepared for most of the acidic drugs and poisons or their metabolites after extractive methylation. This method allowed the detection of nine categories of drugs in urine. The third was published for doping-relevant drugs after enzymatic hydrolysis, SPE, and combined TMS and TFA derivatization. Screening is performed using a mass chromatogram, which may indicate the presence of suspected mass spectra in the full mass spectra stored during GC separation. Positive signals can be confirmed by visual or computerized comparison of the peak full mass with reference spectra. Eight ions per category were individually selected from the mass spectra of the corresponding drugs and their metabolites identified in an authentic biosample. Thirteen common drugs covering relevant retention times were used to optimize the whole procedure. It is not true that all three STA procedures covered all 6,300 mass spectra included in the library. Many compounds need special procedures and they are not detectable under the conditions used or the concentrations of compounds are too low [5].

As a consequence of the development of extraction methods for STA based on mixed-mode SPE columns, as well as of the recent introduction of instruments for the automated sample preparation allowing efficient evaporation and derivatization of the extracts, full automation of STA methods based on GC-MS analysis is also available. It needs GC-MS instruments equipped with an HP PrepStation System. The samples directly injected by the PrepStation are analyzed by full scan GC-MS. Using macrocommands, peak identification and reporting of the results are also automated. Each ion of interest is automatically selected, retention time is calculated, and the peak area is determined. All data are checked for interference, peak selection, and baseline determination.

16.4.2. LC-MS Procedures

Currently no comprehensive libraries have been developed for methods, such as ESI (see Section 2.1.15) and atmospheric pressure chemical ionization (APCI; see Section 2.1.8), used to ionize compounds after LC separation. Generally these types of library are user generated for a specific purpose.

The coupling of MS with LC has long been considered a potential means of increasing the range of compounds amenable to MS. Though ESI and APCI have superseded all the other types of interfaces or ionization sources for LC-MS, they are not compatible with EI (Fig. 16.3), involve a soft ionization process, and generate very few fragments. These limitations can be bypassed by using collision-induced dissociation (CID; see Section 3.2.3), which provides thorough fragmentation of the compounds. Accelerated ions can collide with molecules of a neutral gas either in a specialized collision cell

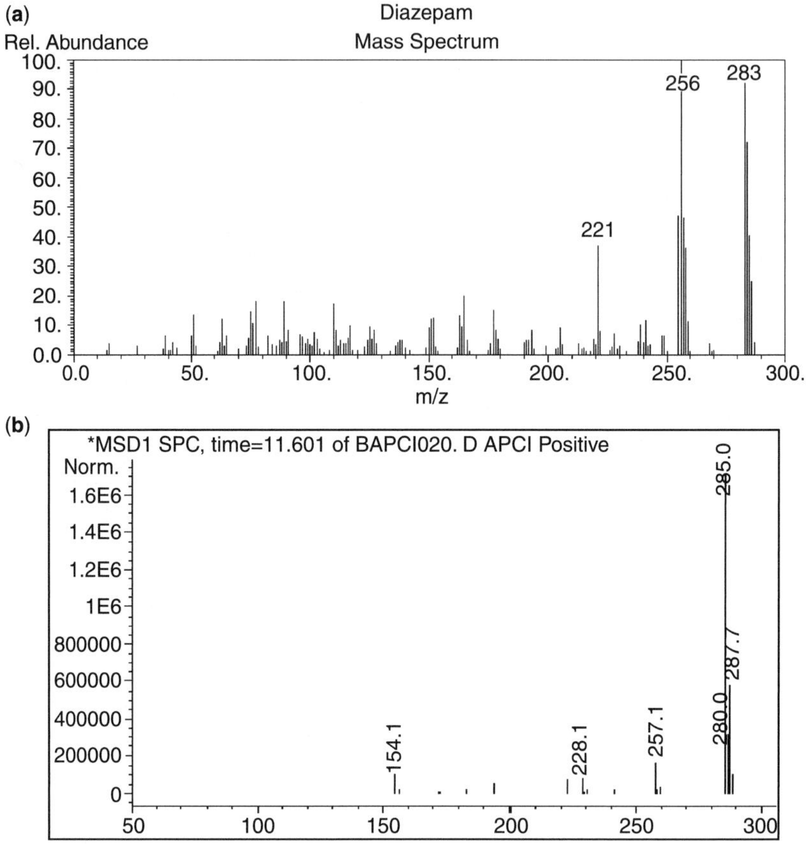

Figure 16.3. Comparison of TIM mass spectra of diazepam obtained by GC-MS/EI (a) and LC-MS/APCI (b). The differences are seen in number, type, and intensity of ions.

(the so-called CID in cell) or in the intermediate pressure part of the MS between the atmospheric pressure source and the high vacuum of the gas analyzer (in-source CID) (see Section 3.2.1).

One of the solutions useful for ion-trap (see Section 2.2.4) or tandem MS of any sort supposes that a limited number of parent ions are selected in the first MS stage and submitted to fragmentation in the collision cell and then the resulting fragments are analyzed in the second MS stage. It can be used easily, with or without chromatographic separation, to confirm the identity of suspected compounds as long as the fragmentation energy is standardized (in terms of nature and pressure of collision gas, and ion kinetic energy), and libraries of mass spectra of compounds of interest are built for each different type of MS.

To be able to use MS/MS spectra library searching for general unknown screening, it is necessary to use an automatic process, called data dependent acquisition or information-dependent acquisition, to select the parent ions of interest, totally unexpected by definition, and to dissociate them and monitor their fragments.

An alternative to the use of MS/MS spectra for general unknown screening is single MS spectra using in-source CID and ionspray interface. It involves an acceleration potential, the name of which varies depending on the manufacturer: orifice voltage for Sciex, cone voltage for Waters–Micromass, octapole offset voltage for Finnigan, and fragmentor voltage for Agilent Technologies. The fragments produced by in-source CID are generally the same as those by CID in the cell of an MS/MS instrument, but not necessarily with the same intensity. For the screening procedure ions formed in the source undergo alternatively a weak or a strong fragmentation (every other scan), induced by in-source CID. Positive and negative ions are then alternatively transmitted to the MS and analyzed in the full-scan mode. The analytical software, completed by macro-commands, allowed analysts to store in the libraries, then to search automatically for, reconstructed mass spectra (sum of spectra obtained in weak and strong conditions), in the positive and in the negative modes separately. Each compound is thus characterized at best by two mass spectra, corresponding to four different physical conditions. In-source CID fragmentation needs to be preceded by an efficient chromatographic separation procedure for good selectivity (no interference), good sensitivity (no ion suppression), and reproducible fragmentation (fragmentation efficiency being dependent on the ion density in the transition zone). This comprehensive screening LC-MS procedure, elaborated by French scientists, covers 1,200 compounds analyzed in plasma, with retention times between 1.5 and 47.4 min, giving about 1,000 positive mass spectra and 500 negative spectra. Recently three MS spectra libraries for LC-MS/MS and ESI-in-source-CID-MS are commercially available. They are for identification of drugs, pesticides and explosives (http://www.chemicalsoft.de/index-ms.htm).

16.5. QUANTITATIVE ANALYSIS

Using MS detectors, with or without GC or LC separation, it is possible to carry out precise and accurate quantification of analytes. MS quantification is more usually based on the peak area for specific ion fragments, called SIM. It is generally necessary to use an internal standard (IS). The ISs labeled with stable isotope (e.g., deuterated) are ideal, since they simulate the analyte very closely, but often a close structural analog (e.g., an alkyl analog) of the analyte is also good enough. The deuterated IS of the drug co-elutes with it from a chromatographic column (sometimes very slightly earlier than the unlabeled compounds) and should have an almost identical response factor. The appropriate IS helps also to minimize matrix effects and to correct for other variables such as slight differences in transfer volumes when using LLE or SPE. The second important matter is the choice of an appropriate calibration method, which is critical to obtaining reliable results. Single-point calibrations are generally unacceptable. Multipoint, five or more, calibration points are preferred. Quantitative results are only acceptable if the analyte concentration lies within the validated concentration range. An assay calibration should be linear and produce a good correlation coefficient (e.g., better than 0.98) (Fig. 16.4). Sometimes the calibration line is nonlinear and may require a quadratic fit. Multiple calibrators are prepared by adding known amounts of the analyte to the drug-free specimen, similar to that being analyzed, but containing

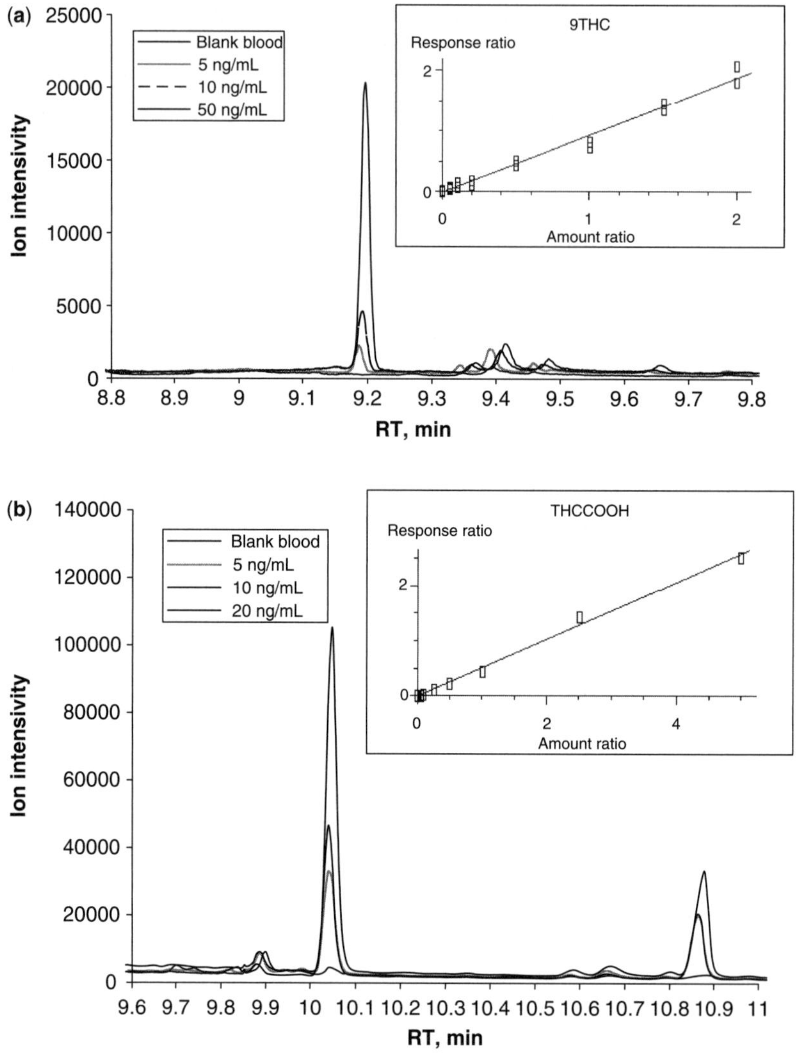

Figure 16.4. Quantification of Δ^9-tetrahydrocannabinol (9THC) after TFA and 11-nor-9-carboxy-Δ^9-tetrahydrocannabinol (THCCOOH) after PFP in blood using GC-MS/NCI. Drug-free blood was spiked with 9THC (a) and THCCOOH (b) to the concentrations of 0, 5, 10, and 50 or 20 ng/mL and with ISs–9THC-D_3 and THCCOOH-D_3 to 20 ng/mL. Monitoring ions were (m/z) 410.3 for 9THC and 572.3 for THCCOOH. The values of validation parameters, expressed in ng/mL, were LOD, 0.25; LOQ, 0.5; limit of linearity, 0.5 to 100 for both analytes [2].

IS. Peak-area or peak-height ratios are calculated for the analyte and IS and plotted against the ratios of known concentration of the analyte and IS.

A quantitative procedure should be validated for selectivity, calibration model, stability, accuracy (bias, precision), linearity, and limit of quantification (LOQ). Additional parameters that might be validated are LOD, recovery, and reproducibility. For accuracy control reference materials and matrix-matched controls are used. All these validation parameters are internationally recommended.

16.6. IDENTIFICATION OF ARSONS

Petroleum products, such as gasoline, kerosene, fuel oils, solvents, and diluters, are often the subjects of criminal investigations. Their wide availability, volatility, and inflammability are the main reasons of their use as accelerants to commit arsons by offenders. A multicomponent mixture of organic compounds isolated from the fire debris significantly differs from the initial one. The process of detection and identification of a petroleum product must consist of two essential stages. The first is separating of analytes from the matrix and concentrating them. The methods most often used are dynamic and static head space; absorbed compounds are then recovered either by solvent extraction (substances with high boiling point) or by heating (low boiling compounds). The second is the identification of substances from isolated mixture.

Chemical analysis of the isolated mixture is carried out by GC-MS equipped with an automated thermal desorber with a module enabling focusing of analytes on a "cold-trap" and a set of suitable adsorption tubes that are used for analyte adsorption and thermodesorption. The accelerant traces are identified by comparison of the chromatograms obtained with standards collected in the computer database and by mass spectra of particular peaks. The spectrum is computer matched against the NIST library spectrum. However, GC-MS can be difficult at high sensitivity when the fragments produced are less than m/z 40. In fire debris traces there is a "lack" of volatile compounds in comparison to the high boiling one, so library matching is very useful in such cases [1].

REFERENCES

1. M. J. Bogusz, (ed.) Forensic Sciences, 2nd ed. Amsterdam: Elsevier, 2008.
2. M. Kala, M. Kochanowski. The Determination of Δ^9-Tetrahydrocannabinol (9THC) and 11-nor-9-Carboxy-Δ^9-Tetrahydrocannabinol (THCCOOH) in Blood and Urine Using Gas Chromatography Negative Ion Chemical Ionisation Mass Spectrometry (GC-MS-NCI), *Chemical Analysis (Warsaw)*, **51**, 2006.
3. P. Marquet. Progress of Liquid Chromatography-Mass Spectrometry in Clinical and Forensic Toxicology, *Therapeutic Drug Monitoring*, **24**, no. 2, (2002).
4. A. C. Moffat, M. D. Osselton, and B. Widdop, (eds.) *Clarke's Analysis of Drugs and Poisons*, 3rd ed. London: Pharmaceutical Press, 2004.
5. K. Pfleger, H. H. Maurer, and A. Weber, *Mass Spectral and GC Data of Drugs, Poisons, Pesticides, Pollutants and their Metabolites*, 2nd ed. Weinheim: VCH, 1992, Part 4, 2000.
6. A. Skulska, M. Kala, A. Parczewski, Fentanyl and its Analogues in Clinical and Forensic Toxicology, *Przeglad Lekarski*, **62**, 2005.

17

NEW APPROACHES TO NEUROCHEMISTRY

Jonas Bergquist, Jerzy Silberring, and Rolf Ekman

17.1. INTRODUCTION

Why bother with all the neurobiology of the human brain, when trying to understand stress? Because the old dogma that our brain remains a stable, unchanging, hardwired black box has to be revised.

Most neurochemists trained in the 1980s and 1990s were comfortable with the value of studying isolated molecular events associated with the nervous system in health and disease. Today we have a growing stock of evidence indicating that the brain can be extensively remodeled throughout the course of life, for example, changes occur in response to the psychosocial environment, life style, and the mind. Events in life sculpture our personality and affect us all the way to the molecules that make us who we are, defining our unique and elusive emotions, memories, consciousness, and ways of coping with life. Thus, understanding the biology and chemistry of the brain is of utmost importance for health outcome from the life perspective of the individual.

The central nervous system, and the brain in particular, is the prime candidate for both being the sensor and normally the major controller of the response to physiological and psychological stimuli. However, under pathological circumstances, such as the influence of sustained stressful, unhealthy psychosocial environments and different damaging lifestyles such as those observed in "unhealthy societies," the brain is altered

Mass Spectrometry. Edited by Ekman, Silberring, Westman-Brinkmalm, and Kraj

resulting in a disturbance of nerve cell chemistry with influence on the immune system and neuroendocrine function.

Significantly, the World Health Organisation has proclaimed that mental disorders may represent a significant share (more than 25%) of the disabilities caused worldwide, with depressive disorders the leading ailment. Prolonged life stress, similar to an unending situation impossible to handle, is the main cause of various diseases, which transcends all borders and is an ever increasing challenge in our daily life, leading to disorders of our modern civilization.

Maybe our ambitions to gain health in general and mental health in particular have failed because we have not realized the importance of the problems that result as the psychosocial stress and our escalating feelings of discontent and annoyance carry an increasing risk to the health of our modern societies. To maintain our goals of physical and mental health throughout life, stress at these levels must be understood and treated.

It has been estimated that in 2020 the number of people suffering from depression and anxiety disorders will be second only to those with ischemic heart disease. With this perspective one obvious question to ask is why is so little research done concerning the relationship between the function of the brain and what makes us content with our life?

17.2. WHY IS THERE SO LITTLE RESEARCH IN THIS AREA?

The simple and straightforward answer to these reflections and questions is that we lack the basic knowledge and hesitate to challenge apparently unsolvable problems. It might also, in part, be due to difficulties of interpreting the complex molecular fingerprint of neuropsychiatric diseases into information of value for diverse treatment strategies. However, it seems likely that today's neurochemists through robust progress with mass spectrometry combined with other methods can now characterize multisite molecular interactions involved in global structural and functional details in the dynamic structures such as synapses (synaptosomes, synaptosomal membranes, presynaptic vesicles, postsynaptic densities), gap junction, receptor complexes, and other neuronal and glial structures of vital importance for life-long mental health.

The communication between neurons occurs at either gap junctions (electrical synapses) or chemical synapses with release of neurotransmitters from a presynaptic neuron and their detection by a postsynaptic nerve cell (Fig. 17.1). Neurotransmitters not used in the synaptic cleft are removed promptly by either uptake into adjacent cells, reuptake in the presynaptic neuron, or are degraded by enzymatic systems.

By extracting new structural as well as functional molecular data on the pathophysiological mechanisms behind major psychiatric diseases it is possible to understand and explain the effects of different treatment strategies, such as those used in traditional medical therapy in combination with physical activities, cognitive therapy, yoga, meditation, and music therapy. It is our hypothesis that basic knowledge in this area, at the molecular level, will be essential to efforts to improve global health and in particular mental health worldwide—a real challenge for our future. Other intriguing points to address are the effects that result from improved social interactions, enthusiastic

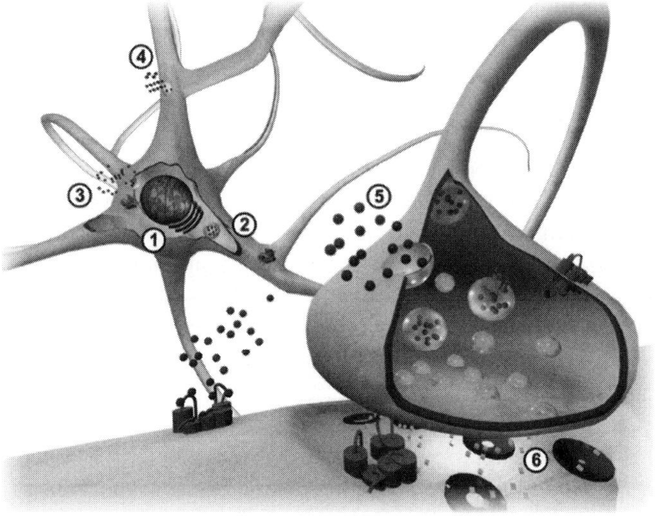

Figure 17.1. Neurotransmission (specific case of peptidergic cells). Production of the peptides in the cell body **(1)**. Packing of the peptides into large dense core vesicles for further transport to the axons **(2)**. Release of neuropeptides from the cell soma **(3)** dendrites **(4)** and outside of the synapse **(5)**. Release of classic neurotransmitters in the synaptic cleft **(6)**. G-protein-coupled type receptors, which act as peptide receptors. (See color insert.)

management and expectations from medical treatment, in other words the placebo effect, still an unsolved problem.

Finally, perhaps the most complex challenge is to understand the different molecular mechanisms and their complex interactions converting psychosocial stress and lifestyle into cellular dysfunction in the brain, endocrine, and immune systems, that is, in the whole human being. Present and future goals are to obtain a thorough understanding of molecular mechanisms underlying how ordinary and persistent psychosocial stressors such as poverty, unstable child–adult/parent relations, and lack of economic, emotional, or social support influence the brain chemistry. How does insulting and prolonged extreme psychosocial stress influence the individual health profile, situations where our brain is both the conductor and the performer?

17.3. PROTEOMICS AND NEUROCHEMISTRY

The rapid progress in proteomics and peptidomics during the last decade offers us new possibilities to study clinical aspects of disorders and diseases related to the brain [1]. These strategies also offer new tools to follow chemical modifications and altered metabolic disturbances that may be indicative of pathophysiological adaptations related to environmental and psychosocial prolonged stress. These techniques can contribute to developments in the diagnostic and therapeutic fields of psychiatric diseases, such as chronic fatigue syndrome, major depression, post-traumatic stress disorders, and neurodegenerative diseases, such as Alzheimer's disease [2-4].

17.3.1. The Synapse

What is a synapse? In the brain, the nerve cells or neurons are connected at special functional junctions called synapses, which depend on many proteins, including large complexes. They participate in basic functions with important roles in coordinating every characteristic of the nervous system, including physiology, emotions, learning, sleep, memory, and pain signal transmission.

Although synapses are composed of almost the same structural components, they change their constituents over time and function. Training determines which synapses will be strengthened. Research has revealed that synapses do not stop functioning without disappearing. If the activity of the synapse goes down it will fade away and, in other words, it is quite appropriate to use the expression "use it or loose it."

It is now possible to describe the protein components of the synapse with proteomics [5]. When individuals, children as well as adults, today in our modern civilization have difficulty in concentrating, are irritable and demonstrate mood changes, it is likely that there are early signs of impaired synaptic homeostasis related to the psychosocial environment and/or their lifestyle. The molecular mechanisms that lead to decreased alertness and poor cognitive performance involve changes in gene expression and proteins. There is a growing need for knowledge about the molecular mechanisms underlying these signs of cognitive disturbances, sleep disturbances, pain, and psychiatric diseases in order to prevent an acceleration of the global burden of health costs from unhealthy societies.

There is now a growing interest in proteomic studies of brain synapses. Recent studies have revealed a high molecular complexity in the pre- and postsynaptic areas, with thousands of proteins [6]. An important investigation for the future is to identify posttranslational modifications, miscoded as well as misfolded proteins, likely to have an impact on different aspects of synaptic function as a response to the environment as well as to the lifestyle. The first challenge is to identify and quantify the presence and variation of different proteins in key structures of the pre- and postsynaptic areas in order to relate protein structures to synaptic function. Recently, a new model has been presented describing the molecular complexity of the synapse with important aspects in emotions, thinking, memory, and consciousness [7] (Fig. 17.2).

17.3.2. Learning and Memory

The developing brain is an exceptionally dynamic organ from the synaptic point of view. It is structured by tradition; training, such as mental challenges for motivation, concentration, and endurance; the psychosocial environment; emotions; and imagination. One basic theory in neuroscience is that changes in synaptic strength and plasticity underlie learning and memory [8]. There seem to be at least two types of changes that take place in the brain during learning, changes in synaptic structure and the formation of new synapses. Initially, newly learned tasks are deposited temporarily in the short-term

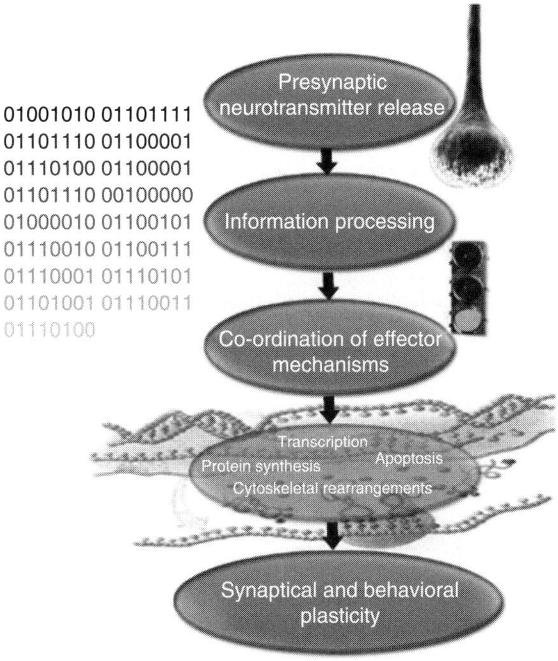

Figure 17.2. A schematic model of clusters of proteins at different cellular levels involved in learning and memory. (See color insert.)

memory. Later, memory consolidates to form more permanent structures residing in long-term memory. It is increasingly clear that the stimulation of synaptic plasticity and the resulting response stimulation vary markedly depending on factors such as brain region, age, and gender, as well as kind of cell and synapse type considered. A comprehensive understanding of learning and memory in the human brain will require a systematic understanding of how synaptic proteins, the interacting lipids, and sugars are involved and how they are related to information processing and storage. Given the vast number of different proteins in neuronal synapses, and the complexity of their interactions and dynamic modifications over time, the plasticity of synapses linked to learning and memory promises to be a new and very exciting research field with pivotal consequences [9].

17.3.3. The Brain and the Immune System

Clearly an important part of understanding stress-related disorders is to better define the chemistry of psycho-neuro-immuno-endocrine response patterns over time, from the healthy and time-limited to unhealthy and sustained individual. There is growing evidence that a physiological communication exists between the brain and the immune system. Several studies have shown that the white blood cells of the immune system, the lymphocytes, may serve as carriers that reflect dynamic changes of modified molecules in the brain as potential markers for ongoing metabolic disturbances.

Sustained stressor exposures, both physical and psychological, have been shown to shift the cytokine profiles of the immune cells toward the promotion of inflammation and allergic response. Neural circuitry underlying stress and emotion can also regulate inflammation. Peripheral inflammatory mediators, in turn, influence mood and cognitive functions. Depressive symptomatology has been associated with the same proinflammatory cytokines that are released during an asthmatic attack or in other forms of severe stressor exposure.

A wide array of growth factors, proinflammatory molecules, including cytokines, prostanoids, and neuropeptides, contributes to the manifestation of inflammatory, neurodegenerative, and metabolic consequences, including increased risk for triggering cell death pathways (Fig. 17.3).

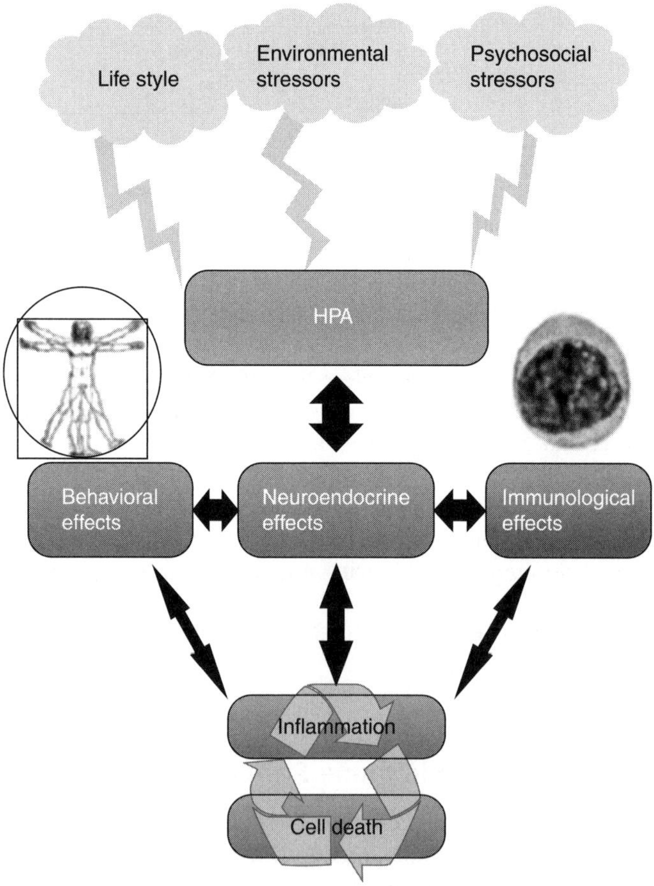

Figure 17.3. The brain in stress (HPA, hypothalamus-pituitary-adrenal axes). Schematic representation of some of the inflammatory response reactions that may even result in cell death as a response to prolonged inflammatory reactions. (See color insert.)

Prolonged or chronic stress in combination with unhealthy lifestyle, a common phenomenon in today's societies, is potentially involved in the development of several different diseases in the following interrelated systems of our organism:

- *The Nervous System*
  - Mental disorders, for example, multiple subtypes of anxiety, chronic fatigue syndrome, depression, sometimes together with chronic pain, posttraumatic stress disorders (PTSD), and schizophrenia
  - Neurodegenerative diseases, for example, early ageing, dementia of Alzheimer type, Multiple Sclerosis, Amyotrophic Lateral Sclerosis, and Parkinson Disease.
- *The Endocrine System*: The metabolic syndromes, including Type 2 diabetes, obesity, and high blood pressure.
- *The Immune System*: Different forms of lymphoma and autoimmune disorders.

Little is known about the molecular mechanisms and complexity converting psychosocial stress into cellular dysfunction in the brain, endocrine, and immune systems. How ordinary and sustained maladapted psychosocial stressors, chronic stress, and an unhealthy lifestyle activate and exert an influence on the biochemistry of the neuro–endocrine–immune axes with implications for future health or disease, is an upcoming innovative research field due to the new and emerging fields of proteomics, metabonomics, and biochip technologies.

Available evidence suggests that prolonged stressor exposure or chronic stress acts on a number of sensitive biological systems, primarily as a consequence of long-term maladaptive changes of the normal function in the brain–endocrine–immune axes resulting from chemical modifications of signaling molecules. It certainly appears likely that these effects will in turn produce molecular metabolic modifications with pathophysiological consequences leading in the long term to the progression of cardiovascular diseases, diabetes, obesity, cancer, dementia, depression, and autoimmunity.

Recent findings point to basic biological mechanisms for molecular sensing of psychosocial, environmental, and lifestyle stressors that induce conditions generating the metabolic perturbations and accumulation of proteins associated with the activation of inflammatory pathways and programmed cell death, apoptosis. It is now more and more evident that oxidative stress generates abnormal, chemically modified proteins as well as protein aggregates that trigger pro-inflammatory processes. In fact, stressors and cytokines are involved in the initiation of depressive disorders through processes involving apoptosis and oxidative stress [10].

17.3.4. Stress and Anxiety

The response of the brain to both acute and chronic stress can be discussed in terms of its capacity to demonstrate its dynamic plasticity. The term plasticity describes almost any change in the brain, from the chemical level to the formation of new neurons and synapses. Prolonged or chronic stress has specific effects on the structure and function of the synapses in different brain regions. The neurons in different regions may show signs of atrophy, cell death, as a result of chronic psychosocial stress, as well as after

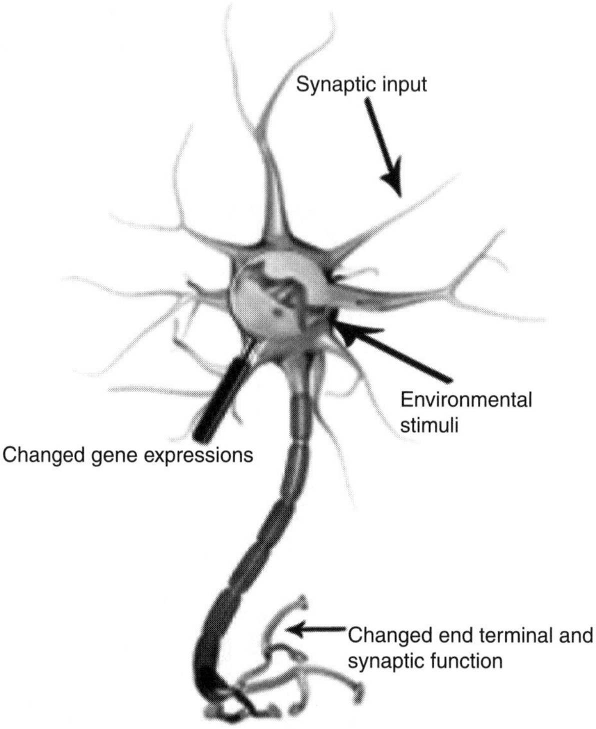

Figure 17.4. Epigenetics in the nervous system. Regulation occurs in response to synaptic inputs and/or other psychosocial-environmental stimuli. The external stimuli result in changes in the transcriptional profile of the neuron and eventually affects neural function(s). Many disorders of human cognition might involve dysfunction of epigenetic tagging. (See color insert.)

severe, traumatic stress. What makes the individual variation a real challenge is that we are ignorant of the effects of the early stress responses (strength and duration), and the fine-tuning of neuronal connections. Some alterations occur through changes at the DNA–RNA level, the epigenetic programming, which will appear later in adult life as some kind of psychiatric vulnerability. There is increasing evidence that complex diseases, such as major psychiatric diseases, may be mediated through epigenetic effects such as DNA acetylation, methylation, and phosphorylation (Fig. 17.4).

Another interesting question is how traumatic accidents and circumstances may underlie a progression of abnormal molecular and structural adjustment, which will reflect the different pathogenesis in several stress-related diseases. What is of particular importance is that the brain cannot be "turned off"; thus dysfunctional circuits are created within regions of the brain, such as the frontal cortex, hippocampus, and amygdale, of relevance for various forms of PTSD. It should be noted that in the daily life of modern urban environments PTSD is now on the rise. In fact, according to recent research, psychosocial stress brings an increasing threat to our modern societies and various physiological response patterns, such as the classic flight and fight reaction, that were vital for survival in evolutionary history now have the potential to create serious diseases [11].

17.3.5. Psychiatric Diseases and Disorders

For several decades neuroresearchers have tried to find biological markers linked to the individual, phenotypic molecular mechanisms related to diseases, such as depression, PTSD, schizophrenia, and dementia. At present, one of the prevailing strategies is proteomic analysis of different brain regions to demonstrate alterations in protein isoforms and other structural modifications that will help in understanding of these serious psychiatric disorders in clinically relevant populations [12]. Furthermore, neuroimaging studies along with postmortem analysis have provided some indications about which brain regions may be involved in these disorders.

However, these strategies do not lead to easy answers. Current approaches that study the material from deceased patients with slow progressive neurodegenerative diseases, by analogy, might be compared to trying to explain what started the war by examination of a battlefield. The use of brain material from deceased patients, an endpoint material, does not reflect different phases of a progressive disease that might occur during a period of 5 to 20 years.

With this background there is an obvious call for novel strategies to follow changes of complex molecular patterns of different stress-related diseases over days, weeks, months, and years as an effect of lifestyle and the psychosocial environment to reflect the effects of unhealthy environments. The molecular interactions between the brain and the immune system in health and disease are reflected in the circulatory system as the white blood cells, the lymphocytes, mimic ongoing activities in the brain. By using lymphocytes from patients with psychosomatic-psychiatric diseases we can find detailed information about protein–peptide translational modifications and transformation essential for the development of new approaches that can prevent and treat major psychiatric diseases.

There is evidence that supports the hypothesis that cytoskeleton and mitochondrial dysfunctions are involved in the neuropathology of the major psychiatric disorders. Furthermore, there is increasing evidence coming from studies of multiprotein complexes, which may provide an insight into molecular interactions influencing the major neuropsychiatric illnesses. Recent advances describe the use of a tandem affinity approach of interest for studies of protein–protein interactions. However, differences in age, gender, and genetic background, as well as the psychosocial environment can reduce the specificity of the biomarker profile information [13].

17.3.6. Chronic Fatigue Syndrome

Fatigue is a common characteristic of a number of somatic and psychiatric diseases and disorders. Patients suffering from syndromes where fatigue is the main constituent of the condition often report that the fatigue has a major impact on their quality of life. In what way does fatigue differ from general tiredness and sleepiness? Fatigue is characterized as extreme and persistent tiredness, weakness, and/or exhaustion. Symptoms are mental, physical, or both.

Fatigue syndromes, such as chronic fatigue syndrome, most probably represent a heterogeneous group of disorders with a multifactorial pathophysiology, including various degrees of disturbances to the neuro-endocrine-immune systems. They are a serious concern affecting millions of patients of all ages, races, and socioeconomic groups worldwide. It is quite clear from the literature that a successful molecular profiling of fatigue syndromes will require integration of genomic and proteomic data with environmental and behavioral data, with fresh approaches in computational biology.

In a recently published article, cerebrospinal fluid was analyzed from patients with the symptoms from the following diseases; chronic fatigue syndrome, Persian Gulf War illness, and fibromyalgia. Identical protein profiles were found from the three groups studied and the authors propose that fatigue-like syndromes may be a syndrome caused by protein misfolding, cerebrovascular amyloidosis. The data from this study points at the importance of future conformation dependent studies in the understanding of molecular structural changes and their correlation to conformational symptoms and diseases [14].

17.3.7. Addiction

Drug dependence remains a serious health problem in developed countries and its etiology is still unknown. However, there appears to be a positive correlation between addiction and the degree of PTSD. Recent research indicates that at least some forms of drug dependence may be inherited. Addiction concerns drug dependence but also, for example, gambling, computer games, uncontrolled shopping, etc. Certainly, many molecules and pathways are involved in addiction processes as these mechanisms vary, depending on the origin of the disease or drug taken. For instance, morphine binds to μ-opioid receptors and cocaine is a dopamine transporter blocker. In contrast, Ecstasy (MDMA) causes massive serotonin release from neurons. Dependence is linked to the reward system, controlled by dopamine, and a limbic system, consisting of several brain structures.

Changes in protein content in these structures are under investigation by several research groups [15]. It is still not recognized how our memory influences the return to the addictive behavior, even after 5 to 10 years of abstinence. Therefore, one approach is to use mass spectrometry for proteome research, to reveal the molecules involved in the general pathways involved mostly in drug addiction, such as morphine dependence (morphinome) or after acute doses of amphetamine. Another field where mass spectrometry is commonly applied in medical practice is control of the methadone program. Methadone is an addictive substance produced synthetically, and is used as a replacement therapy for treatment of heroin addicts. As it is synthetically produced, it is "clean" and does not cause any severe infectious diseases, such as AIDS or hepatitis, as a result of syringe exchanges. But its delivery requires strict control of patients and individual adjustment of the dose injected. Mass spectrometry, GC/MS in particular, helps to maintain control of patients and provides a direct screening process to identify misuse of illegal drugs by patients undergoing methadone-mediated withdrawal from heroin. Another application in this field is the profiling of illegal drugs to identify impurities that are characteristic of a particular clandestine laboratory. This might be helpful in tracing illegal laboratories and routes of drugs transportation even across the continents.

17.3.8. Pain

During the last several decades, pain research has undergone many changes. All pain sensation is subjective and to some degree affected by emotional status, and environmental as well as cultural factors [16]. It is the stress of living with chronic pain that affects the pathophysiological mechanisms. Pharmacological models are still in frequent use to understand pain, but sophisticated methodologies are now being used in search of the molecule(s) of pain. Following such trends, a rapidly growing area of research has been focused mainly on the molecular basis of pain, including proteomics, genomics, molecular biology, etc., to study so-called "molecular pain." The term "pain" is rather diffuse and strongly depends on the actual psychosocial status of the patient and his or her emotions and senses at a given moment. A good definition of pain has recently been given as "the pain is . . . what the patient says it is." There is, however, no absolute measure of pain, except individual feeling. Pain types can be roughly divided into: (1) acute or nociceptive pain, (2) chronic or neuropathic pain, and (3) cancer pain. Each type of pain is controlled by a distinct mechanism(s) and, thus, a thorough knowledge of these mechanisms is necessary to treat various symptoms in a proper way. Current pain management is limited to opioids, such as morphine or oxycodone, and nonsteroidal anti-inflammatory drugs (NSAIDS). The pain treatment procedures follow the scheme called the analgesic ladder (a commonly accepted World Health Organization routine), where a combination of the aforementioned drugs is gradually applied at various "stages" of pain. From this point of view, there is an urgent need for novel strategies to improve pain treatment by individual diagnosis and a more personalized therapy.

Mass spectrometry has been applied mainly in proteome research, but also in discovery and quantitation of neuropeptides that are involved in pain mechanisms, such as nocistatin, substance P, or verification of, for example, the structure of endogenous morphine in the central nervous system. Some proteomics studies of pain are aimed at the search for pain markers in cerebrospinal fluid, as it may reflect changes in brain and spinal cord functioning. Another research area concerns proteome analysis in cancer pain using spinal cord tissue and animal models.

17.3.9. Neurodegenerative Diseases

Alzheimer's disease, Parkinson disease, prion diseases (Creutzfeld-Jacob in humans, scrapie in sheep), Huntington disease, dementia with Levy's bodies, sclerosis multiplex and amyotrophic lateral sclerosis, frontotemporal lobar degeneration, and vascular dementia are the most commonly occurring neurodegenerative diseases, with different (and often unknown) pathophysiology, creating serious health care problems and involving not only medical personnel but also families. The most striking feature of these diseases is that they all develop over many years, in contrast to, for example, juvenile insulin-dependent diabetes, and are most often difficult or impossible to recognize at early stages. Therefore, they are sometimes called a "silent death." Despite extensive research in this field, still little is known about how these diseases develop along the life span, and it is not clear what triggers their onset. Current reports suggest that at least Alzheimer's disease develops constantly during our life and this is, to a certain extent, a natural process. Even the well-recognized markers, such as lack or decreased level of dopamine in Parkinson disease or fibrillary tangles and plaques in Alzheimer's disease, are still not defined as a cause or an effect of the pathophysiology. It is, however, clear that these diseases, as well as many others (if not all) have a much more complicated etiology, thus involving many pathways, perhaps all leading to neurodegeneration. The present clinical treatment for these

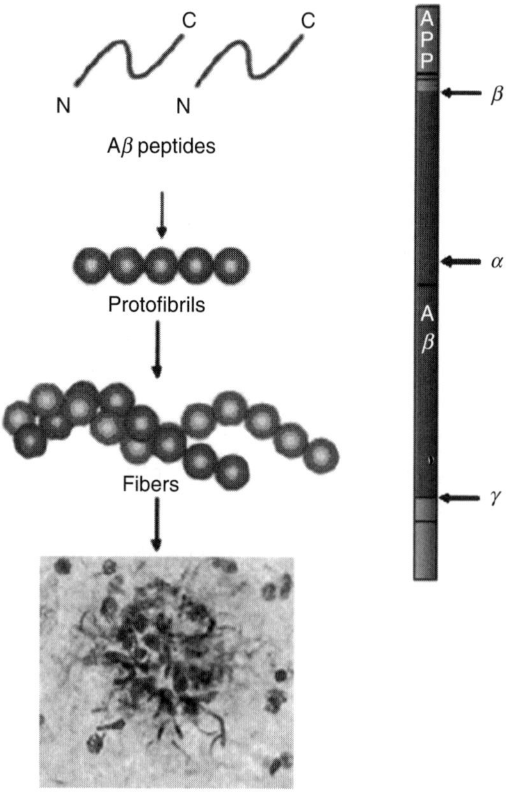

Figure 17.5. The precursor molecule APP and the three different proteases α, β, γ secretase that are involved in the processing of APP to β-amyloid peptide. The aberrant processing of the amyloid precursor protein (APP) leads to accumulation of beta-amyloid fragments, first as protofibrils and then as fibers that aggregate in the senile plaque structures. (See color insert.)

ailments involves supplementary, pharmacological therapy, based on L-Dopa administration for Parkinson patients or therapy with acetylcholinesterase inhibitors for Alzheimer's disease. These compounds, however, do not reverse the diseases to the healthy state, nor do they appear to slow disease progression. The molecular basis, at least at the level of the proteins involved, of a majority of these diseases is still not recognized [17]. Relatively little research in this field has been done with the aid of proteomics approaches, though this strategy seems to be ideal to screen for the entire pathways that might be involved in particular diseases, but a substantial part of the investigations have been performed on specific mechanisms, such as aberrant processing of the amyloid precursor, leading to accumulation of beta-amyloid fragments, as shown in Figure 17.5.

Involvement of several proteolytic enzymes, secretases, is probably crucial for this process but other hypotheses, including, for example, cholinergic transmission or accumulation of metal ions, have also been considered. Future perspectives in this area concern the search for novel pharmaceuticals that cross the blood-brain barrier, without side effects (e.g., the dyskinesias of L-Dopa), or potent and selective inhibitors of improper cleavage of amyloid protein, or even stem cell therapy to restore neuronal cells.

17.4. CONCLUSIONS

In order to understand the mechanisms behind stress, our body's response to stress, its relationship to health and disease, and, ultimately, the treatment and prevention of stress, we need to cross scientific barriers and develop new strategies and new paradigms. We need to think in a transdisciplinary manner. We need to consider how we can move knowledge, not only from the lab bench to the clinic to the society at large, but also how to reverse such information flow. Improving and answering the different issues behind stress-related disorders and diseases (SRDs) will require new diagnostic strategies, such as biopattern technologies and proteomics, that are sensitive to different ways in which SRDs can manifest themselves.

Actually, in a twisted way of thinking, stress might be one of the most fruitful areas in which to apply innovative new thinking and paradigms in order not only to improve mechanistic understandings of disease but also to enhance the ability to implement new knowledge in society to improve overall public health.

ACKNOWLEDGMENTS

We are particular grateful to Jonatan Bergquist and Andreas Dahlin for providing us with the illustrations. Many colleagues have read the manuscript; Andrew G. Ewing did a thorough reading and deserves special mention.

REFERENCES

Mass Spectrometry: Applications in Neurochemistry

1. L. Paulson, R. Persson, G. Karlsson, J. Silberring, A. Bierczynska-Krzysik, R. Ekman, and A. Westman-Brinkmalm. Proteomics and Peptidomics in Neuroscience. Experience of Capabilities and Limitations in a Neurochemical Laboratory. *J. Mass Spectrometry*, **40**(2005): 202–213.

Proteomics in Neuroscience

2. M. Ramström, I. Ivonin, A. Johansson, H. Askmark, K. E. Markides, R. Zubarev, P. Håkansson, S.-M. Aquilonius, and J. Bergquist. Cerebrospinal Fluid Protein Patterns in Neurodegenerative Disease Revealed by Liquid Chromatography Fourier Transform Ion Cyclotron Resonance Mass Spectrometry. *Proteomics*, **4**(2004): 4010–4018.
3. O. N. Jensen. Interpreting the Protein Language Using Proteomics. *Nature Rev. Mol. Cell Biol.*, **7**(2006): 391–403.
4. T. Ekegren, J. Hanrieder, S.-M. Aquilonius, and J. Bergquist. Focused Proteomics in Post-Mortem Human Spinal Cord. *J. Proteome Res.*, **5**(9)(2006): 2364–2371.

Synapse

5. S. G. N. Grant, M. C. Marshall, K.-L. Page, M. A. Cumiskey, and J. D. Armstrong. Synapse Proteomics of Multiprotein Complexes: En Route from Genes to Nervous System Diseases. *Hum. Mol. Gen.*, **14** (Review Issue 2)(2005): R225–R234.
6. K. W. Li, Proteomics of the Synapse. *Anal. Bioanal. Chem.*, **387**(2007): 25–28.
7. A. J. Pocklington, M. Cumiskey, J. D. Armstrong, and S. G. N. Grant. 2006, The Proteomes of Neurotransmitter Receptor Complexes from Modular Networks with Distributed Functionality Underlying Plasticity and Behaviour. www.molecularsystemsbiology.com—Article nr 2006.0023.

Learning and Memory

8. Q. Ding, S. Vaynman, P. Souda, J. P. Whitelegge, and F. Gomez-Pinilla. Exercise Affects Energy Metabolism and Neural Plasticity-Related Proteins in the Hippocampus as Revealed by Proteomics Analysis. *Eur. J. Neurosci.*, **24**(2006): 1265–1276.
9. J. M. Levenson and J. D. Sweatt. Epigenetic Mechanisms in Memory Formation. *Nature Rev. Neurol.*, **6**(2005): 108–118.

The Brain and Immune System

10. R. Ekman and B. Arnetz 2006. The Brain in Stress: Influence of Environment and Lifestyle on Stress-Related Disorders. In B. Arnetz and R. Ekman (eds.) *Stress in Health and Disease*. Wiley-VCH Verlag, 196–213.

Stress and Anxiety

11. B. Arnetz and R. Ekman (eds.) 2006. *Stress in Health and Disease*. Wiley-VCH Verlag.

Psychiatric Diseases and Disorders

12. C. L. Beasley, K. Pennington, A. Behan, R. Wait, M. J. Dunn, and D. Cotter. Proteomic Analysis of the Anterior Cingulated Cortex in the Major Psychiatric Disorders: Evidence for Disease-Associated Changes. *Proteomics*, **6**(2006): 344–3425.
13. C. Rohlff and K. Hollis. Modern Proteomic Strategies in the Study of Complex Neuropsychiatric Disorders. *Biol. Psychiatr.*, **53**(2003): 847–853.

Chronic Fatigue Syndrome

14. J. N. Baraniuk, B. Casado, H. Maibach, D. J. Clauw, L. K. Pannell, and S. S. Hess. A Chronic Fatigue Syndrome-Related Proteome in Human Cerebrospinal Fluid. *BMC Neurology*, **5**(2005): 22.

Addiction

15. N. S. Abul-Husn and L. A. Devi. Neuroproteomics of the Synapse and Drug Addiction. *J. Pharm. Exper. Therapeutics*, **318**(2006): 461–468.

Pain

16. J. Giordano. The Neurobiology of Nociceptive and Anti-Nociceptive Systems. *Pain Physician*, **8**(2005): 277–290.

Neurodegenerative Diseases

17. D. A. Forero, G. Casadesus, G. Perry, and H. Arboleda. Synaptic Dysfunction and Oxidative Stress in Alzheimer's Disease: Emerging Mechanisms. *J. Cell. Mol. Med.*, **10**(2006): 796–805.

PART IV

APPENDIX

Justyna Jarzębińska, Filip Sucharski, and Hana Raoof

TUTORIALS

http://www.spectroscopynow.com

The free online resource for the spectroscopy community, contains a section dedicated to mass spectrometry (Base Peak). Much elementary information, many articles on MS subject matter and related can be found here.

http://www.rzuser.uni-heidelberg.de/~bl5/encyclopedia.html

Little Encyclopedia of Mass Spectrometry includes the main terms from the field of MS, tutorial-like explanations on important topics, illustrated by many photographs, pictures, and spectra. Supplies article links for further reading in the field of interest.

http://www.mc.vanderbilt.edu/msrc/tutorials/

Short monograph concerning mass spectrometry. Full description of technique shown clearly, supplemented by numerous figures. Moreover, a MALDI tutorial in PDF file can be found on this site.

http://www.ionsource.com

Site containing many tutorials, pieces of advice, useful tables, and links connected to MS.

Mass Spectrometry. Edited by Ekman, Silberring, Westman-Brinkmalm, and Kraj

http://www.separationsnow.com

Large portal for separation techniques, including information about the analytical process from sample preparation through separation (HPLC, GC, electrophoresis) up to detection.

http://www.asms.org/whatisms/page_index.html

The web page of the American Society for Mass Spectrometry in the form of FAQ (*frequently asked questions*) contains answers to questions asked especially by beginning users of mass spectrometers.

http://www.vias.org/simulations/simusoft_msscope.html

Simple program simulating a console of a mass spectrometer. One can observe the spectrum directly and modify it by changing resolution or mass range. It is possible to obtain spectra of three compounds.

http://www.eaglabs.com/en-US/research/research.html

Interesting and extensive presentation of SIMS and TOF-SIMS, with a large number of free articles.

http://www.ivv.fhg.de/ms/ms-analyzers.html

Web page that gives attention to different types of analyzers.

http://www.rmjordan.com/tt1.html

TOF analyzer tutorial which contains principles, rules of ion generation, separation steps, and basics of reflectron mode.

http://www.forumsci.co.il/HPLC/lcms_page.html

Description of LC-MS, including free articles. This site has subdivisions such as proteomics, metabonomics, 2D-electrophoresis, and sample preparation.

http://ull.chemistry.uakron.edu/gcms/

A short course on MS in applications for GC.

http://www.gcms.de/index.html

Description of several types of mass spectrometers used as detectors for GC.

http://cot.marine.usf.edu/hems/underwater/

Application of MS in underwater research.

http://www.sift-ms.com/sift.htm

Description of the SIFT (selected ion flow tube) technique, which allows detection of gas traces in real time.

http://www.msterms.com

Mass Spectrometry Wiki, based on Wikipedia Mass Spectrometry Category. It is the IUPAC-sponsored project to update the standard terms and definitions for mass spectrometry. Anyone can edit entries after logging in.

SOFTWARE

http://www.maldi-msi.org/index.php

Site has good software subdivision devoted to matrix-assisted laser desorption/ionization mass spectrometric imaging (MALDI-MSI, also termed imaging MS or MS imaging).

http://sourceforge.net/project/showfiles.php?group_id=90558

Open source framework for LC-MS based proteomics.

http://cabig.nci.nih.gov/

The site of the National Cancer Institute whose aim is to create a network that will connect the entire cancer community. It has a big division that contains many tools and datasets.

http://www.scripps.edu/~cdputnam/protcalc.html

Protein calculator for calculation of, for example, charge, molecular weight, ultraviolet absorption, and some other parameters.

http://idelnx81.hh.se/bioinf/mass_spectro.html

Site contains applications for peak extraction and peak post-processing for both MALDI and ESI.

http://www.bioinfo.no/tools/

Web page offers a large number of tools dedicated mainly to genomic and proteomic research.

http://mmass.biographics.cz/

Site offers mMass program that is an open source package of simple tools, written mainly in Phyton and dedicated to mass spectrometric data analysis. It can also be used for protein sequence handling and other proteomic tasks.

http://www.colby.edu/chemistry/NMR/scripts/peptide.html

Peptide mass finder that can perform calculations, taking into account such factors as ion type, substitutions, and modifications.

DATABASES

Literature

http://www.ncbi.nlm.nih.gov/literature

Search engine of the National Center for Biotechnology Information, including a few databases: PubMed (MEDLINE biomedical literature), PubMed Central (free digital archive of life sciences journal literature), Books, OMIM (Online Mendelian Inheritance in Man, a catalog of genetically linked diseases),

OMIA (Online Mendelian Inheritance in Animals, a catalog of genetic diseases in animals), and others.

http://books.google.com/

Beta version of search engine which allows browsing of full text books to find out where to buy or borrow those of interest.

http://scholar.google.com

Enables easy access to scientific literature, journals, articles, abstracts, citations, and the like.

http://www.scirus.com/

Science-specific search engine. Searches for different types of information (abstracts, articles, books, conferences, theses, and disssertations) in various sources (literature databases, web resources).

http://info.scopus.com/

Abstract and citation database of research literature and quality web resources. Covers titles from Nature, Biomed Central, Springer Verlag, and many others. Updated daily.

Bioinformatics

1. Nucleotide sequence databases

   GeneBank (http://www.ncbi.nlm.nih.gov/Genbank/) database directed by the National Center of Biotechnological Information.

   EMBL (http://www.ebi.ac.uk/embl), database of the European Bioinformatics Institute, a part of the European Molecular Biology Laboratory.

   DDJB (http://www.ddbj.nig.ac.jp/), DNA Data Bank of Japan.

2. Protein databases

   SwissProt (http://www.expasy.org/sprot), database established by the Swiss Institute of Bioinformatics, provides protein sequences with high level of annotation. Highly integrated with other protein databases. All data are verified manually.

   TrEMBL (http://www.expasy.org/sprot), database of the European Bioinformatics Institute, translated EMBL. Generated by computer translation of genetic information from the EMBL database. Automatically annotated.

   PIR (http://pir.georgetown.edu/), Protein Information Resource, located at Georgetown University Medical Center, which has provided the first international Protein Sequence Database.

   UniProt (http://www.expasy.uniprot.org), Universal Protein Resource. Created by joining the Swiss-Prot information, TrEMBL and PIR; contains protein sequence and function.

MSDB (ftp://ftp.ncbi.nih.gov/repository/MSDB), database created especially for MS applications. Contains nonidentical protein sequences obtained from other databases (PIR, TrEMBL, SwissProt). At http://www.matrixscience.com/help/seq_db_setup_msdb.html, a guidebook for MSDB users can be found.

Protein Data Bank (http://www.rcsb.org/pdb), international repository of experimentally resolved structures of biological macromolecules (proteins, nucleic acids, viruses), including anotations.

Prosite (http://www.expasy.org/prosite), database of protein domains, families, and functional sites.

InterPro (http://www.ebi.ac.uk/interpro), database of protein families, domains, and functional sites; allows prediction of the function or structure of a new protein on the basis of its sequence homology to sequences of known proteins.

Merops (http://merops.sanger.ac.uk), database of peptidases and their proteinaceous inhibitors. Includes enzyme classification and nomenclature, external links to literature, and the structure of proteins of interest (if known). Enables one to find the gene coding for a given peptidase or to find the best enzyme to digest a chosen substrate.

3. Others

Entrez (http://www.ncbi.nih.gov/entrez), information retrieval system, integrating NCBI databases of protein and nucleotide sequences, genomes, macromolecular structures, and MEDLINE literature database.

Mascot (http://www.matrixscience.com/), a search engine that uses mass spectrometry data to identify peptides and proteins from primary sequence databases (MSDB, SwissProt, and others).

ProteinProspector (http://prospector.ucsf.edu), database search tools, searching by peptide mass, with MS/MS or Edman data, etc.

ExPASy Proteomics tools (http://expasy.org/tools/), tools and online programs for protein identification and characterization, similarity searches, pattern and profile searches, posttranslational modification prediction, topology prediction, primary structure analysis, or secondary and tertiary structure prediction.

METLIN Metabolite Database (http://metlin.scripps.edu), a repository for mass spectral metabolite data. Each metabolite has a link to the *Kyoto Encyclopedia of Genes and Genomes.*

NIST Chemistry WebBook (http://webbook.nist.gov/), contains thermochemical data for over 7000 organic and small inorganic compounds, reaction thermochemistry data, MS, IR, UV-Vis spectra, gas chromatography data, and more.

SwePep (http://www.swepep.com/), endogenous peptide database to aid their identification by MS.

SignalP (http://www.cbs.dtu.dk/services/SignalP), site of the Center for Biological Sequence Analisys (CBS) for prediction of the presence and location of signal peptides in given amino acid sequences.

PROTOCOLS

http://www.narrador.embl-heidelberg.de/GroupPages/PageLink/activities/protocols.html

A few protocols mainly for analysis of proteins by MS.

http://www.st-andrews.ac.uk/~bmsmspf/protocols.htm#manual

Several protocols including in-gel digestion, preparing gel slices, and some others for MS analysis.

http://www.curie.fr/recherche/themes/equipe-protocoles.cfm/id_equipe/310/lang/_gb.htm

Proteomic protocols, from protein staining to analysis by MS.

http://www.cfgbiotech.com/proteomics/sample_prep.htm

Simple advice on sample preparation.

http://www.chemistry.wustl.edu/~msf/damon/index.html

MALDI tutorial with useful information concerning sample preparation, calibration, and applications.

JOURNALS

Nature
http://www.nature.com

Science
http://www.sciencemag.org

Proceedings of the National Academy of Sciences of the USA
http://www.pnas.org

Science Direct
http://www.sciencedirect.com

BioMed Central
http://www.biomedcentral.com

Springer Link
http://www.springerlink.com

Journal of Mass Spectrometry
http://www3.interscience.wiley.com/cgi-bin/jhome/6043

Mass Spectrometry Reviews
http://www3.interscience.wiley.com/cgi-bin/jhome/49879

Rapid Communications in Mass Spectrometry
http://www3.interscience.wiley.com/cgi-bin/jhome/4849

European Journal of Mass Spectrometry
http://www.impub.co.uk/ems.html

Journal of the American Society for Mass Spectrometry
http://www1.elsevier.com/homepage/saa/webjam

International Journal of Mass Spectrometry
http://ees.elsevier.com/ijms/

Analytical Chemistry
http://pubs.acs.org/journals/ancham

DISCUSSION GROUPS

http://web.chemistry.gatech.edu/~bostwick/stms/

A big forum concerning all MS techniques. It is directed to people dealing with mass spectrometry in everyday life, its character is professional and concerns practical problems.

http://msblog.kermitmurray.com/

MS blog that includes short news, links to related web sites, and other items of interest.

http://www.mspeople.net/

A large number of job offers in industry and academia.

http://groups.google.com/group/sci.techniques.mass-spec/topics

Huge discussion group, directed to MS beginners and professionals.

MASS SPECTROMETRY SOCIETIES

http://www.asms.org/

American Society for Mass Spectrometry

http://www.bsms.be/

Belgium Society for Mass Spectrometry

http://www.bmss.org.uk/

British Mass Spectrometry Society

http://www.csms.inter.ab.ca/

Canadian Society for Mass Spectrometry

http://www.spektroskopie.cz/en/index.htm

Czech Mass Spectrometry Expert Group

http://www.dsms.dk/
Danish Society for Mass Spectrometry

http://www.denvms.nl/
Dutch Society for Mass Spectrometry

http://www.bmb.leeds.ac.uk/esms/
European Society for Mass Spectrometry

http://www.sfsm.info/
French Society for Mass Spectrometry

http://www.dgms-online.de/
German Society for Mass Spectrometry

http://www.ismas.org/
Indian Society for Mass Spectrometry

http://www.mssj.jp/
The Mass Spectrometry Society of Japan

http://www.latrobe.edu.au/anzsms/
The New Zealand and Australian Society for Mass Spectrometry

http://www.nsms.no/
Norwegian Society for Mass Spectrometry

http://ptsm.ibch.poznan.pl/
Polish Mass Spectrometry Society

http://www.ssms.sg/
Singaporean Society for Mass Spectrometry

http://www.saams.up.ac.za/
The South African Association for Mass Spectrometry

http://www.smss.uu.se/
Swedish Mass Spectrometry Society

http://www.sgms.ch/
Swiss Group for Mass Spectrometry

TABLE A.1. Exact Masses and Isotopic Abundances of Selected Elements

Element	Isotope	Relative Abundance (%)	Mass (Da)	Isotope	Relative Abundance (%)	Mass (Da)
Bromine	^{79}Br	50.69	78.918336	^{81}Br	49.31	80.916280
Carbon	^{12}C	98.93	12.000000	^{13}C	1.07	13.003355
Chlorine	^{35}Cl	75.78	34.968853	^{37}Cl	24.22	36.965903
Fluorine	^{19}F	100.00	18.998403			
Hydrogen	^{1}H	99.9885	1.007825	^{2}H	0.0115	2.014102
Iodine	^{127}I	100.00	126.904473			
Nitrogen	^{14}N	99.63	14.003074	^{15}N	0.37	15.000109
Oxygen	^{16}O	99.76	15.994915	^{17}O	0.038	16.999131
	^{18}O	0.21	17.999160			
Phosphorus	^{31}P	100.00	30.973762			
Potassium	^{39}K	93.26	38.963707	^{40}K	0.012	39.963999
	^{41}K	6.73	40.961825			
Silicon	^{28}Si	92.23	27.976927	^{29}Si	4.68	28.976495
	^{30}Si	3.09	29.973771			
Sodium	^{23}Na	100.00	22.989767			
Sulfur	^{32}S	94.93	31.972071	^{33}S	0.76	32.971458
	^{34}S	4.29	33.967867	^{36}S	0.02	35.967081

More can be found at http://www.sisweb.com/referenc/source/exactmas.htm or http://www.webelements.com/.

TABLE A.2. Selected Posttranslational Modifications

Modification	Modification Place	Mass Difference Monoisotopic	Mass Difference Average
Acetylation	N-terminal, Lys	42.010	42.037
Palmitoylation	Cys, Lys, Ser, Thr	238.229	238.408
Farnesylation	Cys	204.188	204.351
Myristoylation	Gly (N-terminal), Lys	210.198	210.355
Biotinylation	N-terminal, Lys	226.078	226.295
Deamidation	Asn, Gln	0.984	0.985
Phosphorylation	Ser, Thr, Tyr	79.966	79.980
Formylation	N-terminal, Lys	27.995	28.010
Glycosylation			
N-glycosylation	Asn-X-Ser(Thr)		>800
N-acetylglucosamine	Asn	203.079	203.192
O-glycosylation	Ser, Thr		> 800
Hydroxylation	Asp, Lys, Asn, Pro	15.995	15.999
Methylation	Cys, His, Lys, Asn, Gln, Arg, N-terminal	14.016	14.027
Ubiquitinylation	Lys	114.042	114.103
Oxidation	His, Met, Trp	15.995	15.999
Sulfonation	Met	31.990	31.999

TABLE A.3. The Masses and Structures of Commonly Occurring Amino Acid Residues

Name	Code	Residue	Mass (Da)	
			Monoisotopic	Average
Alanine	Ala (A)	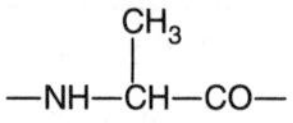	71.03711	71.0779
Arginine	Arg (R)	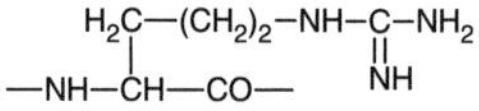	156.10111	156.1857
Asparagine	Asn (N)	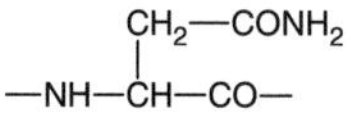	114.04293	114.1026
Aspartic acid	Asp (D)	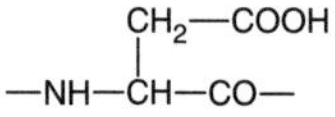	115.02694	115.0874
Cysteine	Cys (C)	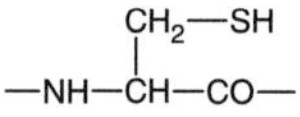	103.00918	103.1429
Glutamic acid	Glu (E)	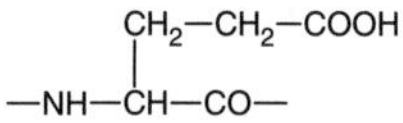	129.04259	129.1140
Glutamine	Gln (Q)	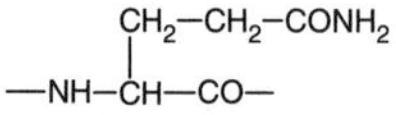	128.05858	128.1292
Glycine	Gly (G)	—NH—CH_2—CO—	57.02146	57.0513
Histidine	His (H)	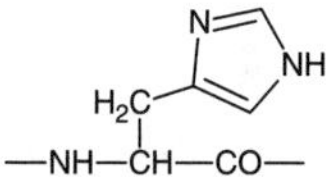	137.05891	137.1393
Isoleucine	Ile (I)	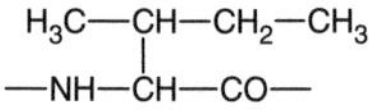	113.08406	113.1576
Leucine	Leu (L)	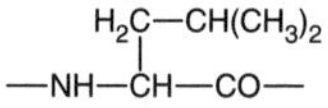	113.08406	113.1576
Lysine	Lys (K)	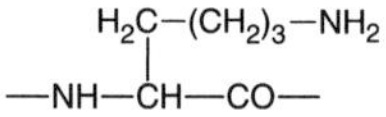	128.09496	128.1723
Methionine	Met (M)	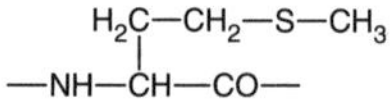	131.04048	131.1961
Phenylalanine	Phe (F)	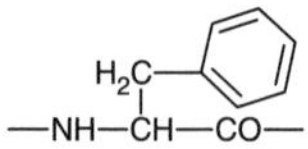	147.06841	147.1739
Proline	Pro (P)	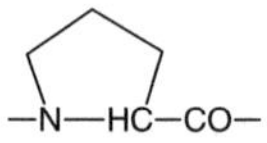	97.05276	97.1152
Serine	Ser (S)	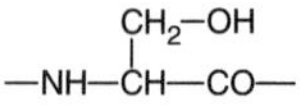	87.03203	87.0773

(*Continued*)

TABLE A.3. *Continued*

			Mass (Da)	
Name	Code	Residue	Monoisotopic	Average
Threonine	Thr (T)	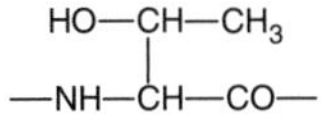	101.04768	101.1039
Tryptophan	Trp (W)	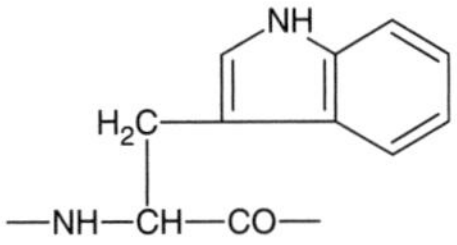	186.07931	186.2099
Tyrosine	Tyr (Y)	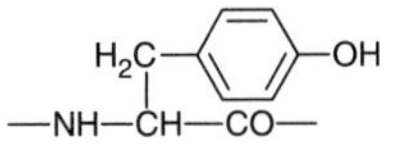	163.06333	163.1733
Valine	Val (V)	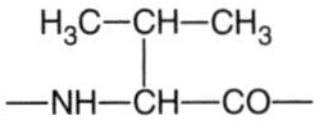	99.06841	99.1311

TABLE A.4. The Masses and Structures of Some Less Common Amino Acid Residues

			Mass (Da)	
Name	Code	Residue	Monoisotopic	Average
2-Aminobutyric acid	2-Aba	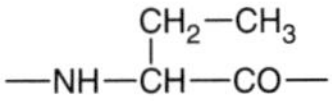	85.05276	85.1045
Aminoethylcysteine	AECys	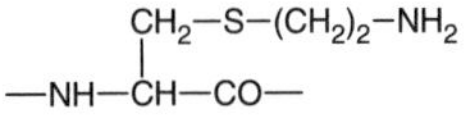	146.05138	146.2107
2-Aminoisobutyric acid	Aib	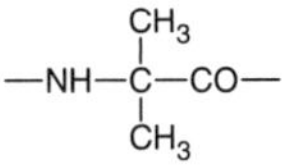	85.05276	85.1045
Carboxymethylcysteine	Cmc	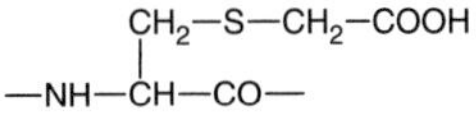	161.01466	161.1790
Cysteic acid	Cys(O_3H)	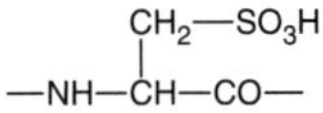	150.99393	151.1411
Dehydroalanine	Dha	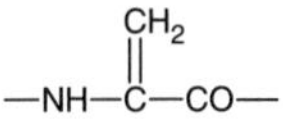	69.02146	69.0620
2-Dehydro-2-aminobutyric acid	Dhb	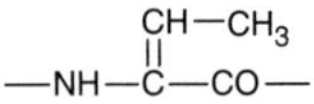	83.03711	83.0886

(Continued)

TABLE A.4. *Continued*

Name	Code	Residue	Mass (Da) Monoisotopic	Mass (Da) Average
4-Carboxyglutamic acid	Gla	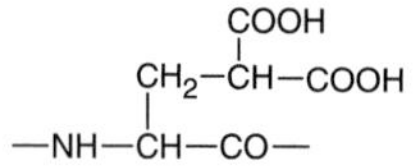	173.03242	173.1235
Homocysteine	Hcy	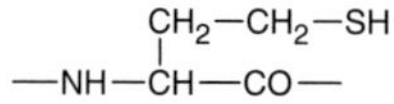	117.02483	117.1695
Homoseryne	Hse	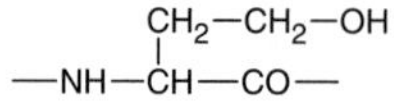	101.04768	101.1039
5-Hydroxylysine	Hyl	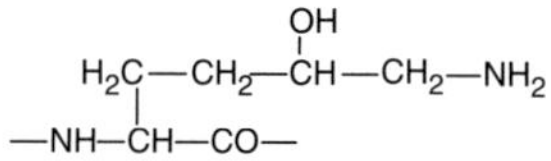	144.08988	144.1717
4-Hydroxyproline	Hyp	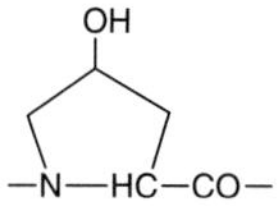	113.04768	113.1146
Isovaline	Iva	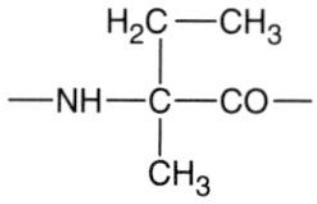	99.06841	99.1311
Norleucine	Nle	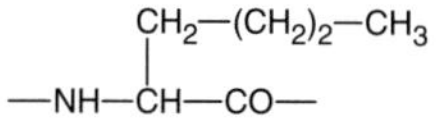	113.08406	113.1576
Norvaline	Nva	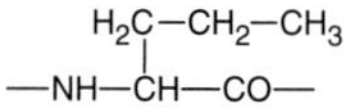	99.06841	99.1311
Ornithine	Orn	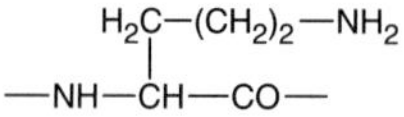	114.07931	114.1457
2-Piperidinecarboxylic acid	Pip	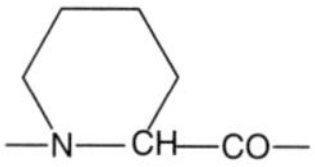	111.06841	111.1418
Pyroglutamic acid	pGlu	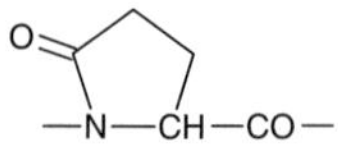	111.03203	111.0987
Sarcosine	Sar	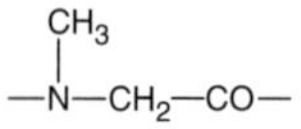	71.03711	71.0779

Monoisotopic masses were calculated on the basis of the atomic masses of the most abundant isotopes of elements: C, 12.000000 Da; H, 1.007825 Da; N, 14.003074 Da; O, 15.994915 Da; S, 31.972070 Da. Average masses were calculated on the basis of the weighted-average atomic masses of elements (taking account of the content of each isotope): C, 12.0107 Da; H, 1.00794 Da; N, 14.0067 Da; O, 15.9994 Da; S, 32.065 Da.

TABLE A.5. The Masses of Some Monosaccharide Residues and Their Derivatives

			Monoisotopic Mass of Derivatives			
Name	Formula	Monoisotopic Mass (Average)	Methylated	Deutero-methylated	Acetylated	Deutero-acetylated
Deoxypentose	$C_5H_{10}O_4$	116,0473	130,0630	133,0818	158,0579	161,0767
Deoxyribose		(116,1167)	(130,1436)	(133,1621)	(158,1540)	(161,1725)
Pentose	$C_5H_{10}O_5$	132,0423	160,0736	166,1112	216,0634	222,1010
Arabinose, ribose, xylose		(132,1161)	(160,1699)	(166,2069)	(216,1907)	(222,2277)
Deoxyhexose	$C_6H_{12}O_5$	146,0579	174,0892	180,1269	230,0790	236,1167
Fucose		(146,1430)	(174,1968)	(180,2337)	(230,2176)	(236,2545)
Hexosamine	$C_6H_{13}NO_5$	161,0688	217,1314*	229,2067*	287,1005	296,1570
Galactosamine		(161,1577)	(217,2652)	(229,3391)	(287,2695)	(296,3250)
Glucosamine						
Hexose	$C_6H_{12}O_6$	162,0528	204,0998	213,1563	288,0845	297,1410
Galactose, glucose, mannose		(162,1424)	(204,2230)	(213,2785)	(288,2542)	(297,3097)
Hexuronic acid	$C_6H_{10}O_7$	176,0321	218,0790	227,1355	260,0532	266,0909
Glucuronic acid		(176,1259)	(218,2066)	(227,2620)	(260,2005)	(266,2375)
Heptose	$C_7H_{14}O_7$	192,0634	248,1260	260,2013	360,1056	372,1810
		(192,1687)	(248,2762)	(260,3501)	(360,3178)	(372,3917)
***N*-Acetylohexosamine**	$C_8H_{15}NO_6$	203,0794	245,1263	254,1828	287,1005	293,1382
N-Acetyloglucosamine		(203,1950)	(245,2756)	(254,3311)	(287,2695)	(293,3065)
N-Acetylogalactosamine						
2-Keto-3-deoxyoctonate (KDO)	$C_8H_{14}O_8$	220,0583	276,1209	288,1962	346,0900	355,1465
		(220,1791)	(276,2866)	(288,3605)	(346,2909)	(355,3464)
Muramic acid	$C_{11}H_{19}NO_8$	275,1005	317,1475	326,2039	317,1111	320,1299
		(275,2585)	(317,3392)	(326,3946)	(317,2958)	(320,3143)

(Continued)

TABLE A.5. *Continued*

Name	Formula	Monoisotopic Mass (Average)	Monoisotopic Mass of Derivatives			
			Methylated	Deutero-methylated	Acetylated	Deutero-acetylated
		Salicylic Acids				
N-Acetylneuraminic acid	$C_{11}H_{19}NO_9$	291,0954 (291,2579)	361,1737 (361,3923)	376,2678 (376,4847)	417,1271 (417,3698)	426,1836 (426,4252)
N-Glyconeuraminic acid	$C_{11}H_{19}NO_{10}$	307,0903 (307,2573)	391,1842 (391,4186)	409,2972 (409,5295)	475,1326 (475,4064)	487,2079 (487,4804)

*Corresponds to dimethylation of amine group. Trimethylation or deuterotrimethylation can also occur, giving ions heavier by 14,0157 (14,0269) or 17,0345 (17,0454), respectively.

INDEX

Abbreviations and units, 9–12
Acceleration potential, 317
Accelerator mass spectrometry (AMS), 38
  MS analyzers, 62–64
  oceanography, 239
  tandem principle, 62–63
Accurate mass
  definition, 4
  measurements, 158
Acetophenone
  deuterated, 167
  intensity calculation, 167
  isotopomers percentage, 168
  molecular ion cluster, 167
Acids. *See* specific type
Acronyms, 9–12
Active Magnetospheric Particle Tracer Experiment (AMPTE), 256
  mission, 260
Acylcarnitines, 293
  precursor ion scan, 293
Addiction. *See also* Doping control
  neurochemistry new approaches, 330
  substance and GC-MS, 330
AE. *See* Appearance energy (AE)
Affinity chromatography, 109
Agroterrorism, 268
Aldehyde mass spectra, 143
Alkyl group preferential loss, 139
Alzheimer's disease, 324
Amherst, Jeffrey, 268
Amine fragmentation, 150
Amino acids, 185
  monoisotopic mass, 190
  MS profiling, 248–251
  MS protein identification, 248–251
  multidimensional chromatography, 248–251
  neutral, 194–201
  tandem MS, 292
  two-dimensional gel electrophoresis, 248–251
Amino hydroxybenzoic acid, 301
Aminopeptidase, 207
AMPTE. *See* Active Magnetospheric Particle Tracer Experiment (AMPTE)
AMS. *See* Accelerator mass spectrometry (AMS)
Analytes
  detected by clinical chemistry, 288
  spectrum MS, 126
Anthrax, 268
Anxiety, 327–328
  neurochemistry new approaches, 327–328
APCI. *See* Atmospheric pressure chemical ionization (APCI)
API. *See* Atmospheric pressure ionization (API)
AP-MALDI. *See* Atmospheric pressure matrix-assisted laser desorption/ionization (AP-MALDI)
Appearance energy (AE), 129, 134
APPI. *See* Atmospheric pressure photoionization (APPI)
Arginine, 201–204
Arson
  GC-MS, 319
  identification, 319
Arylsulfonylazetidin electron ionization mass spectra
  characteristic fragment ions, 177
ASAP. *See* Atmospheric-pressure solids analysis probe (ASAP)
Atmospheric pressure chemical ionization (APCI), 315
  applications, 17
  category, 17
  desorption, 30

Mass Spectrometry. Edited by Ekman, Silberring, Westman-Brinkmalm, and Kraj

Atmospheric pressure chemical ionization (APCI) (*Continued*)
doping control, 231
ion sources, 24
ion type, 17
Atmospheric pressure ionization (API), 24
doping control, 231
Atmospheric pressure matrix-assisted laser desorption/ionization (AP-MALDI)
applications, 18
category, 18
ion sources, 37
ion type, 18
Atmospheric pressure photoionization (APPI)
applications, 17
category, 17
ion sources, 26
ion type, 17
precursor molecule, 332
Atmospheric-pressure solids analysis probe (ASAP), 30
Atomic emission spectrometry, 21
Atomic mass unit
definition, 4
Average mass
definition, 4

Bacillus anthracis, 268
Background ions
peak shape, 156
Background spectrum
MS, 126
Biemann's nomenclature
peptide and protein sequencing, 185–186
Biological markers, 237
Biological oceanography, 235
Biological systems comparison, 212
Biological Weapons Convention of 1972
bioterrorism, 268
Biomarkers, 237
proteins MS identification, 271
Biomolecular mass spectrometry, 246
Biomolecule omics technologies, 244
Bioterrorism, 267–271
agents, 269
Biological Weapons Convention of 1972, 268
biomarker protein MS identification, 270
biothreat agents categories, 268
challenges, 269
counter agents, 269
defined, 267
Geneva Protocol of 1925, 268
historical accounts, 267
immunoproteomics, 271
new therapeutics development, 271
proteomics, 269
vaccine development, 271
Biothreat agents categories, 268
Blackbody infrared radiative dissociation
ion activation methods, 100
Black Death, 268
Blood, 296
Bowen's rules, 144
Brain, 325–326
neurochemistry new approaches, 325–326
stress, 326
Bromine isotope, 159
Bromo-3-chloropropane
EI mass spectrum, 144
Bromoethane
mass spectra, 158

Carboxypeptidase Y, 207
Capillary electrophoresis (CE), 310
Capillary zone electrophoresis (CZE), 112
ESI-MS interface schematic, 112
Carbon, 161
cycle, 166
isotope, 159
isotopic contributions, 162
Carbon dating, 235
oceanography, 239
Carbon disulfide
mass spectra, 158
Carbon 13 nuclear magnetic resonance
inorganic ionic species, 251–252
small molecular weight organic compounds, 251–252
Catalyst
small molecule imaging, 280
TOF-SIMS, 280, 282
Category A agents, 269
Category B agents, 269
CE. *See* Capillary electrophoresis (CE)
Cell death, 326
CFA. *See* Color filter array (CFA)

CF-FAB. *See* Continuous-flow fast atom bombardment (CF-FAB)
Challenger shuttle, 260
Channeltron flying, 260
Chemical bond energy, 144
Chemical elements
  natural isotopic abundances, 159
Chemical images
  acquisition, 276
Chemical ionization (CI)
  applications, 17
  category, 17
  ion sources, 17, 24
Chemical oceanography, 235
Chlorine isotope, 159
Chloroethane mass spectra, 158
Chlorooctane electron ionization mass spectra, 146
Cholesterol TOF secondary ion mass spectrometry images, 279
Chromatography, 106–109
  gas chromatography, 106
  liquid chromatography, 107–108
  MS, 121–127
  organic chemistry, 121–127
  prohibited substances, 227
  screening methods, 227
  shape and area peaks, 127
  supercritical fluid chromatography, 109
Chronic fatigue syndrome, 324, 329–330
  neurochemistry new approaches, 329
CI. *See* Chemical ionization (CI)
CID. *See* Collision induced dissociation (CID)
CIR. *See* Corotating interaction region (CIR)
Claviceps purpurea, 267
Clinical
  defined, 287
Clinical biomarker multianalyte detection, 289–290
Clinical chemistry, 287–298
  analytes detected, 288
  clinical biomarker multianalyte detection, 289–290
  clinical example, 292–293
  isotope dilution and quantification, 295–298
  jelly bean experiment, 295–298
  quantification demonstrations, 294
  quantitative profiling, 291
  tandem MS and sorting, 294
Coins example
  counting and sorting, 295
  MS, 295
Collision activated dissociation
  ion activation methods, 98–99
Collision induced dissociation (CID), 194
  high-energy, 98
  ion activation methods, 98–99
  low-energy, 99
  SORI, 100
  VLE, 100
Color filter array (CFA)
  images, 278
Complementary ions peak intensities, 141
Cone voltage, 317
Confirmatory test, 289
Continuous-dynode electron multiplier, 68
Continuous-flow fast atom bombardment (CF-FAB), 33–34
Corotating interaction region (CIR), 256
Criminalistics, 310
Cryogenic detector, 70
C-terminal carboxyl group, 210
Cyclic peptides
  nomenclature, 188
  and protein sequencing, 187
Cylindrical quadrupole ion trap ions, 54
CZE. *See* Capillary zone electrophoresis (CZE)

Dalton (da), 4
DAPCI. *See* Desorption atmospheric pressure chemical ionization (DAPCI)
DART. *See* Direct analysis in real time (DART)
Database search tools, 206
Daughter ion
  definition, 4
DC. *See* Direct current (DC)
Dead layer, 260
Definitions and explanations, 3–11
Dempster, 16, 19, 23, 45
De novo sequencing, 213
Depression, 324
DESI. *See* Desorption electrospray ionization (DESI)

Desorption atmospheric pressure chemical ionization (DAPCI), 30
Desorption electrospray ionization
ion sources, 29
Desorption electrospray ionization (DESI). *See also* Matrix-assisted laser desorption electrospray ionization (MALDESI)
applications, 18
category, 18
ion type, 18
schematic, 30
Detectors, 15, 65–70
cryogenic detector, 70
electron multipliers, 67–68
Faraday detector, 67
focal plane detector, 69
image current detector, 70
photoplate detector, 65–66
scintillation detector, 69
solid-state detector, 70
Deuterated acetophenone, 167
Diazepam thermal ionization mass spectrometry, 316
Differential gel electrophoresis (DIGE), 249
principle, 250
DIGE. *See* Differential gel electrophoresis (DIGE)
Dimeric ion, 4
Dimethylalkylammonium acetyl (DMAA), 208
Direct analysis in real time (DART)
category, ion type and applications, 18
Direct current (DC)
GD source, 20
Direct inlet
organic chemistry, 121
Discrete-dynode electron multiplier, 67
Displacement electrophoresis, 113
DMAA. *See* Dimethylalkylammonium acetyl (DMAA)
DNA PCR mass arrays, 246–248
Doping control, 225–232
atmospheric pressure chemical ionization (APCI), 231
atmospheric pressure ionization (API), 231
chromatographic screening methods, 227
EPO, 232
gas chromatograph (GC), 228
HBOC, 232
high resolution mass spectrometry, 227
LC-MS-MS, 232
National Measurement Institute screening method, 227
prohibited substances, 227
quadrupole-based MS, 228
selected ion monitoring (SIM), 228
solvent extraction, 227
SPE, 227
stanozolol metabolite detection, 230
tuning ion separation, 229
WADA accredited laboratories, 231
WADA 2008 prohibited list, 226
Doubly-charged precursor
annotated product ion spectra, 202
fragmentation spectrum, 203
product ion spectrum, 198
Drift tube ion mobility spectrometry (DTIMS), 110
Drug testing. *See also* Doping control
gas chromatograph (GC), 228
quadrupole-based MS, 228
selected ion monitoring (SIM), 228
DTIMS. *See* Drift tube ion mobility spectrometry (DTIMS)
Dynamic secondary ion mass spectrometry, 33
Dynode strings, 67

Earth, 255, 256
environment, 255
unknown radiation belt, 254
EC. *See* Electrochromatography (EC)
ECD. *See* Electron capture dissociation (ECD)
Edman degradation, 190, 206
EI. *See* Electron ionization (EI)
ELDI. *See* Electrospray-assisted laser desorption/ionization (ELDI)
Electric analyzer
schematic, 47
Electric-field driven separation, 110–112
electrophoresis, 111–112
ion mobility, 110
Electric sector
MS analyzers, 44–48
Electrochromatography (EC), 113
Electron capture dissociation (ECD), 192
ion activation methods, 101

Electron delocalization, 139
Electron ionization (EI)
  applications, 17
  bromo-3-chloropropane, 144
  category, 17
  compound, 163
  ethylpropionate, 130
  ion sources, 23
  ion type, 17
  labeled molecules, 166
  mass spectrum, 130, 144, 163
  organic chemistry, 129
  schematic, 23
Electron ions Stevenson's rule, 143
Electron multiplier detectors, 67–68
Electron transfer dissociation, 101
Electron volt (eV), 4
Electrophoresis
  electric-field driven separation, 111–112
Electrospray-assisted laser desorption/ ionization (ELDI), 30
Electrospray ionization (ESI), 16, 38, 97. *See also* Desorption electrospray ionization (DESI); Matrix-assisted laser desorption electrospray ionization (MALDESI)
  category, ion type and applications, 17
  ion sources, 27–28
  nanoflow schematic, 108
  schematic, 28
  sequencing, 193
Elemental composition
  fragmentation pathways, 175
  isotopic peaks, 164
Endocrine system, 327
Energetic diagram, 138
Environment
  Earth, 255
Enzymatic digestion, 206
Epigenetics
  nervous system, 328
  tagging, 328
EPO. *See* Erythropoietin (EPO)
Erythropoietin (EPO)
  doping control, 232
ESI. *See* Electrospray ionization (ESI)
ESONET. *See* European Sea Floor Observatory Network (ESONET)
Ethylpropionate
  electron ionization mass spectra, 130
European Sea Floor Observatory Network (ESONET), 241
European Space Agency's *Ulysses* space probe, 256
eV. *See* Electron volt (eV)
Even electron ion rule, 143
Extracted ion chromatogram, 4
Extraterrestrial, 253

FAB. *See* Fast atom bombardment (FAB)
FAIMS. *See* High-field asymmetric waveform ion mobility spectrometry (FAIMS)
False identification results, 215
Faraday cups, 259
Faraday detector, 67
Fast atom bombardment (FAB), 21
  category, ion type and applications, 18
  continuous-flow, 33–34
  ion sources, 33
FD. *See* Field desorption (FD)
Fentanyl analogs
  screening method, 311
FI. *See* Field ionization (FI)
Field asymmetric waveform ion mobility spectrometry, 110
Field desorption (FD)
  category, ion type and applications, 17
  ion sources, 27
Field free region, 4
Field ionization (FI)
  category, ion type and applications, 17
  ion sources, 26
Field's rules, 144
Fingerprint
  image, 283, 284
  lines, 281
Fluorine isotope, 159
Focal plane detector, 69
Forbush decreases, 256
Forensic sciences, 309–318
  arson identification, 319
  definition, 309
  GC-MS procedures, 315
  LC-MS procedures, 315–316
  materials examined, 309–310
  quantitative analysis, 319
  sample preparation, 312
  small molecule imaging, 281

Forensic sciences (*Continued*)
systematic toxicological analysis, 312–317
TOF-SIMS, 281
Forensic toxicology, 310
chemical compound identification, 313
GC-MS, 312
LC-MS, 312
MS, 310
Fourier transform ion cyclotron resonance (FTICR), 38, 57, 96
acquisition speed, 58
analyzer, 96
features, 61
ion chemistry, 61
ion detection, 59
mass calibration, 59
MS analyzers, 58–61
performance parameters, 59
schematic, 59
Four-sector tandem mass spectrometry, 97
Fragmentation, 170
charge center initiation, 150
charge remote organic chemistry, 151
elemental composition, 175
initiation, 149
ion structures, 176
organic chemistry, 175–176
pathways, 175
pattern simplification, 208
peptide and protein sequencing, 208
peptide derivatization, 208
Fragmentation spectrum
doubly-charged precursor, 203
peptides, 184
singly charged peptide, 198
Fragment ion
arylsulfonylazetidin electron ionization mass spectra, 177
definition, 4
IE, 141
organic chemistry, 168–172
Fragmentor voltage, 317
Frank-Condon principle, 132
FTICR. *See* Fourier transform ion cyclotron resonance (FTICR)
Full width at half-maximum (FWHM), 54
definition, 7
FWHM. *See* Full width at half-maximum (FWHM)
Gas chromatograph (GC)
coupled to MS, 251–252
drug testing, 228
inorganic ionic species, 251–252
small molecular weight organic compounds, 251–252
Gas chromatograph mass spectrometry (GC-MS), 315
addictive substance, 330
analysis, 122, 123, 128
arsons, 319
forensic toxicology, 312
organic pollutants, 122, 123, 128
petroleum products, 319
quantitative, 124
systematic toxicological analysis procedures, 315
Gas discharge
applications, 17
category, 17
DC, 20
ion sources, 16
ion type, 17
trapping analyzers, 21
Gas-liquid chromatography (GLC), 106
Gas-solid chromatography (GSC), 107
GC. *See* Gas chromatograph (GC)
GC-MS. *See* Gas chromatograph mass spectrometry (GC-MS)
GD. *See* Gas discharge; Glow discharge
GDMS. *See* Glow discharge mass spectrometry (GDMS)
GE. *See* Gel electrophoresis (GE)
Gel electrophoresis (GE), 111
Gel filtration, 109
Gel permeation chromatography (GPC), 109
GENESIS, 264
Geneva Protocol of 1925
bioterrorism, 268
Genomics applications, 246–247
Geological oceanography, 235
GLC. *See* Gas-liquid chromatography (GLC)
Global climate changes, 240
Global Ocean Observing System (GOOS), 241
Global positioning system (GPS), 240
Glow discharge, 16
category, ion type and applications, 17
ion source schematic, 20

Glow discharge mass spectrometry
(GDMS), 20
Goldstein, 16
GOOS. *See* Global Ocean Observing System
(GOOS)
GPC. *See* Gel permeation chromatography
(GPC)
GPS. *See* Global positioning system (GPS)
GSC. *See* Gas-solid chromatography (GSC)

HBOC. *See* Hemoglobin-based oxygen
carriers (HBOC)
Helium, 106
Hemoglobin-based oxygen carriers (HBOC)
doping control, 232
Hexachlorobiphenyl
MS, 126
HIC. *See* Hydrophobic interaction
chromatography (HIC)
High-energy collision induced
dissociation, 98
Highest occupied molecular orbital (HOMO),
129, 149
High-field asymmetric waveform ion
mobility spectrometry (FAIMS), 110
High ion energies
structural and stereochemical aspects, 145
High mass accuracy analyzers, 206
High performance liquid chromatography
(HPLC), 239
coupled to MS, 251–252
inorganic ionic species, 251–252
oceanography, 239
small molecular weight organic
compounds, 251–252
High resolution mass spectrometry, 157
doping control, 227
organic chemistry, 155–157
HOMO. *See* Highest occupied molecular
orbital (HOMO)
Homologous ion series, 169
organic compounds, 169
HPA. *See* Hypothalamus-pituitary-adrenal
axes (HPA)
HPLC. *See* High performance liquid
chromatography (HPLC)
Hybrid linear ion trap-orbitrap instrument, 57
Hydrogen, 106
isotope, 159
Hydrogen nuclear magnetic resonance
inorganic ionic species, 251–252
small molecular weight organic
compounds, 251–252
Hydrophobic interaction chromatography
(HIC), 108
Hyperthermal regime
collisions, 102
Hypothalamus-pituitary-adrenal axes
(HPA), 326

ICP. *See* Inductively couple plasma (ICP)
ICP-MS. *See* Inductively couple plasma mass
spectrometry (ICP-MS)
ICR. *See* Ion cyclotron resonance (ICR)
Identification strategy, 191
IDMS. *See* Isotope dilution mass
spectrometry (IDMS)
IE. *See* Ionization energy (IE)
IEC. *See* Ion-exchange chromatography (IEC)
IEF. *See* Isoelectric focusing (IEF)
Image current detector, 70
Imaging mass spectrometry (IMS), 275, 276
Imaging method, 275. *See also* specific type
Immonium ions, 185
chemical structure, 185
list, 186
Immune system, 325–326, 327
neurochemistry new approaches,
325–326
Immunogenic proteins
identification, 272
Immunoproteomics
bioterrorism, 271
IMS. *See* Imaging mass spectrometry (IMS)
Inductively couple plasma (ICP), 19
category, ion type and applications, 17
combined with laser ablation, 22
schematic, 22
sources, 16
Inductively couple plasma mass spectrometry
(ICP-MS), 237
Infant screening, 292
Infrared (IR), 119
Infrared multiphoton dissociation
(IRMPD), 100
Inlet systems, 121
Inorganic ionic species, 251–252
In-source dissociation, 192

Integrated Ocean Observing System (IOOS), 241
Intense peaks mass spectra, 170
Intensity calculation
  acetophenone molecular ion cluster, 167
  ions, 151
International Olympic Committee (IOC), 225
IOC. *See* International Olympic Committee (IOC)
Iodine isotope, 159
Ion
  chemistry FTICR analyzer, 61
  cylindrical QIT, 54
  densities solar wind, 257
  electric-field driven separation, 110
  elemental composition, 158–163
  formed in different instruments, 189
  fragmentation schemes, 176
  generation methods, 17
  isotopic peaks, 158–163
  mobility, 110
  organic chemistry, 158–163
  peak intensities, 133, 141
  structures, 174, 176
Ion activation methods, 97–101
  blackbody infrared radiative dissociation, 100
  collision induced/activated dissociation, 98–99
  electron capture dissociation, 101
  electron transfer dissociation, 101
  in-source decay, 97
  photodissociation, 100
  post-source decay, 98
  surfaced-induced dissociation, 102
  tandem MS, 97–101
Ion cyclotron resonance (ICR), 99. *See also* Fourier transform ion cyclotron resonance (FTICR)
  cells, 56
  principle, 58–59
Ion-exchange chromatography (IEC), 108
Ionization
  efficiency, 4
  hard *vs.* soft, 304
  methods, 277
Ionization energy (IE), 129, 134
  fragments, 141
Ion sources, 15–37
  atmospheric pressure chemical ionization, 24
  atmospheric pressure matrix-assisted laser desorption/ionization, 37
  atmospheric pressure photoionization, 26
  chemical ionization, 24
  desorption electrospray ionization, 29
  electron ionization, 23
  electrospray ionization, 27–28
  fast atom bombardment, 33
  field desorption, 27
  field ionization, 26
  gas discharge, 16
  GD, 20
  inductively coupled plasma, 21–22
  laser desorption/ionization, 34
  matrix-assisted laser desorption/ionization, 35–36
  multiphoton ionization, 25
  new generation, 278
  photoionization, 25
  plasma desorption, 34
  real time direct analysis, 30
  role, 15
  secondary ion MS, 31–32
  spark source, 19
  thermal ionization, 16–18
  thermospray ionization, 27
IOOS. *See* Integrated Ocean Observing System (IOOS)
IR. *See* Infrared (IR)
IRMPD. *See* Infrared multiphoton dissociation (IRMPD)
IRMS. *See* Isotope ratio mass spectrometry (IRMS)
Isobaric interferences, 237
Isochronous TOF mass spectrometry
  space MS history, 262–263
Isoelectric focusing (IEF), 111
Isoleucine, 185, 186
Isotachophoresis (ITP), 113
Isotope, 159
  dilution and quantification, 295–298
Isotope dilution mass spectrometry (IDMS), 291
  definition, 4
Isotope ratio mass spectrometry (IRMS), 4–5
Isotopic abundances, 159

Isotopic mass spectrometry, 166
Isotopic peaks
  deuterated acetophenone, 167
  elemental composition, 164
  intensity, 161
  theoretic intensity, 163
ITP. *See* Isotachophoresis (ITP)

Jelly bean example
  clinical chemistry, 295–298

Kingdon trap, 55

LA-ICP. *See* Laser ablation combined with inductively couple plasma (LA-ICP)
Laser ablation combined with inductively couple plasma (LA-ICP), 22
Laser desorption/ionization (LDI), 21
  category, ion type and applications, 18
  ion sources, 34
LC. *See* Liquid chromatography (LC)
LC-MS. *See* Liquid chromatography mass spectrometry (LC-MS)
LDI. *See* Laser desorption/ionization (LDI)
LEO. *See* Low Earth orbit (LEO)
Leucine, 185, 186
Library sequencing, 213
Library spectrum, 126
Light scattering (LS)
  polymer analysis, 304
Light stable isotope, 238
Limit of detection (LOD), 313
Linear ion traps, 56
Linear quadrupole ion trap, 54
Linear TOF mass spectrometry, 260–261
Liquid chromatography (LC), 106
  chromatography, 107–108
  ESI-MS interface schematic, 108
Liquid chromatography mass spectrometry (LC-MS), 315
  forensic science procedures, 315–316
  forensic toxicology, 312
  quantitative, 124
  systematic toxicological analysis, 315–316
Liquid chromatography mass spectrometry mass spectrometry (LC-MS-MS)
  doping control, 232
Liquid-liquid extraction (LLE), 312
Liquid metal ion source (LMIS), 277
Liquid secondary ion mass spectrometry (LSIMS), 33
LLE. *See* Liquid-liquid extraction (LLE)
LMIS. *See* Liquid metal ion source (LMIS)
LOD. *See* Limit of detection (LOD)
Low Earth orbit (LEO), 254
Low-energy collision induced dissociation, 99
Low mass region, 205–206
LS. *See* Light scattering (LS)
LSD. *See* Lysergic acid (LSD)
LSIMS. *See* Liquid secondary ion mass spectrometry (LSIMS)
Lyman-alpha radiation, 253
Lysergic acid (LSD), 267

Magnetic sector analyzer
  MS, 44–48
  operating modes, 46–47
  performance parameters, 48–49
  principle, 46
Magnetosphere, 255
Makarov, 55
MALDESI. *See* Matrix-assisted laser desorption electrospray ionization (MALDESI)
MALDI. *See* Matrix-assisted laser desorption/ionization (MALDI)
MALDI-MS. *See* Matrix-assisted laser desorption/ionization mass spectrometry (MALDI-MS)
MALDI-TOF. *See* Matrix-assisted laser desorption/ionization time of flight (MALDI-TOF)
Marine organisms
  water column chemicals, 239
Mark-Huwink-Sakurada equation, 304
Mass, 5
  accuracy, 5
  calibration, 5
  charge, 6
  defect, 5
  fragmentography, 124
  limit, 5
  nominal, 8
  number, 5
  peak width, 5
  precision, 5
  range, 5

Mass (*Continued*)
selection, 6
separation, 6
Mass-analyzed ion kinetic energy spectrometry (MIKES), 97
Mass analyzer, 15
actions, 5
features, 39
Mass array
primer binding, 247
primer extension, 247
primer terminates, 247
Mass resolution
definition, 6
establishing, 7
peak width definition, 6
ten percent valley definition, 6
Mass resolving power
definition, 6
establishing, 7
Mass selection mass spectrometry, 6
Mass spectra
aldehydes, 143
bromoethane, 158
carbon disulfide, 158
chloroethane, 158
compound, 120
definition, 6
identify compound, 120
intense peaks, 170
monofunctional aliphatic compounds, 172
monofunctional compounds, 171
para-fluorobenzophenone, 125
synthetic peptide, 180
Mass spectrometry (MS), 1
amino acids, 248–251
biomarker proteins, 271
blood, 296
building blocks, 15–70
coin counting and sorting, 295
definition, 6
detection limit, 219
dynamic range, 219
effect, 219
forensic toxicology, 310
fragmentation processes, 130–134
identification, 271
interpretation, 137–176
ionization methods, 277
ion sources, 15–37
libraries, 173
MMD, 304
oceanography, 235–236
organic chemistry, 128–136, 137–176
pain treatment, 331
parts, 15
peptide, 248–251
peptide fragmentation, 208
peptide identification, 213–217
physical bases, 128–136
polymer analysis, 304
profiling, 248–251
protein identification, 213–217, 248–251
proteins, 248–251
proteome analysis, 211–220
quantitation, 212
quantitative comparison, 212
salbutamol, 280
scheme, 120
sensitivity and specificity optimization, 211–220
serum profiles, 251
signals, 212
success rate and relative dynamic range, 218–219
Mass spectrometry (MS) analyzers, 38–64
accelerator MS, 62–64
Fourier transform ion cyclotron resonance, 58–61
magnetic/electric sector, 44–48
orbitrap, 55–57
quadrupole ion trap, 51–54
quadrupole mass filter, 49–50
time-of-flight, 40–44
Mass spectrometry/mass spectrometry, 6
Mass-to-charge ratio, 6
Matrix-assisted laser desorption electrospray ionization (MALDESI), 38
Matrix-assisted laser desorption/ionization (MALDI), 24
applications, 18
category, 18
in-source dissociation, 192
ion sources, 35–36
ion type, 18
molar masses, 303
PC, 305
poly(bisphenolA carbonate) (PC), 301

polycarbonate, 302
polymers, 300
PSD, 93
schematic, 36
spectra, 302, 305
Matrix-assisted laser desorption/ionization mass spectrometry (MALDI-MS), 37
Matrix-assisted laser desorption/ionization tandem time of flight (MALDI-TOF-TOF)
schematics, 94
Matrix-assisted laser desorption/ionization time of flight (MALDI-TOF)
schematics, 94
Matrix-assisted laser desorption/ionization time of flight mass spectrometry (MALDI-TOF-MS), 37
schematic, 40
McLafferty rearrangement, 142, 150, 151
MCP. *See* Microchannel plate (MCP)
MECA. *See* Multiple excitation collisional activation (MECA)
Medical observation, 287
MEKC. *See* Micellar electrokinetic chromatography (MEKC)
Membrane introduction mass spectrometry (MIMS), 240
Metabolites
analytical access, 252
classes, 252
Metabolome profiling, 252
Metastable ion (MI spectra), 136
definition, 7
registration, 136
Methadone, 330
Mi. *See* Millikan (Mi)
Micellar electrokinetic chromatography (MEKC), 113
Microchannel plate (MCP), 44
schematic, 68
MIKES. *See* Mass-analyzed ion kinetic energy spectrometry (MIKES)
Millikan (Mi), 5
MIMS. *See* Membrane introduction mass spectrometry (MIMS)
MI spectra. *See* Metastable ion (MI spectra)
MM. *See* Molar mass (MM)
MMD. *See* Molar mass distribution (MMD)
Modified Knight/Kingdon trap, 55
Molar mass (MM)
definition, 7
measure, 303
Molar mass distribution (MMD)
measure, 303
MS, 304
polymers, 303–304
Molecular ion
definition, 7
detected, 154
organic chemistry, 152–154
Molecular weight
definition, 7
Monoenergetic UV photon beam, 25
Monofunctional aliphatic compounds
mass spectra, 172
Monofunctional compounds
mass spectra, 171
Monoisotopic mass
definition, 8
Moon, 255, 256
MPI. *See* Multiphoton ionization (MPI)
MPI-MS. *See* Multiphoton ionization mass spectrometry (MPI-MS)
M/Q analysis *vs.* SWICS M, 262
MRM. *See* Multiple reaction monitoring (MRM)
MS. *See* Mass spectrometry (MS)
MS/MS
definition, 6
information, 214
multistage, 89
MS^n. *See* Multiple-stage mass spectrometry (MS^n)
Müller, E.W., 26
Multidimensional chromatography
amino acids, 248–251
peptide, 248–251
proteins, 248–251
Multiphoton ionization (MPI)
applications, 17
category, 17
ion sources, 25
ion type, 17
Multiphoton ionization mass spectrometry (MPI-MS), 25
Multiple excitation collisional activation (MECA), 100
Multiple reaction monitoring (MRM), 8, 89

Multiple-stage mass spectrometry (MS^n), 8, 89
Multizonal electrophoresis, 113

Nano-electrospray ionization, 28
Nanoflow liquid chromatography
  ESI-MS interface schematic, 108
Nanospray, 28
National Measurement Institute screening method, 227
National Ocean Sciences Accelerator Mass Spectrometry Facility (NOSAMS), 239
Navier-Stokes hydrodynamics, 255
Nervous system, 327
  epigenetics, 328
Neurochemistry, 321–333
  addiction, 330
  brain and immune system, 325–326
  chronic fatigue syndrome, 329
  learning and memory, 324
  neurodegenerative diseases, 331–332
  new approaches, 323–331
  pain, 331
  proteomics and neurochemistry, 323–331
  psychiatric diseases and disorders, 329
  stress and anxiety, 327–328
  synapse, 324
Neurodegenerative diseases, 324
  neurochemistry new approaches, 331–332
Neurotransmission, 323
Neutral amino acids, 194–201
Neutral loss, 89, 170
  definition, 8
  principles, 90
  purpose, 91
9THC. *See* Tetrahydrocannabinol (9THC)
Nitrogen
  isotope, 159
  rule, 164–165
NMR. *See* Nuclear magnetic resonance (NMR)
Nominal mass, 8
Normal-phase chromatography (NPC), 108
NOSAMS. *See* National Ocean Sciences Accelerator Mass Spectrometry Facility (NOSAMS)
Nozzle-skimmer dissociation (NSD), 97
NPC. *See* Normal-phase chromatography (NPC)
NSD. *See* Nozzle-skimmer dissociation (NSD)
N-terminal amino group, 209
Nuclear magnetic resonance (NMR), 119
  inorganic ionic species, 251–252
  polymer analysis, 304
  small molecular weight organic compounds, 251–252

Oceanography, 235–240
  carbon dating, 239
  ESONET, 241
  global climate changes, 240
  GOOS, 241
  IOOS, 241
  isobaric interferences, 237
  MIMS, 240
  MS, 235–236
  NOSAMS, 239
  PDMS, 240
  radiocarbon analysis, 239
  soft ionization method, 240
  trace elements, 237
  water column chemicals, 239
  Woods Hole Oceanographic Institution, 239
Octapole offset voltage, 317
Oligonucleotides gene microarrays, 246–248
  PCR mass arrays, 246–248
Omes, 244
Omics applications, 243–251
  defined, 244
  genomics, 246–247
  metabolomics, 251–252
  proteomics, 248–250
  research objects, 245
  transcriptomics, 246–247
Omification, 244
Orbitrap, 38
  analyzer schematic, 56
  MS analyzers, 55–57
  performance parameters, 57
  principle, 55
Organic chemistry, 119–176
  charged and neutral particle stability, 137–147
  charge remote fragmentation, 151
  chromatography-MS, 121–127
  direct inlet, 121

electron ionization, 129
fragmentation scheme, 175–176
fragment ions, 168–172
high-resolution MS, 155–157
inlet systems, 121
ion elemental composition, 158–163
ion isotopic peaks, 158–163
molecular ion, 152–154
MS fragmentation processes, 130–134
MS interpretation, 137–176
MS libraries, 173
MS physical bases, 128–136
natural samples carbon 13 isotope content, 166
nitrogen rule, 164–165
sample isotopic purity calculation, 166–167
unpaired electron localization and charge, 148–150
Organic compounds, 169
Organic molecules, 236
Organic pollutants analysis, 122, 123, 128
Orifice voltage, 317
Ortho-effect, 147
Osmometry polymer analysis, 304
Overlapping product ions, 201
Oxygen, 161
isotope, 159

Pain
definition, 331
MS, 331
neurochemistry new approaches, 331
treatment, 331
Parabolic electric field, 263
Para-fluorobenzophenone
mass spectra, 125
Paul trap, 52, 56
PC. *See* Poly(bisphenolA carbonate) (PC)
PCR mass arrays
DNA, RNA, oligonucleotides gene microarrays, 246–248
PD. *See* Plasma desorption (PD)
PDMS. *See* Polydimethyl siloxane mass spectrometry (PDMS)
Peaks
definition, 8
intense, 170
intensity definition, 8
ion intensities, 133
ion isotopic, 158–164, 167
mass resolution, 6
width, 5, 6
Peptide(s)
charge state, 180
chemical structure, 99
fragmentation nomenclature, 183–187
fragmentation spectrum, 184
identification, 213–217
mass spectrometry fragmentation, 208
matching, 215
MS profiling, 248–251
MS protein identification, 248–251
MS proteome analysis, 213–217
multidimensional chromatography, 248–251
nomenclature, 184, 188
peptide sequencing, 183–187
production, 323
protein sequencing, 183–187
sensitivity and specificity optimization, 213–217
two-dimensional gel electrophoresis, 248–251
Peptide derivatization, 207
fragmentation pattern simplification, 208
peptide and protein sequencing, 207–209
stable isotope labeling, 209
Peptide mass fingerprinting (PMF), 191, 213
identification, 216
Saccharomyces cerevisiae proteins, 216
Peptidergic cells, 323
Peptide sequencing, 179–209
Biemann's nomenclature, 185–186
cyclic peptides, 187
data acquisition, 193
de novo sequencing, 192
fragmentation pattern simplification, 208
peptide and protein tandem MS, 181–182
peptide derivatization, 207–209
peptide fragmentation nomenclature, 183–187
rationale, 190–191
Roepstorff's nomenclature, 183–184
sequencing procedure examples, 194–204
stable isotope labeling, 209
technical aspects and fragmentation rules, 188–189
tips and tricks, 205–206

Peptide tandem mass spectrometry
peptide and protein sequencing, 181–182
Pfleger-Maurer-Weber library, 315
Pharmaceutical cellular monitoring, 278
Phe, 290
Phenylalanine, 185, 289, 290
Phosphate, 279
Phospholine phosphate, 279
Phosphorus isotope, 159
Photodissociation ion activation methods, 100
Photoionization (PI)
applications, 17
category, 17
ion sources, 25
ion type, 17
Photoplate detector, 65–66
schematic, 66
Physical oceanography, 235
PI. *See* Photoionization (PI)
PKU, 290
Planetology, 253
Plasma desorption (PD)
applications, 18
category, 18
ion sources, 34
ion type, 18
PLOT. *See* Porous layer open tubular (PLOT) columns
PMF. *See* Peptide mass fingerprinting (PMF)
Poison group analysis, 314
POLAR/TIMAS schematic, 258
Poly(bisphenolA carbonate) (PC), 301
MALDI spectra, 305
SEC trace, 305
Polycarbonate, 302
Polydimethyl siloxane mass spectrometry (PDMS), 240
Polymers, 299–305
analysis, 301–304
average molar mass determination, 303–304
instrumentation, 300
LS, 304
MALDI, 300
matrices, 300
MS, 304
NMR, 304
osmometry, 304
sample preparation, 300
ultrapure polymer samples, 301
POPOP, 301
Porous layer open tubular (PLOT) columns, 106
Post-source decay (PSD)
ion activation methods, 98
MALDI, 93
Posttraumatic stress disorders (PTSD), 324, 328
Precursor ion
definition, 8
mode, 89
purpose, 91
scanning principles, 90
Precursor molecule atmospheric pressure photoionization, 332
Primer
binding, 247
extension, 247
mass array, 247
terminates, 247
Product ion
chemical structures, 185
definition, 9
mode, 89
purpose, 91
scanning principles, 90
spectrum, 195
Progeny ions, 9
Prohibited substances screening methods, 227
Protective antigen identification, 272
Protein
expression, 249
genetic blueprint, 249
identification, 213–217
learning and memory, 325
MS profiling, 248–251
MS protein identification, 248–251
MS proteome analysis, 213–217
multidimensional chromatography, 248–251
sensitivity and specificity optimization, 213–217
separation effect, 219
study, 248
two-dimensional gel electrophoresis, 248–251

Protein sequencing, 179–209
  Biemann's nomenclature, 185–186
  cyclic peptides, 187
  data acquisition, 193
  de novo sequencing, 192
  fragmentation pattern simplification, 208
  peptide and protein tandem MS, 181–182
  peptide derivatization, 207–209
  peptide fragmentation nomenclature, 183–187
  rationale, 190–191
  Roepstorff's nomenclature, 183–184
  sequencing procedure examples, 194–204
  stable isotope labeling, 209
  technical aspects and fragmentation rules, 188–189
  tips and tricks, 205–206
Protein tandem mass spectrometry, 181–182
Proteomics
  bioterrorism, 269
  neurochemistry new approaches, 323–331
  omics applications, 248–250
  SRD, 333
  studies, 211
  workflow, 250
Protonated molecule, 9
PSD. *See* Post-source decay (PSD)
Psychiatric diseases and disorders, 329
Psycho-neuro-immuno-endocrine response patterns, 325
PTSD. *See* Posttraumatic stress disorders (PTSD)

QIT. *See* Quadrupole ion trap (QIT)
Qq. *See* Quadrupole mass filter (Qq)
QqQ. *See* Triple quadrupole (QqQ)
Qq-TOF. *See* Quadrupole mass filter TOF (Qq-TOF)
Quadrupole-based mass spectrometry
  drug testing, 228
Quadrupole ion trap (QIT), 38, 52, 95–96
  mass spectrometry cross-section schematic, 52
  MS analyzers, 51–54
  performance parameters, 54–55
  principle, 52–54
  schematic, 53
Quadrupole mass filter (Qq), 38
  MS analyzers, 49–50
  principle, 50–51
Quadrupole mass filter TOF (Qq-TOF), 95
  mass spectrometry schematic, 40
Quantitative analysis
  forensic sciences, 319
Quantitative gas chromatograph mass spectrometry, 124
Quantitative liquid chromatography mass spectrometry, 124
Quantitative profiling
  clinical chemistry, 291

Radiation belt of Earth, 254
Radical site reaction initiation, 149
Radiocarbon analysis, 239
Radio frequency (RF) potential, 19
RDR. *See* Relative dynamic range (RDR)
Reaction coordinate potential energy, 134
Real time direct analysis, 30
Rearrangement process
  types, 150
Reflector, 93
  single-stage, 42
Relative dynamic range (RDR), 211, 219
  definition, 218
REMPI. *See* Resonance-enhanced multiphoton ionization (REMPI)
Research, 287
Resonance-enhanced multiphoton ionization (REMPI), 25
Resonance ionization (RI). *See* Resonance-enhanced multiphoton ionization (REMPI)
Resonant ejection, 53
Reversed-phase chromatography (RPC), 108
RF. *See* Radio frequency (RF) potential
RI. *See* Resonance-enhanced multiphoton ionization (REMPI)
RNA, 246–248
Roepstorff-Fohlmann's conventions, 187
Roepstorff's nomenclature
  peptide and protein sequencing, 183–184
RPC. *See* Reversed-phase chromatography (RPC)
RTOF. *See* Reflector
Russian ion traps, 259

Saccharomyces cerevisiae proteins, 216
Salbutamol
  MS, 280
  TOF-SIMS image, 280
Satellite instrumentation, 257
Satellite ions, 185
  analysis, 186
Schwarz, H., 147
Scintillation detector, 69
Score distribution, 214, 215
Screening methods
  infants, 292
  National Measurement Institute, 227
SEC. *See* Size exclusion chromatography (SEC)
Secondary ion mass spectrometry (SIMS), 21
  applications, 18
  category, 18
  ion sources, 31–32
  ion type, 18
  schematic, 31
  small molecule imaging, 277
Secondary neutral mass spectrometry (SNMS), 33
Sector analyzers, 45
Selected ion monitoring (SIM)
  drug testing, 228
Selected reaction monitoring (SRM), 89
  definition, 9
  principles, 90
  purpose, 91
Selectivity, 211, 216, 217
Semiconductors, 282
Sensitivity, 211, 216, 217
Separation methods, 105–111
  chromatography, 106–109
  electric-field driven separation, 110–112
Sequence collection searching, 213
  protein identification, 214
Sequencing
  de novo, 213
  specific enzymes, 207
SFC. *See* Supercritical fluid chromatography (SFC)
Short tandem repeats (STR), 246
Significance testing, 214
Silent death, 332
Silicon
  isotope, 159
  wafer TOF-SIMS images, 285
SIM. *See* Selected ion monitoring (SIM)
SIMS. *See* Secondary ion mass spectrometry (SIMS)
Single nucleotide polymorphisms (SNP), 246
Single reaction monitoring, 89
Single-stage reflector, 42
Singly charged peptide, 198
Singly charged precursor, 200
Size exclusion chromatography (SEC), 109
  PC trace, 305
Small molecular weight organic compounds, 251–252
Small molecule imaging, 275–283
  biological applications, 278–279
  catalysis, 280
  forensics, 281
  future, 283
  semiconductors, 282
  SIMS imaging, 277
SNMS. *See* Secondary neutral mass spectrometry (SNMS)
SNP. *See* Single nucleotide polymorphisms (SNP)
Soft ionization method, 27
  coupling, 240
Soft ion sources, 16
Solar wind, 255
  ion densities, 257
  types, 256
Solid phase extraction (SPE)
  doping control, 227
Solid state detector (SSD), 70, 259
  flying, 260
Solvent extraction, 227
SORI. *See* Sustained off resonance (SORI)
Space charge effect, 9
Space mass spectrometry
  beginnings, 259
  history, 259–261
  isochronous TOF-MS, 262–263
  linear TOF-MS, 260–261
  space sciences, 257–258
Space sciences, 253–264
  beginnings, 259
  dynamics, 256
  GENESIS and future, 264

isochronous TOF-MS, 262–263
linear TOF-MS, 260–261
origins, 254–255
space MS history, 259–263
space MS paradox, 257–258
Spark ionization. *See* Spark source (SS)
Spark source (SS)
applications, 17
category, 17
ion sources, 19
ion type, 17
schematic, 20
Spark source mass spectrometry (SSMS), 19
SPE. *See* Solid phase extraction (SPE)
Specificity, 211
Spectral interpretation algorithm, 176
Spectral resolution, 155
Sport, 225. *See also* Doping control
SRD. *See* Stress-related disorders and diseases (SRD)
SRM. *See* Selected reaction monitoring (SRM)
SS. *See* Spark source (SS)
SSD. *See* Solid state detector (SSD)
SSMS. *See* Spark source mass spectrometry (SSMS)
Stable isotope, 238
labeling, 209
peptide derivatization, 209
protein sequencing, 209
Stanozolol metabolite detection, 230
Stevenson's rule (Stevenson-Audier), 141, 142, 150, 151
even electron ions, 143
STJ. *See* Superconducting tunnel junction (STJ)
STR. *See* Short tandem repeats (STR)
Stress, 327–328
Stress-related disorders and diseases (SRD)
biopattern technologies and proteomics, 333
Success rate, 211
definition, 218
Sulphur isotope, 159
Sun, 253, 254, 255
Superconducting tunnel junction (STJ), 70
Supercritical fluid chromatography (SFC), 106, 109
Surfaced-induced dissociation, 102
Surface ionization. *See* Thermal ionization mass spectrometry (TIMS)
Sustained off resonance (SORI), 100
SWICS M *vs.* M/Q analysis, 262
Synapse, 324
Synthetic peptide mass spectra, 180
Systematic toxicological analysis, 312–317
forensic sciences, 312–317
GC-MS procedures, 315
LC-MS procedures, 315–316
poison groups, 314

Tandem accelerator
mass spectrometry principle, 62–63
performance parameters, 64–65
schematic, 63
Tandem in space, 188
MS analyzer combinations, 91–94
Tandem in time, 188
MS analyzer combinations, 95
Tandem mass spectrometry, 89–101, 290
amino acids, 292
analyzer combinations, 91–96
clinical chemistry, 294
definition, 9
four-sector, 97
ion activation methods, 97–101
metabolites, 294
purpose, 91
real-life example, 182
separation methods, 105–111
sorting, 294
stages, 181
tandem in space, 91–94
tandem in time, 95
Tandem time of flight (TOF-TOF), 93
Tate, J.T.
laboratory, 23
Taylor cone, 27
TDC. *See* Time-to-digital converter (TDC)
Ten percent valley definition
mass resolution, 6
Tetrabromobiphenyl
partial spectra, 160
Tetrachlorobiphenyl
partial spectra, 160
Tetrahydrocannabinol (9THC)
quantification, 318

Th. *See* Thomson (Th)
THC. *See* Tetrahydrocannabinol (9THC)
Thermal ionization (TI), 21
  category, ion type and applications, 17
  definition, 19
  ion sources, 16–18
  schematic, 18
Thermal ionization mass spectrometry (TIMS), 16
  diazepam, 316
Thermospray ionization (TSI)
  category, ion type and applications, 17
  ion sources, 27
Thomson (Th), 5, 16, 45
  definition, 9
TI. *See* Thermal ionization (TI)
TIC. *See* Total ion current (TIC)
Time focusing devices
  TOF analyzer, 40
Time-lag focusing
  principle, 43
Time of flight (TOF) analyzer, 21, 32–33, 38
  mass range, 44
  MS, 40–44
  performance parameters, 43–44
  principle, 40
  time focusing devices, 40
Time of flight mass spectrometry (TOF-MS) reflectrons, 262
Time of flight secondary ion mass spectrometry (TOF-SIMS), 277, 278
  advantages, 278
  catalyst, 282
  catalysts, 280
  cholesterol, 279
  forensic science, 281
  instruments, 45
  pharmaceutical cellular monitoring, 278
  salbutamol, 280
  silicon wafer, 285
  surface image, 281
Time-to-digital converter (TDC), 44
TIMS. *See* Thermal ionization mass spectrometry (TIMS)
TOF. *See* Time of flight (TOF) analyzer
TOF-MS. *See* Time of flight mass spectrometry (TOF-MS) reflectrons
TOF-SIMS. *See* Time of flight secondary ion mass spectrometry (TOF-SIMS)
TOF-TOF. *See* Tandem time of flight (TOF-TOF)
Torr
  definition, 9
Total ion current (TIC)
  chromatogram, 9
  definition, 9
Toxicological analysis of nonbiological materials, 311–312
Trace elements, 237
Trace evidence, 310
Transcriptomics applications, 246–247
Transmission, 9
Trapping analyzers, 21
Trimethoxyphenyl phosphonium acetate, 208
Triple quadrupole (QqQ), 92
TSI. *See* Thermospray ionization (TSI)
Tulp, Nicolaes, 223
Tuning ion separation, 229
2-DGE. *See* Two-dimensional gel electrophoresis (2-DGE)
Two-dimensional gel electrophoresis (2-DGE), 105
  amino acids, 248–251
  peptide, 248–251
  proteins, 248–251
Tyr, 290

u. *See* Unified atomic mass unit (u)
Ultra performance liquid chromatography (UPLC), 108
Ultrapure polymers, 301
Ultraviolet (UV), 119
  photodissociation, 100
Ulysses launch, 260
Ulysses space probe
  European Space Agency's, 256
Ulysses/SWICS
  schematic, 261
Unified atomic mass unit (u), 9
Units, 9–12
Universal detector, 289
UPLC. *See* Ultra performance liquid chromatography (UPLC)
Urey, Harold, 238
UV. *See* Ultraviolet (UV)

Vaccine development to combat
bioterrorism, 271
Vacuum spark. *See* Spark source (SS)
Van Allen's surprise, 254
Very low energy (VLE)
CID, 100
Viscometry
molar masses, 303
Visible photodissociation, 100
VLE. *See* Very low energy (VLE)

WADA. *See* World Anti-Doping Agency
(WADA)
Wahrgaftig diagram, 131
Wall coated open tubular (WCOT)
columns, 106
Water, 122, 123, 128
WCOT. *See* Wall coated open tubular
(WCOT) columns
Wien, 16
WIND/MASS schematic, 263
Winkler, Hans, 243
Woods Hole Oceanographic Institution
(Massachusetts), 239
Woodword-Hoffman theory, 150
World Anti-Doping Agency
(WADA), 225
accredited laboratories for doping
control, 231
2008 prohibited list, 226
World Anti-Doping
Code, 225

WILEY-INTERSCIENCE SERIES IN MASS SPECTROMETRY

Series Editors

Dominic M. Desiderio
Departments of Neurology and Biochemistry
University of Tennessee Health Science Center

Nico M. M. Nibbering
Vrije Universiteit Amsterdam, The Netherlands

John R. de Laeter • *Applications of Inorganic Mass Spectrometry*
Michael Kinter and Nicholas E. Sherman • *Protein Sequencing and Identification Using Tandem Mass Spectrometry*
Chhabil Dass • *Principles and Practice of Biological Mass Spectrometry*
Mike S. Lee • *LC/MS Applications in Drug Development*
Jerzy Silberring and Rolf Eckman • *Mass Spectrometry and Hyphenated Techniques in Neuropeptide Research*
J. Wayne Rabalais • *Principles and Applications of Ion Scattering Spectrometry: Surface Chemical and Structural Analysis*
Mahmoud Hamdan and Pier Giorgio Righetti • *Proteomics Today: Protein Assessment and Biomarkers Using Mass Spectrometry, 2D Electrophoresis, and Microarray Technology*
Igor A. Kaltashov and Stephen J. Eyles • *Mass Spectrometry in Biophysics: Confirmation and Dynamics of Biomolecules*
Isabella Dalle-Donne, Andrea Scaloni, and D. Allan Butterfield • *Redox Proteomics: From Protein Modifications to Cellular Dysfunction and Diseases*
Silas G. Villas-Boas, Ute Roessner, Michael A.E. Hansen, Jorn Smedsgaard, and Jens Nielsen • *Metabolome Analysis: An Introduction*
Mahmoud H. Hamdan • *Cancer Biomarkers: Analytical Techniques for Discovery*
Chabbil Dass • *Fundamentals of Contemporary Mass Spectrometry*
Kevin M. Downard (Editor) • *Mass Spectrometry of Protein Interactions*
Nobuhiro Takahashi and Toshiaki Isobe • *Proteomic Biology Using LC-MS: Large Scale Analysis of Cellular Dynamics and Function*
Agnieszka Kraj and Jerzy Silberring (Editors) • *Proteomics: Introduction to Methods and Applications*
Ganesh Kumar Agrawal and Randeep Rakwal (Editors) • *Plant Proteomics: Technologies, Strategies, and Applications*
Rolf Ekman, Jerzy Silberring, Ann M. Westman-Brinkmalm, and Agnieszka Kraj (Editors) • *Mass Spectrometry: Instrumentation, Interpretation, and Applications*